THE SIGNIFICANCE OF
NONLINEARITY IN THE
NATURAL SCIENCES

Studies in the Natural Sciences

A Series from the Center for Theoretical Studies
University of Miami, Coral Gables, Florida

Recent Volumes in this Series

A Continuation Order Plan is available for this series. A continuation order will bring
delivery of each new volume immediately upon publication. Volumes are billed only upon
actual shipment. For further information please contact the publisher.

ORBIS SCIENTIAE

THE SIGNIFICANCE OF NONLINEARITY IN THE NATURAL SCIENCES

Chairman
Behram Kursunoglu

Editors
Arnold Perlmutter
Linda F. Scott

Scientific Secretaries
Mou-Shan Chen
Joseph Hubbard
Michel Mille
Mario Rasetti

Center for Theoretical Studies
University of Miami
Coral Gables, Florida

SPRINGER SCIENCE+BUSINESS MEDIA, LLC

Library of Congress Cataloging in Publication Data

Orbis Scientiae, University of Miami, 1977.
    The significance of nonlinearity in the natural sciences.

    (Studies in the natural sciences; v. 13)
    "Held by the Center for Theoretical Studies, University of Miami, Coral Gables,
Florida, January 17-21, 1977."
    Includes index.
    1. Nonlinear theories—Congresses. 2. Mathematical physics—Congresses. 3. Fluid
dynamics—Congresses. 4. Biology—Congresses. I. Kursunoğlu, Behram, 1922-   II.
Perlmutter, Arnold, 1928-   III. Scott, Linda F. IV. Miami, University of,
Coral Gables, Fla. Center for Theoretical Studies. V. Title. VI. Series.
QA401.068 1977   500.1'01'515252   77-9630

ISBN 978-1-4684-7226-4   ISBN 978-1-4684-7224-0 (eBook)
DOI 10.1007/978-1-4684-7224-0

Proceedings of Orbis Scientiae 1977 on The Significance of Nonlinearity in the Natural Sciences
held by the Center for Theoretical Studies, University of Miami,
Coral Gables, Florida, January 17-21, 1977
© 1977 Springer Science+Business Media New York
Originally published by Plenum Pres, New York 1977
Softcover reprint of the hardcover 1st edition 1977

PREFACE

In accordance with the established tradition of these annual meetings under the aegis of Orbis Scientiae we have, this year, included the very important field of "The Significance of Nonlinearity in the Natural Sciences." We are pleased to join many scientists in recognizing the nonlinearity arising from the underlying interaction of all natural phenomena. It is tempting to say that in the long run things are nonlinear and that we shall have to design new techniques and methods to solve nonlinear equations. This year's Orbis Scientiae did include four sessions on nonlinear equations pertaining to elementary particle physics, molecular physics, fluid dynamics, and also to biology. Our Center intends to pursue the inclusions of these topics in its future Orbis Scientiae.

Appreciation is extended to Mrs. Helga S. Billings, Mrs. Elva Brady, and Ms. Yvonne L. Leber for their skillful typing of the proceedings, which they have performed with great enthusiasm and dedication.

The Editors

ORBIS SCIENTIAE 1977 PARTICIPANTS

CONTENTS

SOME RECENT DEVELOPMENTS ON SOLITONS IN TWO-DIMENSIONAL
FIELD THEORIES

Andre Neveu*

Institute for Advanced Study

Princeton, New Jersey 08540

This review is written for both mathematical
physicists and applied mathematicians. It discusses
some recent results, conjectures, and open problems of
classical and semiclassical mechanics of field theories
in one space and one time dimension. We concentrate on
theories with either Lorentz or Galilean invariance,
dividing the review into three parts: One on exactly
soluble models, one on a model which we conjecture is
exactly soluble, and one on approximate methods in non-
exactly soluble models.

The exactly soluble model which has been most in-
tensely investigated from a quantum mechanical point of
view is the sine-Gordon theory. This is because it exhibits relativistic invariance together with interesting
classical solutions with particle-like properties. Its
Lagrangian density is[1]

---

*Permanent address after April 1977: Laboratoire de
 Physique Theorique, Ecole Normale Superieure, Paris, France

$$L = \frac{m^2}{\lambda} \left[-\frac{1}{2} (\partial_\mu \phi)^2 + \cos\phi - 1 \right] \quad , \tag{1}$$

where

$$(\partial_\mu \phi)^2 = (\partial_x \phi)^2 - (\partial_t \phi)^2 \quad .$$

The velocity of light has been set equal to one: λ/m^2 is the dimensionless coupling constant, m the mass of the fundamental field (with $\hbar$ set equal to 1).

The classical mechanics of the sine-Gordon theory has been solved by Ablowitz et al.[2] in light-cone coordinates and by Takhtadzhyan and Faddeev[3] in laboratory coordinates, using inverse scattering methods. Classically, there is a discrete set of equivalent vacua, corresponding to the space-time independent field configurations $\phi = 2n\pi$. The soliton and antisoliton connect two vacua with adjacent values of n. In their rest frame, they correspond to the field configurations

$$\phi = 4 \text{ Arc tan exp} \pm x \quad . \tag{2}$$

The classical rest mass of the soliton is $M_s = 8m^3/\lambda$. The only other particle-like classical solution of the sine-Gordon theory is the doublet, or breather mode. It is a soliton-antisoliton bound state which, in its rest frame, corresponds to

$$\phi(x,t) = 4 \text{ Arc tan} \frac{\epsilon \sin[t(1+\epsilon^2)^{-\frac{1}{2}}]}{\cosh[\epsilon x(1+\epsilon^2)^{-\frac{1}{2}}]} \quad , \tag{3}$$

where ϵ is any real (positive) number. The rest energy of the doublet is

$$E(\epsilon) = 2M_s \sin\theta \quad , \tag{4}$$

with $$\theta = \text{Arctan } \varepsilon \quad . \tag{5}$$

The quantum mechanics of (1) was first considered in ref. (4), using semiclassical methods. The result is that the soliton and antisoliton become particles which are heavy in perturbation theory ($\lambda/m^2 \ll 1$); one can compute corrections to M_s due to quantum corrections in a systematic fashion[5]; the breather yields a finite set of energy levels. The angle θ of Eq. (5) comes out quantized in the leading semiclassical approximation according to the formula

$$\theta_n = n \frac{\lambda}{16m^2(1 - \frac{\lambda}{8\pi m^2})} = n\frac{\gamma'}{16} \quad , \tag{6}$$

where n is any positive integer such that $0 < \theta_n < \pi/2$.

The quantum mechanical sine-Gordon theory is equivalent to the massive Thirring model,[6] which is defined by the Lagrangian

$$L = i\bar{\psi}\partial\!\!\!/\psi - M\bar{\psi}\psi - \frac{1}{2}g(\bar{\psi}\gamma_\mu\psi)^2 \quad , \tag{7}$$

where ψ is a fermion field with mass M coupled to itself by a contact vector interaction. The analysis of ref. (6) implies among other things that

$$\frac{\lambda}{4\pi m^2} = \frac{1}{1 + \frac{g}{\pi}} \tag{8}$$

and that λ/m^2 is restricted to the range $0 \le \lambda/m^2 < 8\pi$. In the language of the massive Thirring model, the soliton (antisoliton) becomes the fundamental fermion (antifermion). Breather states are fermion- antifermion bound states.

Interestingly enough, a lattice version of the
massive Thirring model had been solved exactly some
time ago by Baxter and Johnson et al.[7] Using these results, Luther[8] has been able to prove that the semiclassical quantization formula (6) for the masses of the
soliton- antisoliton bound states is actually exact.
This is very much analogous to the nonrelativistic
hydrogen atom, where the Bohr formula is also exact.

The S-matrix for soliton-antisoliton scattering was
first considered by Korepin et al.[9] They proposed the
following interesting conjecture for the soliton antisoliton S-matrix element

$$S(s) = e^{in\pi} \prod_n \frac{e^{\omega - i\frac{n\pi}{N}} + 1}{e^{\omega} + e^{-i\pi\frac{n}{N}}} \quad , \qquad (9)$$

where s is the center of mass energy,

$$\omega = \ln \frac{s - 2M_s^2 + [s(s - 4M_s^2)]^{\frac{1}{2}}}{2M_s^2} \qquad (10)$$

and γ' is equal to $8\pi/N$; N is an integer. This restric-
tion on the value of the coupling constant is needed for
eq. (9) to have the correct crossing properties. In
this conjecture, soliton-antisoliton reflection vanishes
identically. This vanishing is already present at the
classical level. However, reflection is not forbidden
by the existence of the infinite set of conservative
laws of (1). Indeed, using functional techniques,
Korepin[10] has computed the semiclassical reflection coefficient in soliton-antisoliton scattering, and found
that it vanishes precisely for $\gamma' = 8\pi/N$. Putting all
this together, Zamolodchikov[11] has recently generalized
the conjecture (9) to arbitrary values of γ, demanding

only the correct unitarity crossing and analyticity
properties of the S-matrix. Although explicit, his
formula is too lengthy to be reproduced here.

Semiclassical calculations on the S-matrix[12]
support the above-mentioned conjectures.

With all these developments that have taken place
in the last year or so, it does not seem that much is
left to squeeze out of Lagrangian (1).

Another classically exactly soluble model with
particle-like solitons is the nonlinear Schrödinger
equation[1], whose Lagrangian is

$$L = i\phi^*\partial_t\phi - \frac{1}{2}|\partial_x\phi|^2 + \frac{\kappa}{2}|\phi|^4 \ . \qquad (11)$$

It can be considered as the nonrelativistic weak field
limit of the sine-Gordon theory. In its rest frame, it
has a soliton solution

$$\phi = \sqrt{\frac{2\omega}{\kappa}}\ e^{i\omega t}\ \text{sech}\ (x\sqrt{2\omega})\ , \qquad (12)$$

whose classical energy is

$$E_{c\ell} = -\frac{2\sqrt{2}}{3\kappa}\ \omega^{3/2} \qquad . \qquad (13)$$

Suitable interpretation of the semiclassical quantization[13,14] gives the following discrete energy levels

$$E_n = \frac{-\kappa^2}{24}(n^3-n)\ ,\ n = 1,2,3\ldots \qquad . \qquad (14)$$

The quantum version of (11) has been solved exactly.[15] It describes the behavior of n one dimensional
nonrelativistic particles of unit mass interacting via δ
function potentials of strength κ. Such a system has

only one bound state (if $\kappa > 0$), with energy given by
(14). Note that for small λ/m^2, the expansion of the
sine in eq. (4) will give, combined with eq. (6), the
n^3 term of eq. (14).

We now come to a model which is not known to be
soluble, but possesses so many nice features (in the
form of exact, explicit static and time dependent
solutions) that it might very well be so, and in any
case, deserves further study. It is the model defined
by the Lagrangian[16]

$$L = i\bar{\psi}\partial\psi - g\sigma\bar{\psi}\psi - \frac{Z}{2}\sigma^2 \quad . \tag{15}$$

Here, ψ is actually a short hand notation for a large
number N of fermion species, all interacting with the
same scalar field $\sigma(x,t)$. When $N\to\infty$ at fixed g^2N, the
Lagrangian (15) can be replaced by another effective
Lagrangian, obtained by integrating out the fermion
fields. This can be done formally because (15) is
quadratic in ψ. The net result is to consider $-\frac{1}{2}$
$Z\sigma^2$ plus the one fermion loop as an effective classical
Lagrangian for σ. It is shown in ref. (17) where this
effective Lagrangian is studied, to correspond to the
following Hartree-Fock like system of equations of motion. First, given a $\sigma(x,t)$ one must solve for all the
solutions of the Dirac equation

$$(i\partial - g\sigma)\psi = 0 \tag{16}$$

(with appropriate boundary conditions on ψ, see ref.
(17)). In general, this equation will have both scattering
and bound states. Then, one must form the quantity
$-g \sum_{states} \bar{\psi}\psi(x,t)$, where the sum over states extends over

all occupied fermion states (in general all the negative energy Fermi sea, plus some bound states). One
will have found a variational minimum of the effective
action if this sum over states precisely equals σ of
eq. (16) up to the fixed renormalization constant Z.
Hence, the other equation of motion is

$$Z\sigma = -g \sum_{\substack{\text{occupied} \\ \text{states}}} \bar{\psi}\psi \quad . \tag{17}$$

This set of equations (16,17) has been studied in
ref. (17). A very large class of exact, analytic solutions has been found, both space and space-time dependent. The situation resembles the sine-Gordon situation
in many respects. The symmetry $\sigma \rightarrow -\sigma$, $\psi \rightarrow \gamma_5 \psi$ is spontaneously broken. There are two space-time independent
vacua, $\sigma = \pm \sigma_0$, where σ_0 which can be chosen at will,
is related to the renormalization point and to g. A
soliton[17]

$$\sigma = \sigma_0 \tanh(g\sigma_0 x) \tag{18}$$

in its rest frame, connects these two vacua. This is
presumably a real soliton, because the solution corresponding to soliton-antisoliton scattering has been
found analytically in ref. (17).

Besides the soliton, there is another set of time
independent solutions to the system (16-17). They
correspond to the field configurations

$$\sigma = \sigma_0 + \sigma_0 y \tanh \left(g\sigma_0 yx - \frac{1}{4} \ell n \frac{1+y}{1-y}\right)$$

$$-\sigma_0 y \tanh\left(g\sigma_0 yx + \frac{1}{4} \ell n \frac{1+y}{1-y}\right) , \tag{19}$$

where
$$y = \sin \frac{\pi}{2} \frac{n_f}{N} \quad . \tag{20}$$

n_f is the number of fermions in the only bound state of the potential $\sigma(x)$ described by (18). Hence, strictly speaking, different values of n_f correspond to slightly different forms of eq. (17) since in eq. (17) the weight of the bound state $\bar{\psi}\psi$ is precisely n_f. Interestingly enough, the energy of the field configuration (19) is precisely

$$E(n_f) = 2M_s \sin \frac{\pi}{2} \frac{n_f}{N} \quad , \tag{21}$$

where M_s is the mass of the soliton (18). The field configuration (19) is also presumably a soliton in the strict sense of the word.

Finally, a whole class of time-dependent solutions of (16-17) has been found explicitly in ref. (17). These solutions correspond to either scattering of soliton and antisoliton, or to excitations of the field configuration (19), which can also be considered as soliton-antisoliton bound states. These time dependent solutions are analytically very much similar to the sine-Gordon breather mode, or to the sine-Gordon soliton-antisoliton scattering solution. Also their semi-classical quantization yields exactly the same formula as (21) and a large degeneracy is present in the theory. All this makes it a very intriguing and potentially very interesting system.

Finally, for completeness, one should envisage higher order corrections to this semiclassical treatment of the Lagrangian (15). These corrections are of order 1/N. Here, 1/N plays the role of λ/m^2 in the sine-Gordon Lagrangian and the semiclassical treatment,

eq. (16) and (17) is only valid for large N. Indeed,
for N=1, (15) represents after Fierz rearrangement the
massless Thirring model; for N = 2, (15) has been
shown[18] to be equivalent to two decoupled sine-Gordon
equations with a special value of λ/m^2. For other
values of N, no exact result is known.

We now come to models which are not exactly
soluble and describe some approximation methods and
some partial results. Let us begin with the ϕ^4 theory
in two space-time dimensions, whose Lagrangian is

$$L = \frac{m^4}{\lambda} \left[-\frac{1}{2}(\partial_\mu \phi)^2 + \frac{1}{2}\phi^2 - \frac{1}{4}\phi^4 \right] \quad . \qquad (22)$$

The parameters of the Lagrangian are chosen in such a
way that there are two classical vacua, $\phi \pm 1$, which
spontaneously break the $\phi \to -\phi$, symmetry of (22). The
kink (and antikink) that connects these two vacua
corresponds to the static field

$$\phi(x) = \pm \tanh \frac{x}{\sqrt{2}} \quad . \qquad (23)$$

The classical mass of this kink is $\frac{2}{3}\sqrt{2}\frac{m^3}{\lambda}$. It is
corrected by quantum fluctuations. There can also be
excited states of the kink[4] which correspond to kink-
meson bound states or resonances. Such states do not
exist in the sine-Gordon theory.

Computer experiments show that the kink and antikink (23) do not behave as solitons in a collision.
Some radiation is produced and they can even fall into
a bound state. This bound state is similar to the
sine-Gordon breather and can actually be approximated
analytically[4] when its amplitude is small enough. Indeed, setting $\phi = 1 + z$, one has to solve the equation

$$\ddot{z} - z'' + 2z + 3z^2 + z^3 = 0 \quad . \tag{24}$$

Introducing a small parameter ε related to the amplitude of the doublet, one expands z, by analogy with eq. (3), simultaneously in powers of ε and in harmonics of the fundamental frequency

$$z(x,t) = \varepsilon^2 g_1(\xi) + \sum_{n=0}^{\infty} [\varepsilon^{2n+1} f_{2n+1}(\xi) \sin(2n+1)\tau$$

$$+ \varepsilon^{2n+2} g_{2n+2}(\xi) \cos(2n+2)\tau] , \tag{25}$$

where

$$\tau = \frac{t\sqrt{2}}{\sqrt{1+\varepsilon^2}} \quad \xi = \frac{\varepsilon x \sqrt{2}}{\sqrt{1+\varepsilon^2}} \quad .$$

From eq. (24), one can then solve for the f's and g's in powers of ε. For example, one finds

$$f_1 = \frac{2}{\sqrt{3}} (1 + \frac{20}{g} \varepsilon^2 + \ldots) \frac{1}{\cosh \xi} - \frac{103}{18\sqrt{3}} \frac{\varepsilon^2}{\cosh^3 \xi} + 0(\varepsilon^2). \tag{26}$$

One can also compute the classical energy $E(\varepsilon)$ and action $W(\varepsilon)$ around one period for such a solution

$$E = \frac{m^3}{\lambda} \frac{2\sqrt{2}}{3} (2\varepsilon + \frac{37}{27} \varepsilon^3) + 0(\varepsilon^5) , \tag{27}$$

$$W = \int \dot{z}^2 \, dx \, dt = \frac{4\pi}{3} \frac{m^2}{\lambda} (2\varepsilon + \frac{46}{27} \varepsilon^3) + 0(\varepsilon^5) . \tag{28}$$

If one eliminates ε between these two equations, one obtains

$$E(W) = m\sqrt{2} \frac{W}{2\pi} - \frac{3\sqrt{2}}{32} (\frac{W}{2\pi})^3 \frac{\lambda^2}{m^3} + 0(W^5) \quad . \tag{29}$$

Identifying the first two terms of this series with the expansion of a sine

$$E(W) = \frac{4}{3} \sqrt{2} \frac{m^3}{\lambda} \sin \frac{3\lambda W}{8\pi m^2} + \ldots \quad . \quad (29)$$

Remarkably enough[19] the coefficient of the sine is exactly equal to twice the mass of the kink (23). Certainly, eq. (29) does not hold in higher orders in W, but this coincidence is very intriguing. It is not excluded that $E(W)$ could be a simple function while $z(x,t)$ would be very complicated.

As mentioned in ref. (4), it is also surprising to remark that in lowest order in ε, g_2, f_3, g_4, f_5, etc., are the terms of a geometric series. Although the expansion (24) is specifically designed for small values of ε, one can, by analogy with sine-Gordon, set $\varepsilon = i/v$, $0 < v < 1$, and sum all the terms of this geometric series. One then obtains, for $\phi = 1 + z$ and large $|t|$

$$\phi \approx \frac{\frac{1}{4} \sqrt{3} \, v \cosh \frac{x}{\sqrt{1-v^2}} - \cosh \frac{vt}{\sqrt{1-v^2}}}{\frac{1}{4} \sqrt{3} \, v \cosh \frac{x}{\sqrt{1-v^2}} + \cosh \frac{vt}{\sqrt{1-v^2}}} \quad . \quad (30)$$

This looks like two well separated kinks, moving with velocities $\pm$ v. It is not clear what the $\approx$ sign in this equation actually means. There are no obvious small parameters in which to expand.

The periodic motion (25) can be quantized semiclassically. In the domain of validity of the approximation, it yields a series of nonrelativistic bound states of n particles interacting with δ-function potentials. The binding energy agrees with that found by

summation of Feynman diagrams. Presumably these states
can also be considered as kink-antikink bound states.

A method analogous to the expansion (25) can also
be used to derive higher conservation laws for the
Lagrangian (22): ϕ, its canonical momentum $\pi = \dot{\phi}$, and
a space derivative are all considered to be of the same
order in ε as suggested by (25). For example, the
analog for (22) of the exact conserved quantity

$$P_1(\text{sine-Gordon}) = \int dx[\phi'''\pi - \tfrac{3}{4}\,\pi\phi'\cos\phi + \tfrac{1}{8}(\phi'\pi^3 + \phi'^3\pi]$$

$$(31)$$

is found to be

$$P_1(\phi^4) = \int dx[\tfrac{4}{9}\,\phi'''\pi - \pi\phi'\phi^2 + \tfrac{1}{3}(\phi'^3\pi + \phi'\pi^3)$$

$$- \tfrac{1}{3}\,\pi^3\phi^2\phi' - \tfrac{1}{10}\,\pi^5\phi' + O(\varepsilon^9)]\ . \quad (32)$$

Another technique for handling approximately the
differential equations of interacting field theories
makes use of the trace identities and the inverse
scattering method in a new fashion. The idea has only
been applied to time independent solutions. Hence, one
has only to deal with a one dimensional scattering
problem. Instead of functionally varying the Hamiltonian
with respect to the field, one instead considers the
field as the potential of an appropriate scattering
problem, such that the Hamiltonian can be expressed as
a combination of its trace identities. One can then replace the functional variation with respect to the field
with a variation with respect to the scattering data.
This generally reduces the problem to an algebraic equation to determine the bound state energies of the

potential which turns out to be reflectionless. This
technique has been developed in ref. (20) to study the
time independent solutions of the two-dimensional nonlinear sigma model whose Lagrangian is

$$L = \bar{\psi} [i\not{\partial} - g(\sigma+i\pi\gamma_5)]\psi -\frac{1}{2}[(\partial_\mu\sigma)^2+(\partial_\mu\pi)^2]-\frac{\lambda}{4}(\sigma^2+\pi^2-1)^2.$$

In the special case $\lambda = 2g^2$, it turns out that the last
two terms of (33) are precisely the second trace identity for the Dirac equation obtained from the first term.

Scattering theory and inverse scattering is not in
an advanced enough stage to permit extension of such
techniques to two and higher space dimensions.

REFERENCES

1. For a review of exactly soluble two-dimensional classical field theories, see: A.C. Scott, F.Y.F. Chu and D.W. McLaughlin, Proc. IEEE 61, 1443 (1973).

2. M. Ablowitz, D. Kaup, A. Nevell and H. Segur, Phys. Rev. Lett. 30, 1262 (1973).

3. L. Takhtadzhyan and L.D. Faddeev, Theor. Math. Phys. 21, 160 (1974).

4. R.F. Dashen, B. Hasslacher and A. Neveu, Phys. Rev. D11, 3424 (1975).

5. See A. Jevicki's report at this conference and references therein.

6. S. Coleman, Phys. Rev. D11, 2088 (1975).

7. R.J. Baxter, Phys. Rev. Lett. 26, 834 (1971), Ann. Phys. 70, 323 (1972), J.D. Johnson, S. Krinsky and B.M. McCoy, Phys. Rev. A8, 2525 (1973).

8. A. Luther, Phys. Rev. B14, 2153 (1976).

9. V.E. Korepin, P.P. Kulish and L.D. Faddeev, JETP Lett. 21, 138 (1975).

10. V.E. Korepin, JETP Lett. 23, 201 (1976).

11. A.B. Zamolodchikov, ITEP preprint, Moscow (1976).

12. J.L. Gervais and A. Jevicki, Nucl. Phys. B110, 113 (1976).

13. A. Klein and F. Krejs, Phys. Rev. D13, 3282 (1976).

14. C. Nohl, Ph.D. Thesis, Princeton University 1976.

15. C.N. Yang, Phys. Rev. 168, 1920 (1968). Yu. C. Tyupkin, V.A. Fateev and A.S. Schwartz, Sov. J. Nucl. Phys. 22, 321 (1975).

16. D.J. Gross and A. Neveu, Phys. Rev. D10, 3235 (1974).

17. R.F. Dashen, B. Hasslacher and A. Neveu, Phys. Rev. D12, 2443 (1975).

18. A. Luther, private communication.

19. This was noticed by A. Klein, whom the author thanks for a most interesting conversation on this subject at the 1976 Orbis Scientiae Conference.

20. D.K. Campbell and Y.T. Liao, Phys. Rev. $\underline{D14}$, 2431 (1976).

PATH INTEGRAL QUANTIZATION OF SOLITONS

A. Jevicki[*]

Institute for Advanced Study

Princeton, New Jersey 08540

ABSTRACT

A review of some recent work on the collective coordinate approach to perturbation expansion about soliton solutions is presented. Consistent formulation of the method in framework of the Feynman path integral quantization procedure is described. As an illustrative application we discuss in detail a perturbation expansion for scattering of solitons.

INTRODUCTION·

After the relevance of classical soliton solutions to the quantum theory was established at the semi-classical level[1], the main problem remaining was to develop a systematic perturbation expansion about these solutions The straightforward way of doing this, which consists of shifting the field by the classical solution, runs into serious difficulties. Namely, by the shift various symmetries (like translational invariance) get broken

---
[*] Research sponsored by the Energy Research and Development Administration Grant No. E(11-1)-2220.

and consequently zero frequency modes appear, leading
to an infrared problem. The most effective way of
eliminating this difficulty is given by the collective
coordinate method[2] in which a proper treatment of
symmetry degrees of freedom is achieved. In this approach
it is then easy to develop a consistent set of Feynman
rules and to formulate a systematic perturbation
expansion.[3]

In what follows we will review some further work
on collective coordinates and their application to
soliton quantization, most of which was done during the
last year in collaboration with J. L. Gervais. First,
in sec. I we give a brief summary of the canonical
collective coordinate method. Then in sec. II a
nontrivial application is presented on the example of
soliton scattering. Finally in sec. III we describe an
alternative, noncanonical treatment of zero frequency
modes.

I. THE CANONICAL COLLECTIVE COORDINATE METHOD

In the initial works dealing with the one soliton
sector perturbation expansion[2,4,5] the center of mass
position was treated as a collective variable since it
was necessary to preserve the translational invariance
when shifting by a space dependent classical solution.
To treat more complicated soliton solutions we have
formulated in collaboration with J. L. Gervais and
B. Sakita a general method of collective coordinates in
ref. (6). The basic idea consists in promoting the
symmetry parameters X_α, which appear in certain classical
solutions $\phi_{c\ell}(x,t;X_\alpha)$ to dynamical variables (collective
coordinates). There conjugate variables $p_\alpha(t)$ are then
precisely the generators $P_\alpha[\Pi, \phi]$ of the symmetry group.

This new set of canonical variables is introduced into
the path integral expression for transition amplitude

$$T_{fi} = \int D\Pi D\phi \psi_f^* \psi_i \, \exp\{iA[\Pi, \phi]\} \tag{1.1}$$

by using the identities

$$\int \prod_\alpha D \, p_\alpha \prod_t \delta \, (p_\alpha(t) - P_\alpha[\Pi, \phi]) = 1 \, ,$$

$$\int \prod_\alpha D \, X_\alpha \prod_t \delta \, (Q_\alpha[\phi_{[X(t)]}]) \, J = 1 \, , \tag{1.2}$$

$$J = \det \left\| \frac{\partial Q_\alpha}{\partial X_\beta} \right\| \quad .$$

Here $\phi_{[X]}$ denotes the transform of ϕ and the Q_α's
represent arbitrary symmetry fixing conditions. The
collective coordinates can be transferred into the
action by performing a canonical transformation

$$\phi(x,t) \rightarrow \phi_{[-X(t)]}(x,t), \quad \Pi(x,t) \rightarrow \Pi_{[-X(t)]}(x,t) \, . \tag{1.3}$$

Furthermore if the initial and final states represent
eigenstates of the conserved quantities P_α it is possible
to integrate out the collective variables and obtain

$$T_{fi} = \prod_\alpha (2\pi) \delta(p'_\alpha - p_\alpha) \int D\pi D\phi \psi_f^* \psi_i \prod_{\alpha,t} \delta(p_\alpha - p_\alpha) \delta(Q_\alpha) J e^{iA} \, . \tag{1.4}$$

This result is formal since in deriving it we made a
canonical transformation by formally changing variables
in the path integral.

In general it is not known how to perform canonical
transformations in quantum theory. We have made in ref.
(7) an investigation of point transformations in the path
integral since they are the ones most often used in
application of the collective coordinate method. Let us

consider a general system with n-degrees of freedom

$$L(\dot{q},q) = \frac{1}{2} \sum_{a=1}^{n} \dot{q}_a^2 - V(q)_J$$

in the presence of an external source J. The generating functional is then given by

$$Z[J] = \int \prod_{a=1}^{n} \mathcal{D}q_a \exp\{i \int dt\ L(\dot{q},q)\} \ . \qquad (1.5)$$

Performing the point transformations $q_a = F^a(Q)$ we obtain formally

$$\int \prod_{i=1}^{n} \mathcal{D}Q_i\ g(Q)^{1/2} \exp\{i \int dt\ L(F_i,\dot{Q}_i,F(Q))\} \ , \qquad (1.6)$$

where

$$F^a_{,i} = \frac{\partial}{\partial Q_i} F^a(Q) \ ,$$

$$g_{ij} = \sum_{a=1}^{n} F^a_{,i}\ F^a_{,j} \ , \qquad (1.7)$$

$$g = \det g_{ij} \ .$$

However, by making use of the diagram technique one can show that this expression differs from the original generating functional (1.5). This means that a formal change of integration variables does not represent a correct step in the path integral.

To derive the correct answer we use a more precise form for the path integral

$$Z[J] = \lim_{N \to \infty} \int \prod_{a=1}^{n} \prod_{k=1}^{N-1} \frac{d q_a(k)}{(2\pi i \epsilon \hbar)^{1/2}} \prod_{k=0}^{N-1} \exp\{\frac{i}{\hbar} A(q(k+1),q(k))\}, \qquad (1.8)$$

obtained by dividing the time interval into N equal intervals $\epsilon = T/N$ and denoting $t_K = \epsilon k$, $q(k) \equiv q(k+1)$. Then after the change of variables $q_a(k) = F^a(Q(k))$, the new short-time action becomes

$$\tilde{A}(Q(k+1), Q(k)) = \frac{1}{2\epsilon} \sum_{a=1}^{n} (F^a(Q(k+1))-F^a(Q(k)))^2 -\epsilon V(F(Q(k))). \quad (1.9)$$

Now it is crucial to observe that it is not correct to approximate this expression by using

$$F^a(Q(k+1)) - F^a(Q(k) \approx F^a_{,i} \Delta Q(k) ,$$

since in the path integral $\Delta Q = O(\epsilon^{1/2})$ and not $O(\epsilon)$. Deciding to expand about the midpoint $\overline{Q}(k) = \frac{1}{2}(Q(k+1)+Q(k))$ the correct approximation for the short-time action reads

$$\tilde{A} \approx \frac{1}{2\epsilon} \Sigma [g_{ij}(\overline{Q})\Delta Q_i \Delta Q_j + \frac{1}{12} F^a_{,i} F^a_{,j\ell m}\Delta Q_i \Delta Q_j \Delta Q_\ell \Delta Q_m] - \epsilon V(F(\overline{Q})). \quad (1.10)$$

Similarly for the measure we obtain

$$g(\overline{Q})^{1/2} [1 + \frac{1}{16}(g^{ij} g_{ij,\ell m} + g^{ij}_{,\ell} g_{ij,m})\Delta Q_\ell \Delta Q_m] . \quad (1.11)$$

Next it is possible to substitute the additional terms appearing in (1.10) and (1.11) by an equivalent potential term. The form of this additional potential is

$$\Delta V(\overline{Q}) = \frac{\hbar^2}{8} \Gamma^i_{j\ell}(\overline{Q})\Gamma^j_{im}(\overline{Q})g^{\ell m}(\overline{Q}) , \quad (1.12)$$

where $g^{\ell m}$ represents the inverse of g_{ij} and $\Gamma^i_{j\ell}$ are the Christoffel symbols.

To summarize, we have found that a careful treatment of point transformations in the path integral leads to:

(a)   An additional set of vertices coming from $\Delta V(Q)$.

(b)   Specification of the otherwise ambiguous
      contractions at the same point as implied by the
      midpoint path integral.

Using this general result it is then simple to derive
the correct Feynman rules in the path integral collective
coordinate method. We can also mention that it is not
difficult to demonstrate the equivalence of this method
with the parallel operator approach.[8] However there
are certain advantages to the path integral approach,
since first of all it leads directly to the Feynman
rules and furthermore it appears to be more powerful in
discussing more complicated processes, like, for example,
scattering of solitons.

II. AN ILLUSTRATIVE APPLICATION: PERTURBATION EXPANSION FOR SOLITION SCATTERING

By now there is an extensive literature on applications of the collective coordinate method to develop
perturbation expansions in the one soliton sector for
various field theories. In contrast, the multisoliton
sector appears to be much less studied since it is
considerably more difficult to treat. So instead of going
over the various one soliton sector examples we choose
to discuss in this section the problem of soliton
scattering.[9] It is here that the full power of the
path integral approach becomes evident.

We consider the sine-Gordon model

$$L(\phi) = \frac{1}{2} (\partial\phi)^2 - V(\phi) \ ,$$

$$V(\phi) = \frac{1}{\gamma} (1-\cos(\sqrt{\gamma}\ \phi)) \ ,$$

since for this case exact multisoliton solutions are
known. The static soliton and antisoliton solutions
read

$$\phi_s(x) = \frac{4}{\sqrt{\gamma}} \, tg^{-1} e^x \quad ,$$

$$\phi_a(x) = \frac{4}{\sqrt{\gamma}} \, tg^{-1} e^{-x} \quad ,$$

and their mass is $M_0 = 8/\gamma$. Our treatment of the soliton-
antisoliton scattering is based on the exact classical
solution $\phi_{sa}(x,t;u_1,u_2)$, which describes the transition
of soliton and the antisoliton through each other. It
has the following asymptotic behavior

$$\lim_{t \to \pm\infty} \phi_{sa}(x,t) = \phi_s\left(\frac{x-u_1 t - X_1^\pm}{(1-u_1^2)^{1/2}}\right) + \phi_a\left(\frac{x-u_2 t - X_2^\pm}{(1-u_2^2)^{1/2}}\right),$$

$$(2.1)$$

so that after the collision only the positions become
changed

$$\Delta X_1 = X_1^+ - X_1^- = (1-u_1^2)^{1/2} \ell n \left(1 - \frac{4M_0^2}{s}\right),$$

$$\Delta X_2 = X_2^+ - X_2^- = -(1-u_2^2)^{1/2} \ell n \left(1 - \frac{4M_0^2}{s}\right), \quad (2.2)$$

where s is the usual Mandelstam variable.

We now consider the s-a scattering amplitude written
in the path integral representation

$$S_{fi} = \lim_{\substack{t_i \to -\infty \\ t_f \to +\infty}} e^{i(E_f t_f - E_i t_i)} \int \! \mathcal{D}\pi \mathcal{D}\phi \; \Psi^*_{p_1' p_2'}(t_f) \Psi_{p_1 p_2}(t_i) e^{iA[\pi,\phi]}$$

$$(2.3)$$

with the initial and final wave functionals defined by

$$\Psi_{p_1 p_2} [\phi] = < \phi | p_1, p_2 > \quad .$$

This two-particle wave functional is simply obtained from separate soliton and antisoliton functionals. The latter are easily derived if we use the fact that for the one soliton sector

$$\hat{\phi}(x) = \phi_s \left(\frac{x - \hat{X}}{(1-u^2)^{1/2}} \right) + \hat{\eta}(x) \quad ,$$

$$\int dx \; \phi_s' \; \hat{\eta} = 0 \quad ,$$

where the hats denote operators in the Schrödinger picture. Now

$$\Psi_p [\phi] = < X, \eta | p > = e^{ipX} \Psi_0 [\eta] \big|_{\int dx \phi_s' \eta = 0}$$

where Ψ_0 stands for a Gaussian wave functional representing the ground state. From this we conclude that

$$\Psi_{p_1 p_2} = \prod_{\ell=1}^{2} \int dX_\ell \, e^{ip_\ell X_\ell} \delta \left(\int dx \phi'_\ell \left(\frac{x - X_\ell}{(1-u_\ell^2)^{1/2}} \right) \phi(x) \right) \Psi_0 [\phi - \phi_s - \phi_a].$$

$$(2.4)$$

As the next step we apply the collective coordinate method to separate and integrate out the variables corresponding to space and time translations. The result reads

$$S_{fi} = (2\pi)^2 \delta(p_f - p_i) \delta(E_f - E_i) \lim_{\substack{t_i \to -\infty \\ t_f \to +\infty}} e^{i(E_f t_f - E_i t_i)} \int D\pi D\phi \Psi_f^* \Psi_i$$

$$\cdot \delta(Q) \delta(\overline{Q}) \delta(p_i - P) \delta(E_i - H) e^{iA[\pi, \phi]} \quad . \qquad (2.5)$$

Here P and H denote the total momentum and the hamiltonian while Q and $\bar{Q}$ represent arbitrary gauge conditions. We make the choice

$$Q[\phi] = \int dx \, \phi'_{sa}(\phi - \phi_{sa}) \quad ,$$

$$\bar{Q}[\phi] = \int dx \, \dot{\phi}_{sa}(\phi - \phi_{sa}) \quad , \tag{2.6}$$

which leads to an elegant elimination of zero frequency modes which are associated with space and time translations. At the initial and final times the conditions (2.6) imply

$$\int dx \, \phi'_\ell \left(\frac{x - u_\ell t - X_\ell^\pm}{(1 - u_\ell^2)^{1/2}} \right) \phi(x,t) = 0 \quad \ell = 1,2 \quad .$$

Therefore only the values

$$X_\ell = u_\ell t_i + X_\ell^- \quad ,$$
$$\ell = 1,2 \tag{2.7}$$
$$X'_\ell = u_\ell t_f + X_\ell^+ \quad ,$$

contribute to the integrals which appear in the intial and final wave functionals.

At this point we are ready to do perturbation about the stationary point ϕ_{sa} by making the shift

$$\phi = \phi_{sa} + \eta \quad ,$$

$$\Pi = \phi_{sa} + \Pi_\eta \quad .$$

We will be concerned only with the first quantum correction to the classical answer and in this approximation the path

integral expression for the s-a scattering phase shift
reads

$$e^{2i\delta_{sa}} = \exp\{i \sum_{\ell=1}^{2} (E(p_\ell)T - p_\ell u_\ell T - p_\ell \Delta X_\ell) + i A_{sa}\}$$

$$\cdot \int D\Phi^{+} D\Phi \delta(Q) \delta(\overline{Q}) \delta(i \int dx \Phi^{+} I \Phi'_{sa}) \delta(i \int dx \Phi^{+} I \dot{\Phi}_{sa}) \Psi_f \Psi_i e^{iA^{(2)}},$$

$$(2.8)$$

with the notation

$$\Phi = (_{\pi_\eta}{}^{\eta}) \ , \ \Phi_{sa} = \begin{pmatrix} \phi_{sa} \\ \dot{\phi}_{sa} \end{pmatrix}, \ I = (_i^0 \ {}^{-i}_0) \ ,$$

and the quadratic action given by

$$A^{(2)} = \frac{1}{2} \int dt \ dx \ \Phi^{+} H_2 I \Phi \ ,$$

$$H_2 = i \begin{pmatrix} \partial/\partial t & -1 \\ -\dfrac{\partial^2}{\partial x^2} + V^{(2)}(\phi_{sa}) & \partial/\partial t \end{pmatrix}. \qquad (2.9)$$

From the expression (2.8) we see that the leading
contribution reads

$$2 \ \delta_{sa}^{c\ell} = \sum_{\ell=1} \left(TM_0 (1-u_\ell^2)^{1/2} - \frac{M_0 u_\ell}{(1-u_\ell^2)^{1/2}} \Delta X_\ell \right) + A_{sa} \ ,$$

$$(2.10)$$

which is the classical phase shift[4,5,10]. We point out
here that the precise specification of initial and final
wave functionals appeared to be necessary in order to
obtain this answer. Otherwise one would have to supply

the first two terms in (2.10) essentially by hand.

To compute the first quantum correction we need to evalute the Gaussian functional integral with constraints. This is done by finding the normal modes as follows. First, we solve for the complete set of solutions to the equation of motion $H_2\Phi = 0$, which consists of the discrete modes Φ'_{sa}, $\dot{\Phi}_{sa}$ and the continuum $\{\Phi_k\}$. Imposing periodic boundary conditions we calculate the stability angles[11] ν_n and obtain a set of orthonormal eigenfunctions

$$f_n(x,t) = e^{i\frac{\nu_n t}{2T}} \Phi_{k_n}(x,t) \; ,$$

$$H_2 f_n = -\frac{\nu_n}{2T} f_n \; , \qquad\qquad (2.11)$$

with the inner product defined by

$$(f,g) = \int dx \; f^+(x) \; I \; g(x) \; .$$

Now we still have to consider the zero frequency modes which in fact have zero norms. However, in order to satisfy the first two subsidiary conditions in (2.8) we need to define a new basis by making linear combinations

$$\tilde{f}_n = (\Phi_{k_n} - \alpha_n \Phi'_{sa} - \beta_n \dot{\Phi}_{sa})e^{i\frac{\nu_n t}{2T}} \; , \qquad (2.12)$$

and furthermore the zero frequency modes give identically zero.

Next, by expanding the field

$$\Phi(x,t) = \sum_n (\tilde{f}_n(x,t) \; a_n(t) + c.c.) \; ,$$

we reduce the action to a set of decoupled harmonic
oscillators

$$A^{(2)} = \int_{t_i}^{t_f} dt \sum_n \left[\frac{1}{2i} (\overset{\cdot}{a}_n^* a_n - a_n^* \overset{\cdot}{a}_n) - \frac{\nu_n}{2T} a_n^* a_n \right] \quad , \quad (2.13)$$

while the initial and final wave functionals give
precisely the ground state of this oscillator.
Consequently the expression for the first quantum
correction reads

$$2\delta_{sa}^q = M^{(1)} T \sum_{\ell=1}^2 (1 - u_\ell^2)^{1/2} - \sum_n \frac{1}{4} \nu_n \quad , \quad (2.14)$$

where $M^{(1)}$ denotes the quantum correction to the soliton
mass.   After renormalization and some calculation we
obtain in the C.M. frame the following answer

$$\delta_{sa}^q = -\frac{2}{\pi} \int_0^u dz \frac{\ell n \ z}{1 - z^2} + \frac{2u}{\pi} \ell n \ u + \frac{1}{4} \pi \quad . \quad (2.15)$$

For comparison, the classical phase shift is given by

$$\delta^{c\ell} = \frac{16}{\gamma} \int_0^u dz \frac{\ell n z}{1 - z^2} + \frac{4\pi^2}{\gamma} \quad . \quad (2.16)$$

An interesting conjecture concerning the exact
soliton-antisoliton scattering amplitude has been made
by Faddeev, Korepin and Kulish.   Namely, knowing the
classical result (2.16) and also the bound state spectrum,
these authors conjectured that for special values of the
coupling constant $\delta' = \dfrac{\delta}{1 - \dfrac{\delta}{8\pi}}$, given by

$$\gamma' = \frac{8\pi}{N} \ , \ N = 1, 2, \ldots \quad , \quad (2.17)$$

in this theory there is no reflection and that

$$S_{sa} = \prod_{n=1}^{N-1} \left(-\frac{\xi + e^{i\frac{\pi n}{N}}}{1 + \xi e^{i\frac{\pi n}{N}}} \right), \qquad (2.18)$$

where $\xi = \dfrac{s - 2M^2 + \sqrt{s(s - 4M^2)}}{2M^2}$. Expanding now this expression in the coupling constant γ one indeed obtains for the first and second terms precisely (2.16) and (2.15).

At the end of this section a comment is in order concerning the fact that up to now we were discussing only the transition part of the full s-a amplitude. The question is whether the reflection part can be treated in a similar fashion. One may doubt this since at the classical level there exists no solution describing reflection and it is a purely quantum effect (like tunnelling in quantum mechanics). However, it has been shown by Korepin in ref. (12) that by employing complex time trajectories in the path integral it is possible to give a semiclassical treatment of reflection. Namely, if in the usual soliton-antisoliton classical solution

$$\phi_{sa}(x,t) = \frac{4}{\sqrt{\gamma}} tg^{-1} \left(\frac{1}{u} \frac{sh\dfrac{ut}{(1-u^2)^{1/2}}}{ch\dfrac{x}{(1-u^2)^{1/2}}} \right), \qquad (2.19)$$

we choose the following time contour

$$Im\ t = 0 \qquad\qquad , \quad t < 0$$

$$Im\ t = \frac{\pi(1-u^2)^{1/2}}{u} \quad , \quad t > 0 \qquad (2.20)$$

then this solution describes reflection. Taking into
account the two possible ways of approaching the turning
point $t_0 = i\pi(1-u^2)^{1/2}/2u$ one derives an expression for
the reflection part of the s-a scattering amplitude.
Interestingly enough, for $\gamma=8\pi/N$ this amplitude indeed
vanishes as conjectured before. We mention that the
quantum correction can be computed along the lines of
our computation in case of the transition.

III. NONCANONICAL COLLECTIVE COORDINATES

We will now describe an alternative treatment of
zero frequency modes[13] in which the symmetry parameters
are not promoted to new canonical variables, but are
rather treated as ordinary integration variables in the
path integral. Consequently this method is not canonical
but it appears to be simpler than the one described in
section I.

For illustration we consider a two-dimensional
scalar field theory

$$L(\phi) = \frac{1}{2}(\partial\phi)^2 - V(\phi) \quad ,$$

possessing a static classical solution $\phi_0(x)$. After the
shift $\phi(x,t) = \phi_0(x) + \eta(x,t)$ the quadratic part of the
action reads

$$A^{(2)}[\eta] = \int dt\ dx\ \frac{1}{2}\eta\ \hat{M}\eta \quad ,$$

$$\hat{M} = -\frac{\partial^2}{\partial t^2} + \frac{\partial^2}{\partial x^2} - V^{(2)}(\phi_0(x)) \quad . \quad (3.1)$$

The equation satisfied by the "naive" propagator is then

$$\hat{M}D(x,t;x',t') = i\delta(t-t')\delta(x-x') \quad , \quad (3.2)$$

and the corresponding eigenvalue problem in the box

$$\hat{M} \, f_{mn}(x,t) = \lambda_{mn} f_{mn}(x,t) \quad ,$$

is solved by

$$f_{mn}(x,t) = \frac{e^{i\nu_m t}}{\sqrt{T}} \, \psi_n(x) \quad ,$$

$$\lambda_{mn} = \nu_m^2 - \omega_n^2 \, , \quad \nu_m = \frac{2\pi m}{T} \quad . \qquad (3.3)$$

Here $\{\psi_n\}$ represents a complete set of eigenfunctions of

$$. \, [- \frac{d^2}{dx^2} + V^{(2)}(\phi_0)] \psi_n(x) = \omega_n^2 \psi_n(x) \, , \qquad (3.4)$$

and we assume that it consists of the translational zero frequency mode $\psi_0(x) = \frac{1}{\sqrt{N}} \phi_0'(x)$, $\omega_0 = 0$ and the continuum $\{\psi_k(x)\}$.

Now the "naive" expression for the propagator which reads

$$\sum_{m,n} f_{mn}(x,t) \, \frac{i}{\lambda_{mn} - i\epsilon} \, f_{mn}^*(x',t') \qquad (3.5)$$

contains an infrared problem due to the m=0, n=0, mode, since then $\lambda_{oo} = 0$. However, in the path integral it is possible to preserve the translational symmetry and introduce a subsidiary condition of the form

$$\int dt dx \, f_{oo}(x,t) \, \eta(x,t) = 0 \qquad (3.6)$$

by the argument due to Faddeev and Popov, which we describe in detail later. Consequently we have the more sensible expression for the propagator

$$D(x,t;x',t') = \sum_k \int\frac{d\nu}{2\pi} e^{i\nu(t-t')}\frac{i}{\nu^2-\omega(k)^2+i\varepsilon}\psi_k(x)\psi_k^*(x')$$

$$+ \sum_{m\neq 0} \frac{e^{i\nu_m(t-t')}}{T} \cdot \frac{i}{\nu_m^2} \psi_0(x)\psi_0(x') \ . \quad (3.7)$$

Here the first part represents the contribution of
continuum modes and it is the usual propagator of the
canonical collective coordinate method denoted by
$G(x,t;x',t')$. But, now in addition we have the
contribution coming from the zero mode which can be
summed up to read

$$G_0(x,t;x',t') = i[\frac{T}{12} - \frac{1}{2}|t-t'| + \frac{1}{2T}(t-t')^2]\psi_0(x)\psi_0(x') \ .$$
$$(3.8)$$

Now, we show how the subsidiary condition (3.6) is
introduced. Namely, into the path integral expression
for the one soliton sector transition amplitude

$$<p'|e^{-iHT}|p> = \int D\phi \ \Psi_{p'}[\phi(T)]\Psi_p[\phi(0)]e^{iA[\phi]} \quad (3.9)$$

we insert the identity

$$\int da \ \delta(\int dtdx \ f(x,t)\phi(x-a,t))J = 1 \ ,$$

$$J = \int dtdx \ f(x,t) \frac{\partial}{\partial a} \phi(x-a,t) \ . \quad (3.10)$$

Next, changing variables $\phi(x,t) \to \phi(x+a,t)$ which implies

$$\Psi_p[\phi] \to e^{ipa}\Psi_p[\phi], \ \Psi_{p'}[\phi] \to e^{-ip'a}\Psi_{p'}[\phi] \ ,$$

and performing the integration over a we obtain

$$<p'|e^{-iHT}|p> = (2\pi)\delta(p'-p) F(p) ,$$

$$F(p) = \int D \phi \ \Psi_p^* \Psi_p \delta(\int dt dx \ f(x,t)\phi(x,t))Je^{iA[\phi]} . \quad (3.11)$$

Here $f(x,t)$ is arbitrary and observe that the subsidiary
condition in this path integral is not canonical. Now
for the zero momentum case one simply chooses
$f(x,t) = f_{oo}(x,t)$. In general for arbitrary momenta we
have to use the time dependent solution instead of the
static one. Furthermore it is possible to introduce
even a more general translation invariance fixing term
than the δ-function in (3.11). By the well-known
argument one derives for example

$$F(0) = \int D \eta \ J \ exp\{iA[\phi_o+\eta] + \frac{i}{2\alpha} \ (\int dt dx \psi_o(x)\eta(x,t))^2\} ,$$
$$(3.12)$$

and then the zero mode part of the propagator reads

$$G_{o\alpha}(x,t;x',t') = i[\alpha - \frac{1}{2}|t-t'|]\psi_o(x)\psi_o(x') . \quad (3.13)$$

Here α is arbitrary and the dependence on this parameter
cancels out in the calculation.[14] The Feynman rules
are completed with the vertices coming from the higher
than quadratic part of the action and also the Jacobian.

The advantage of the method described above is in
its simplicity. First of all, it appears that the transformation which has been performed to introduce the
collective coordinate does not require any special
treatment in contrast to the case of canonical transformations discussed in Section I. This virtue is

especially useful in developing for example the
perturbation expansion for soliton scattering. Applying
the method of this section to that case one easily
derives a simple set of Feynman rules (now the propagator
contains the contribution of two zero modes), which can
then be used to perform even higher loop calculations.[15]

Furthermore, being noncanonical this method is
also appropriate for treating the zero-frequency modes
which appear in perturbation expansion about instanton
solutions. For example, let us consider the SU(2)
invariant Euclidean Yang-Mills field theory

$$L = - \frac{1}{4} F_{\mu\nu}^a F^{a\mu\nu} \, ,$$

$$F_{\mu\nu}^a = \partial_\mu A_\nu^a - \partial_\nu A_\mu^a + g\varepsilon_{abc} A_\mu^b A_\nu^c \, , \tag{3.14}$$

possessing an exact classical solution

$$A_\mu^a(x;\rho,z) = \frac{2}{g} \frac{\eta_{a\mu\nu}(x^\nu - z^\nu)}{(x-z)^2 + (1+\rho)^2} \, , \tag{3.15}$$

found by Belavin et al.[17] Here z^ν and ρ are arbitrary
constants associated with translation and scale
invariance. The corresponding zero-frequency eigenfunctions now read

$$\phi_\mu^a(x) = \frac{\partial}{\partial\rho} A_\mu^a(x;\rho,z)_{cl}\Big|_{\rho=0,z=0} \, ,$$

$$\phi_\mu^a(x;\nu) = \frac{\partial}{\partial x_\nu} A_\mu^a(x;\rho,z)_{cl}\Big|_{\rho=0,z=0} \, . \tag{3.16}$$

The zero-frequency modes associated with the SU(2)
internal symmetry read

$$\psi_\mu^a(x;\alpha) = \frac{\partial A_\mu^a(x;\Lambda)_{cl}}{\partial \Lambda_\alpha}\bigg|_{\Lambda=0} \quad , \tag{3.17}$$

where we denote

$$A_\mu(x;\Lambda)=U^{-1}(\Lambda)A_\mu(x)U(\Lambda), \quad U(\Lambda)=\exp\{ig\Lambda^\alpha\tau^\alpha/2\},$$

$$A_\mu(x) = A_\mu^a\tau^a/2 \quad . \tag{3.18}$$

The gauge is specified by $C^a(x) = \partial^\mu A_\mu^a(x)$ and the vacuum to vacuum amplitude which we consider then has the following path integral form:

$$_{out}{<}0|0{>}_{in} = \int DA \prod_{x,a}\delta(C^a(x))\Delta_{FP}\exp\{i\int d^4x L(A(x))\} \cdot \tag{3.19}$$

To proceed we need to introduce the notion of invariant SU(2) group integration

$$\int dU(\Lambda) = \int u(g\Lambda) \prod_{\alpha=1}^{3} (gd\Lambda_\alpha) \quad , \tag{3.20}$$

and the following condensed notation for the zero-frequency modes

$$\phi_\mu^a(x) = f_\mu^a(x;1) \quad ,$$

$$\phi_\mu^a(x;\nu)=f^a(x;\nu+1) \quad , \quad \nu = 1,2,3,4,$$

$$\psi_\mu^a(x;\alpha) = f^a(x;\alpha+5), \quad \alpha = 1,2,3, \tag{3.21}$$

To eliminate these modes we insert the identity

$$\int d\rho \prod_{\nu=1}^{4} dz_\nu \int dU(\Lambda) \prod_{i=1}^{8} \delta(\frac{1}{\sqrt{N}_i} \int d^4 x f_\mu^a(x;i) A_\mu^a(x;\rho,z,\Lambda) J=1,$$

$$(3.22)$$

into the path integral. Now changing the field
integration variables $A_\mu^a(x;\rho,z,\Lambda) \rightarrow A_\mu^a(x)$ we obtain

$$_{out}<0|0>_{in} = \int d\rho \prod_{\nu=1}^{4} dz_\nu \int dU(\Lambda) \int DA \prod_{x,a} \delta(C^a(x)) \Delta_{FP}$$

$$J \prod_{i=1}^{8} \delta(\frac{1}{\sqrt{N}_i} \int d^4 x f^a(x;i) A^a(x)) exp\{i\int d^4 x L\} \qquad (3.23)$$

Here the Jacobian J is to be evaluated from (3.22) using
infinitesimal symmetry transformations and is given by

$$J = \frac{1}{g^3} det M_{ij},$$

$$M_{ij} = \frac{1}{\sqrt{N}_i} \int d^4 x f_\mu^a(x;i) \frac{\partial A^a(x;a)}{\partial a_j}\Big|_{\bar{a}=0}, \qquad (3.24)$$

where $\bar{a} \equiv (\rho,z^\nu,\Lambda_\alpha)$. For explicit one-loop calculations
the reader is referred to ref. (16). We also mention
that it is possible to give a canonical description of
vacuum tunnelling[18] in theories with instantons by
employing the canonical collective coordinate method
discussed in Sect. I.

REFERENCES

1. R. Dashen, B. Hasslacher and A. Neveu, Phys. Rev. D 10, 4114 (1974); 10, 4130 (1974).

2. J. L. Gervais and B. Sakita, Phys. Rev. D 11, 2943 (1975) J. L. Gervais, A. Jevicki and B. Sakita, Phys. Rev. D. 12, 1038 (1975).

3. For an entirely different approach see: J. Goldstone and R. Jackiw, Phys. Rev. D 11, 1486 (1975).

4. C. Callan and D. Gross, Nucl. Phys. B 93, 29 (1975).

5. L. D. Faddeev, V. E. Korepin and P. P. Kulish, JETP Lett. 21, 260 (1975).

6. J. L. Gervais, A. Jevicki and B. Sakita, Phys. Reports 23, 237 (1976); see also A. Hosoya and K. Kikkawa Nucl. Phys. B 101, 271 (1976).

7. J. L. Gervais and A. Jevicki, Nucl. Phys. B 110, 93 (1976).

8. E. Tomboulis, Phys. Rev. D 12, 1678 (1975), N. Christ and T. D. Lee, Phys. Rev. D 12, 1606 (1975).

9. J. L. Gervais and A. Jevicki, Nucl. Phys. B 110, 113 (1976); for the discussion of N-soliton scattering see A. Jaeckel, Ecole Normal preprint (1976).

10. R. Jackiw and G. Woo, Phys. Rev. D 12, 1643 (1975).

11. R. Dashen, B. Hasslacher and A. Neveu, Phys. Rev. D 11, 3424 (1975).

12. V. C. Korepin, JETP Lett. 23, 224 (1976); for the "complex time" method see D. McLaughlin, J. Math. Phys. 13, 1099 (1972).

13. A. Jevicki, Nucl. Phys. B 117, 365 (1976).

14. L. D. Faddeev and V. E. Korepin, Phys. Lett. 63B, 435 (1976); similar propagators appear also in the classical perturbation expansion for solitons developed by J. Keener and D. McLaughlin (preprint 1976).

15. The only higher loop calculation performed so far
     is the two-loop correction to soliton mass in the
     Sine-Gordon theory; H. deVega, Nucl. Phys. B $\underline{115}$,
     411 (1975).

16. G. 't Hooft, Phys. Rev. D $\underline{14}$, 3432 (1976);
     A. Polyakov Nordita preprint (1976); E. Brezin,
     J. C. Le Guillou and J. Zinn-Justin, Saclay
     preprint (1976).

17. A. A. Belavin, A. M. Polyakov, A. S. Schwartz and
     Yu. S. Tyupkin, Phys. Lett. $\underline{59B}$, 85 (1975).

18. J. L. Gervais and B. Sakita, CCNY preprint (1976).

NONTOPOLOGICAL SOLITONS[*]

R. Friedberg

Barnard College and Columbia University

New York, New York 10027

I shall be describing mainly the work carried out over the last few years by T.D. Lee and a number of collaborators, including G.C. Wick, N.H. Christ, M. Margulies, A. Sirlin, and myself.

This work is guided by the premise that we already know enough basic physics to explain the extended structure of hadrons in terms of a conventional local field theory. On the evidence of (e,N) and (ν,N) reactions, we expect local field theory to hold good down to $\sim 10^{-16}$ cm. Therefore the hadrons extending over 10^{-13} should be bound states of a more elementary object. In view of the successes of group theory, we expect the elementary objects to be colored quarks, on the basis of deep in-elastic e-scattering, the quarks should have no internal structure larger than 10^{-16} cm and can therefore be regarded as the elementary excitations of a Fermi field.

---

[*]This research was supported in part by the U.S. Energy Research and Development Administration.

If everything is this easy, what stands in the way?
Since the quarks have not been seen in isolation, they
must be supposed either extremely massive or unstable;
in the latter case they are at least as massive as the
proton. Thus a three-quark nucleon or two-quark meson
must be strongly bound, with relativistic effects
prominent. It is the difficulty of handling bound
states in relativistic quantum field theory that has
hampered development of a simple conventional model of
hadrons.

The study of solitons offers a new departure for
this problem, by suggesting a different 0th-order
approximation from the one most often studied in the
past. The principle is illustrated by the case of a
quantum mechanical particle moving in two dimensions
under the centrally symmetric and quasi-harmonic
potential $V = \frac{1}{2} r^2 [(1 - gr)^2 + \Delta^2 g^2]$. In Fig. 1 we
graph the effective radial potential $V_\ell = V + \ell^2/2r^2$
against r^2 for several values of ℓ. In general there
may be two minima, one at small r^2 and one at large r^2.

When ℓ exceeds a critical value ℓ_c, the absolute
minimum of V_ℓ lies at the larger r^2. The ground state
wave function is then concentrated at this radius,
which is very large if g is small. But the 0th-order
of perturbation theory, which consists of taking g = 0,
gives a ground state concentrated at the much smaller
r^2 (minimum of dotted curve) and is therefore a very
poor approximation. On the other hand, a very good
approximation is obtained by using classical theory with
the exact g. This gives a circular trajectory at the
right radius, on which quantum fluctuations may be super-
imposed as corrections.

We note that as $g \to 0+$ at fixed Δ, $\ell_c \to \Delta^2$ and so

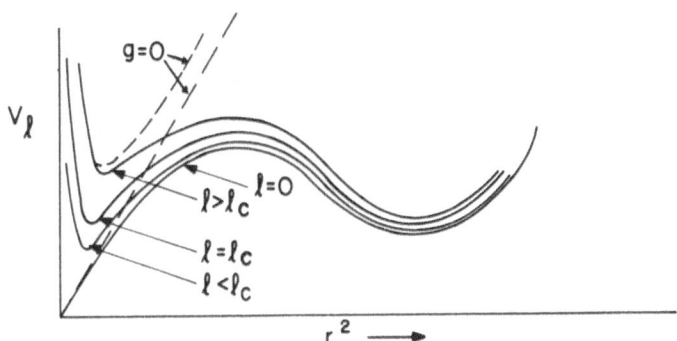

Figure 1. $V_\ell = V + \dfrac{\ell^2}{2r^2}$ against r^2 in simple 1st quantized example.

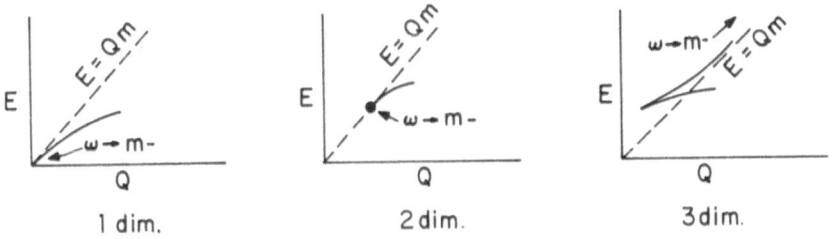

Figure 2. E vs Q in upper range of ω.

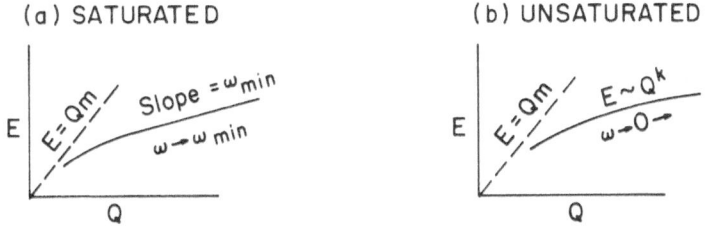

Figure 3. E vs Q in lower range of ω.

if we make $\Delta^2 < 1$ the situation described above occurs
even for small quantum numbers.

We postulate that a similar situation arises from
the interaction between quarks and gluons: that even
for low quark numbers, the quantum bound state is better
approximated by a steady solution of the classical field
equations than by the free-particle state of the quantum
theory without interactions. The stable bound state of
the classical theory is what we call a soliton.

Our solitons are stable only in isolation; they are
what traditional usage has called "solitary waves".
The traditional meaning of "soliton" is that of a configuration stable even against collisions; we propose
to call such configurations "indestructible solitons",
but do not anticipate any role for them in particle
physics. We prefer to use the simpler term for the
more applicable concept.

Our investigations are confined to nontopological
solitons. This does not rule out a theory with degenerate vacua; it does mean that in such a theory we
assume a single choice of vacuum at large distances,
independent of direction.

Before these ideas are applied to a model of
hadrons, two questions should be investigated. First,
one wishes to know how widespread is the occurrence of
solitons (i.e., stable, localized solutions of field
equations) in classical field theories, and what the
characteristic features are. Second, one may ask
whether the classical soliton is indeed a good approximation to the exact quantum ground state in those few
limiting cases where the latter is known.

The first principle in looking for nontopological
solitons in a classical theory of commuting fields is

that stability cannot be achieved with completely time-independent solutions. This is because the energy could always be decreased by shrinking the volume of the soliton while maintaining the peak field amplitude. To achieve stable solitons, we introduce a conserved "charge" associated with a particle whose free mass we denote by m.

If the "charged" particle is a boson, we may call the corresponding field Φ. We pick a Lagrangian that depends on Φ only through $\Phi^{\dagger}\Phi$, and look for solutions in which $\Phi = e^{-i\omega t}\phi$, where ϕ and all other fields are time-independent. (More generally, we have $\partial\Phi/\partial t = \omega(G\Phi)$ where G is a generator of some group on Φ that leaves the Lagrangian unchanged.) The "charge" Q is then defined by $Q = \frac{\partial L}{\partial\omega}$. Q is automatically conserved, and in the quantized theory it must be an integer. It plays the same role here as the angular momentum ℓ in the first-quantized illustration given previously. The energy E may now be minimized for fixed Q, since shrinking the volume would necessitate intensifying the field Φ. After minimization we have $\frac{dE}{dQ} = \omega$ when Q is taken as a continuous variable.

If the "charged" particle is a fermion, we denote its field by ψ and simply let $N = \int \psi^{\dagger}\psi$ be the conserved fermion number. N then plays the part of Q, and ε (the single-fermion energy in the presence of the soliton fields) plays that of ω. After arriving at the field equations we can formally allow N to vary continuously; then we have $\frac{dE}{dN} = \varepsilon$.

The principal information we seek is the graph of E vs. Q (or N). This curve is to be compared with the straight line $E = Qm$, which represents the energy of an aggregate of free particles bearing the same total charge.

For stable solitons to exist, the curve must lie below
the straight line.

We recall that the slope of this curve is ω, and
it is easily seen from the field equations at large
distances that $\omega \lesssim m$. The behavior of E vs Q when
$\xi \equiv \sqrt{m^2 - \omega^2} \ll 1$ depends only on the dimensionality of
space and not on the particular model. It is generally
true in this limit that $\phi \sim \xi$ and the radius $\sim \xi^{-1}$;
hence $Q \sim \xi^2 \xi^{-d}$ in d space dimensions. The resulting
curves for 1, 2 and 3 space dimensions are shown in
Fig. 2; the low -ω part of the curve is omitted. In 1
or 2 dimensions, the curve can only lie under the
straight line. In 3 dimensions the curve lies under the
line when Q exceeds some value Q_S. In more detailed
investigations we have found that the parameters can be
chosen to make $Q_S < 1$ if desired.

As ω decreases to its lower limit ω_{min}, both the
radius and Q become large. Here we encounter two kinds
of behavior, depending on the specific model. In some
cases $\omega_{min} > 0$; the solution is then <u>saturated</u> in the
sense that the amplitude of the charged field is roughly
constant throughout the interior and does not increase
as the radius is made large. The energy then grows as
$E \sim Q\omega_{min}$. (See Fig. 3a.) In other cases $\omega_{min} = 0$.
This leads to an <u>unsaturated</u> solution in which the
charged field grows without limit as the radius increases.
Generally we then have $E \sim Q^k$, where $0 < k < 1$. (See
Fig. 3b.)

In 1 dimension solitons can arise from a single
field. We put

$$L = \dot{\phi}^\dagger \dot{\phi} - \phi'^\dagger \phi' - U(\phi^\dagger \phi) + \text{counterterms},$$

$$m^2 = \left| \frac{dU(\phi^2)}{d\phi^2} \right|_{\phi=0} ,$$

$$\Phi = e^{-i\omega t}\phi(x) \; , \; \phi'^2 = U - \omega^2\phi^2 \; ,$$

$$Q = \frac{\partial}{\partial\omega} \int L \, dx = 2\omega \int \phi^2 dx \; , \; E = \int (\phi'^2 + U + \omega^2\phi^2) dx \; .$$

At the turning point $\phi' = 0$, we must have $U = \omega^2\phi^2$. Therefore the curve $U = U(\phi^2)$ must intersect the straight line $U = \omega^2\phi^2$. Since $\omega^2 < m^2 = U'(0)$, the curve $U(\phi^2)$ must have a portion concave down. Several soluble examples are known:

$$\text{(A)} \quad U = m^2\phi^2 [1 - 2g^2\phi^2 + g^4(1 + \epsilon^2)\phi^4] . \overset{(1)}{\underset{\text{(Fig.4a)}}{}}$$

Here the curve cannot intersect the straight line unless $\omega > \omega_{min}$ where ω^2_{min} is the slope of the tangent shown in Fig. (4ai). Hence the behavior is saturated at large Q (Fig. 4aii).

$$\text{(B)} \quad U = \frac{1}{4}\frac{m^2}{g^2} (\sqrt{1 + 8g^2\phi^2} - 1) . ^{(1)} \qquad \text{(Fig.4b)}$$

Here ω can be as close to 0 as desired (Fig. 4bi). The behavior at large Q is unsaturated with $E \sim Q^{3/4}$ (Fig.4bii).

$$\text{(C)} \quad U = m^2\phi^2 \, e^{-2g^2\phi^2} \qquad . \qquad \text{(Fig.4c)}$$

Another unsaturated case, but here U decreases for large ϕ (Fig. 4ci) and hence the energy is concentrated at the boundary of the soliton. When $\omega \to 0$, the radius grows large but both Q and E remain finite (Fig. 4cii)-

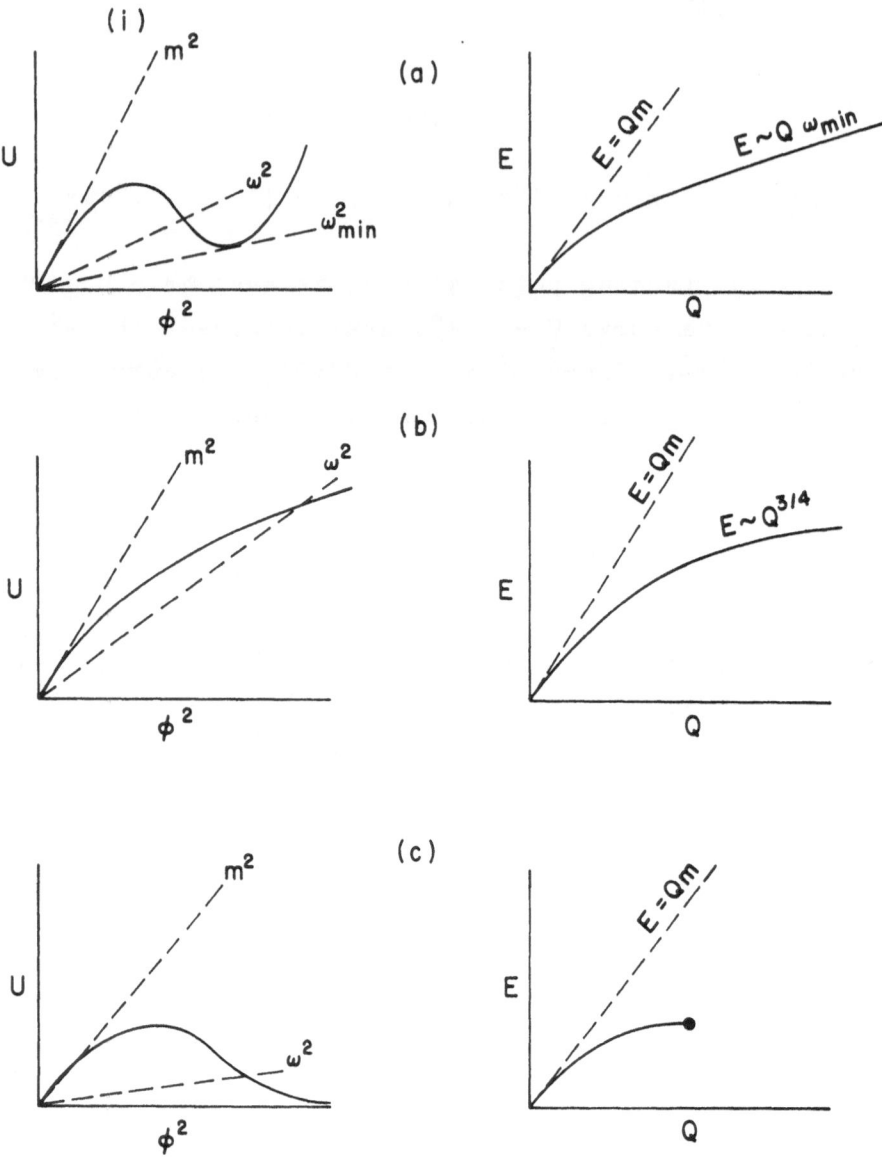

Figure 4.   Solitons in 1 dimension.

an exception possible only in 1 dimension.

In 3 dimensions we restrict ourselves to 4th-order polynomials for the sake of renormalization. This rules out positive definite functions $U(\phi^2)$ with a downward concavity as in the previous examples. But solitons are still possible with two fields.

We have studied two unsaturated examples:

(D) $\Phi \rightarrow e^{-i\omega t} \phi$, "charged" scalar boson,

χ , neutral scalar boson,

$$L = - \partial_\mu \phi^\dagger \partial_\mu \phi - \frac{1}{2}(\partial_\mu \chi)^2 - f^2 \chi^2 \phi^\dagger \phi - U(\chi) + \text{counter-terms,} \qquad (2)$$

where U has a minimum at $\chi = \chi_{vac} = m/f$. We then have

$$Q = 2\omega \int \phi^2 \, d^3r \quad ,$$

$$E = \int [(\nabla\phi)^2 + \frac{1}{2}(\nabla\chi)^2 + f^2\chi^2\phi^2 + U(\chi) + \omega^2\phi^2]d^3r$$

and by minimizing E at fixed Q we obtain coupled non-linear 2nd-order differential equations in ϕ and χ. The E-Q curve (Fig.5) has the spike characteristic of 3 dimensions. The stability threshold Q_S can be made <1 by suitably choosing the ratio of free masses. For large Q we have $E \sim Q^{3/4}$ unless U has a special form making the vacuum degenerate with a second minimum at $\chi = 0$. Then the energy is surface-dominated and $E \sim Q^{2/3}$ at large Q.

(E) ψ , "charged" spinor fermion

σ , neutral scalar boson.[3], [4]

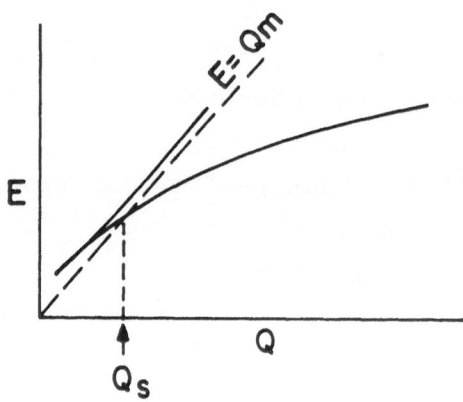

Figure 5.   Unsaturated 3-dimensional examples.

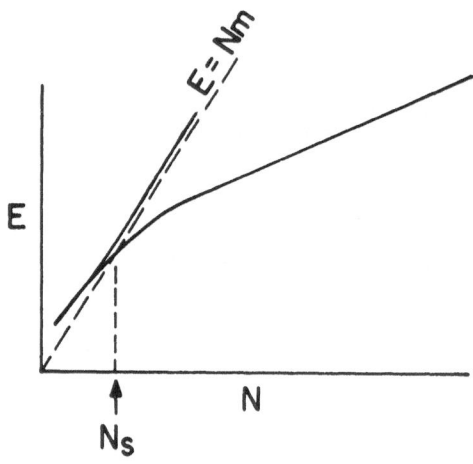

Figure 6.   Saturated 3-dimensional examples.

Here we adopt a slightly different formal approach.
We start from the Hamiltonian density

$$H = H_\sigma + \sum_{k=1}^{n} \psi^{k\dagger} H_D \psi^k + \text{counterterms} ,$$

where

$$H_\sigma = \frac{1}{2} \pi^2 + \frac{1}{2}(\nabla\sigma)^2 + U(\sigma) , \quad \pi \text{ conjugate to } \sigma ,$$

$$H_D = \frac{1}{i} \vec{\alpha}.\vec{\nabla} + \beta(m + g\sigma)$$

and k is a "color" index which permits us to put up to
n quarks in the same orbital. We then approximate the
whole Hamiltonian by the "quasiclassical" Hamiltonian

$$H_{qcl} = \int[\frac{1}{2}(\nabla\sigma_c)^2 + U(\sigma_c)]d^3r + \sum_k^n \sum_\ell^\infty \epsilon_\ell (a_\ell^{k\dagger} a_\ell^k + b_\ell^{k\dagger} b_\ell^k) ,$$

where the index ℓ ranges over the eigenfunctions of H_D
with $\sigma = \sigma_c$, and ϵ_ℓ, a_ℓ, b_ℓ are the corresponding
eigenvalues and annihilation operators for fermion and
antifermion. Here σ_c is a c-number function, but the
fermion field is still quantized (the a and b operators
obey anticommutation rules).

The fermion number $N = \sum_k \sum_\ell (a_\ell^{k\dagger} a_\ell^k - b_\ell^{k\dagger} b_\ell^k)$ is
conserved. We give it some fixed value $\leqslant$ n; then the
lowest eigenvalue of H_{qcl} for a given σ_c is

$$E = N\epsilon + \int[\frac{1}{2}(\nabla\sigma_c)^2 + U(\sigma_c)]d^3r ,$$

where ϵ is the lowest positive solution of

$$\epsilon\psi_c = [\frac{1}{i} \vec{\alpha}.\vec{\nabla} + \beta(m+g\sigma_c)]\psi_c .$$

We minimize E with respect to σ_c and obtain

$$\nabla^2 \sigma_c - U'(\sigma_c) = g \, N\psi_c^\dagger \beta\psi_c \quad .$$

The qualitative properties of this soliton are very similar to those of the previous example using only boson fields. In Fig. 5, simply replace Q by N.

We have also studied saturated examples in 3 dimensions.

(F) Same as (E) but with N>>n .

This case has been studied by Lee and Wick[5] and by Lee and Margulies[6] as a model of nuclear matter. Here the nucleon density remains constant as N grows large, because of the Pauli principle. Both N and E, in that limit, are simply proportional to volume. (See Fig. 6.)

(G) $\vec{V}_\mu$, Yang-Mills vector field

ϕ , Higgs scalar, isospinor.[7]

The Lagrangian has $SU_2 \otimes SU_2$ symmetry, with only the first group gauged. After the Higgs symmetry breaking, there remains one global SU_2 group under which ϕ is scalar and $\vec{V}_\mu$ is vector. Writing $\vec{V}_\mu = (\vec{V}_k, V_0)$, the time-dependence of the soliton is given by

$$\frac{\partial}{\partial t} \vec{V}_k = \vec{\omega} \times \vec{V}_k \ , \quad \frac{\partial}{\partial t} \vec{V}_0 = 0 \ , \quad \frac{\partial}{\partial t} \phi = 0$$

The "charged" field is thus $\vec{V}_k$, and the scalar "glue" is provided by ϕ. The 4th component $\vec{V}_0$ injects a new element: it mediates a self-repulsion of $\vec{V}_k$ which leads to saturation as in Fig. 6.

I now turn to the second question: how good is the

quasiclassical approximation? Let us define the binding
energy $E_B = NE(1) - E(N)$ and the ratio $R = E_B(\text{exact})/$
$E_B(\text{qcl})$ where $E_B(\text{qcl})$ is the value given by our soliton
approximations.

For relativistic theories, exact results are known
only when $g^2 \ll 1$. When $N \gg 1$, in all three cases (D),
(E), and (F), we find $R = 1$. For $N = 2$, the exact result can be found by solving the Coulomb problem in
center-of-mass coordinates, and $R = .768$.[3]

For nonrelativistic "charged" particles we find
also $R = 1$ in the limit $g^2 \gg 1$ for any N,[3] as well as
for $N \gg 1$ and any g^2. Again, when $N = 2$ and $g^2 \ll 1$, we
have $R = .768$.

Either large g^2 or large N tends to make the
quantum corrections negligible, and even in the expected "worst" case $N = 2$, $g^2 \ll 1$, the approximation is
not so bad. Thus we are encouraged to think that in a
model of hadrons ($N = 2$ or 3, $g^2 \gg 1$ but with relativistic
binding) a quasiclassical treatment will be fairly
accurate.

The hadron model we have in mind involves an
$SU_3 \otimes SU_3$ half-gauged symmetry leading to global SU_3
after Higgs breaking, and four fields as shown in Table
1.

The ψ field represents the quark; its 3-fold symmetry is color. We have not yet injected flavor. The
σ field is the scalar "glue". The vector field V_μ, as
suggested by Nambu, mediates a repulsive interaction
raising the energy of color nonsinglets; that explains
why they are not seen. The Higgs field ϕ is then needed
to give the $\vec{V}_\mu$ a mass in the usual way. All the free
masses are assumed large, so that the observed spectrum
is merely that of the color-singlet bound states.

Table 1						
Field	Spin	SU_3^{gauged}	X	SU_3	$\rightarrow$	SU_3
ψ	$\frac{1}{2}$	3	X	1		3
σ	0	1	X	1		1
V_μ	1	8	X	1		8
ϕ	0	3	X	3		1,1,8

In the soliton approximation, a color-singlet state has $\vec{V}_\mu = 0$ everywhere. The ϕ field does not interact directly with ψ ($\phi^2\psi^\dagger\psi$ is not renormalizable, while $\phi\psi^\dagger\psi$ breaks the pre-Higgs symmetry) and its effect on σ can presumably be incorporated into $U(\sigma)$. Thus the bound-state problem simplifies to that of example (E) above, with just ψ and σ, and N = 2 or 3. It is the same model that led to the SLAC bag,[8] and with different parameters was used by Creutz and Soh[9] to give the MIT bag[10] in a certain limit.

There are four main styles of bag, characterized by (a) either MIT- or SLAC- type quark wave function (see Fig. 7) and (b) either volume- or surface- dominated σ energy. See Table 2.

To obtain I or II from the field model, one sets (with Creutz and Soh) $m + g\sigma_0 = 0$ where $U(\sigma)$ has its second minimum at $\sigma = \sigma_0 < 0$. To obtain III or IV, one sets $m + g\sigma_0 < 0$. For I or III, one must have $U(\sigma_0) > U(0)$, for II or IV $U(\sigma_0) = U(0)$. The published MIT bag is I, the SLAC bag is IV.

If $m + g\sigma = 0$, before passing to the MIT limit, one

Table 2		
I	MIT - type quarks, E_σ ~ volume	
II	MIT - type quarks, E_σ ~ surface	
III	SLAC - type quarks, E_σ ~ volume	
IV	SLAC - type quarks, E_σ ~ surface	

can also obtain hybrid quark wave functions as in Fig. 8. These arise from a tendency for σ to overshoot the value σ_0 so as to give the quark the lower SLAC-type energy. This effect can be seen in the curves of Rafelski for the case $U \sim \sigma^2$. If an SU_3 flavor group with unbroken symmetry is introduced, the model gives 36 degenerate mesons, and 56 baryons.

Some preliminary numerical comparisons are shown in Table 3. (Some of these numbers are already known.) Letting

$$m = \text{free quark mass}$$

$$\mu = \text{free gluon mass}$$

$$m_B = \text{average } \textcircled{56} \text{ baryons mass}$$

$$= 1.40 m_N$$

$$B_- = U(\sigma_0) \to B \text{ of MIT bag}$$

$$B_+ = \max U(\sigma) \text{ for } 0 > \sigma > \sigma_0$$

$$\xi = (\frac{m_B}{m})^2 \quad , \quad \eta = \frac{m_B}{\mu} \quad , \quad \zeta = \frac{B_-}{B_+} \quad ,$$

we assume ξ, η, ζ all $\ll 1$ and obtain case

I   when $\xi \gg \eta \gg \zeta$

II   when $\xi \gg \zeta \gg \eta$

III   when $\xi \gg (\xi , \eta)$

IV   when $\eta \gg (\xi , \zeta)$.

Table 3					
Quantity	Expt.	I	II	III	IV
$m_N \langle r^2 \rangle^{1/2}$	3.86	4.25	4.78	2.86	3.21
$2m_N \, \mu_p$	2.79	2.36	2.66	1.90	2.14
μ_n / μ_p	-.685	$\longleftarrow$		$-\,2/3$	$\longrightarrow$
g_A / g_V	1.2	$\longleftarrow$ 1.09 $\longrightarrow$		$\longleftarrow$ 5/9 $\longrightarrow$	

The main result of the comparison is that all the numbers are in the general vicinity of experiment, although the MIT-type models seem somewhat closer. We mean to investigate the hybrid models as well.

Further questions: How do we calculate form factors from such a model? How should we introduce flavor breaking? Does PCAC come in at all?

To conclude: This approach seems the most conservative of all avenues to the study of hadron structure. At the same time it incorporates most of the features definitely indicated by the research of the last fifteen years. Since a preliminary comparison with experiment leads to no glaring contradiction, the approach should be pursued until it leads either to a significant success or to an informative failure.

REFERENCES

1. T.D. Lee, at the Symposium on Frontier Problems in High Energy Physics, Pisa, 1976 [Columbia University preprint No. CO-2271-76].

2. R. Friedberg, T.D. Lee and A. Sirlin, Phys. Rev. D13, 2739 (1976).

3. R. Friedberg and T.D. Lee, Phys. Rev. D15, (1977).

4. A. Nishimura, University of Tokyo preprint UT-275.

5. T.D. Lee and G.C. Wick, Phys. Rev. D9, 2291 (1974).

6. T.D. Lee and M. Margulies, Phys. Rev. D11, 1591 (1975).

7. R. Friedberg, T.D. Lee and A. Sirlin, Nucl. Phys. B115, 1, 32 (1976).

8. W.A. Bardeen, M.S. Chanowitz, S.D. Drell, M. Weinstein and T.M. Yan, Phys. Rev. D11, 1094 (1975).

9. M. Creutz and K.S. Soh, Phys. Rev. D12, 443 (1975).

10. A. Chodos, R.L. Jaffe, K. Johnson, C.B. Thorn and V.F. Weisskopf, Phys. Rev. D9, 3471 (1974).

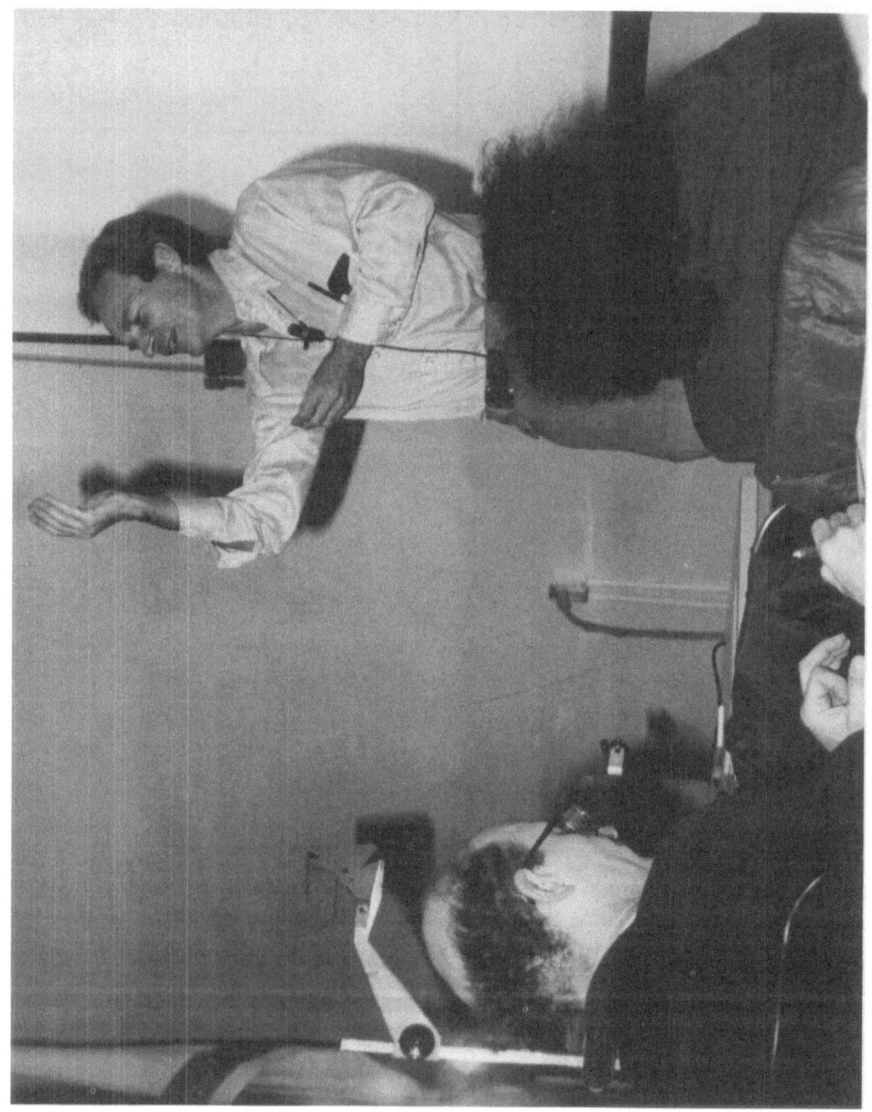

RICHARD FEYNMAN "MAKING A POINT"

VACUUM BUBBLE INSTANTONS*

P. H. Frampton

Department of Physics

University of California, Los Angeles, CA. 90024

In these remarks, we discuss the existence of certain instanton solutions[1] occurring in both non-gauge and gauge field theories. They are not associated with the mapping of a non-Abelian gauge group on S_3, have no conserved topological charge and are hence unstable. They provide probably the principal decay mechanism of an unstable vacuum and standard instanton methods[2-4] can be exploited to estimate such a vacuum decay lifetime.

These vacuum-bubble instantons may be of interest for the following reasons:

(1) they illuminate the limitations of a well-known mathematical theorem that time-independent solitons and instantons are impossible in a four-dimensional field theory with only scalar fields;

(2) they seem to be of pedagogic value in clarifying the assumptions made in the

---
*Work supported in part by the National Science Foundation.

usual identification of the vacuum state
in a quantum field theory;

(3)   they lead to the idea that the physical
       vacuum <u>might</u> be weakly unstable. That this
       apparently bizarre possibility may be enter-
       tained seriously is due to the exceedingly
       small probability for instanton processes in
       a weak interaction theory; it could have
       interesting consequences.

Suppose   that the expectation value of the energy
density, i.e. the effective potential, has a shape like:

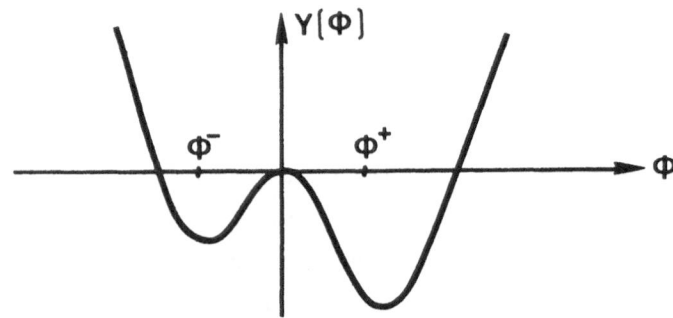

To find the decay probability of the unstable
vacuum at $\phi = \phi^-$ we consider the Euclidean instanton
solution of the classical field equations. This is
interpreted as quantum tunnelling through the potential
barrier to form a vacuum bubble with null four-momentum[5]
and whose interior is built on the lower-energy vacuum
$\phi = \phi^+$ (c.f. bubble formation in a superheated liquid)
Defining a 4-dimensional radius ρ by

$$\rho^2 = \underline{x}^2 + T^2 \qquad (T = it), \qquad (1)$$

the field equation, for spherical symmetry, is

$$\left(\frac{\partial^2}{\partial\rho^2} + \frac{3}{\rho}\frac{\partial}{\partial\rho}\right)\phi = \frac{d}{d\phi}V(\phi). \qquad (2)$$

The instanton with spherical symmetry has the minimum
classical action and hence the greatest tunnelling
probability.[6] Assuming that the surface thickness is
very small compared to the radius the classical action
for radius R is

$$A = -\frac{1}{2} \pi^2 R^4 \varepsilon + 2\pi^2 R^3 S \quad . \tag{3}$$

This is stationary for $R_m = 3S/\varepsilon$ with

$$A_m = \frac{27}{2} \pi^2 S^4/\varepsilon^3 \quad . \tag{4}$$

In these expressions ε is the volume energy density
given by

$$\varepsilon = V(\phi_-) - V(\phi_+) \tag{5}$$

and S is the action per unit hypersurface or equivalently
the surface energy density. In the leading (WKB)
approximation it is given by

$$S = \int_{\phi_-}^{\phi'} d\phi \sqrt{2(V(\phi) - V(\phi_-))} \quad . \tag{6}$$

The probable number of bubbles formed in a space-time
volume V is then (assuming A_m is very large)

$$N = \frac{V}{L^4} \exp(-A_m) \quad , \tag{7}$$

in which L is the typical length scale of the process.
The precise value of L can be calculated as a quantum
correction to the exponent. Such a calculation is being
done by Callan and Coleman;[7] unfortunately neither is
here to report on their progress. It will be sufficiently
accurate for the present to set $L = R_m$ since all that we
miss is an overall numerical factor together with possible

powers of π.

The vacuum-bubble instantons can occur in purely scalar theories. With a <u>stable</u> vacuum there is a well-known theorem that in 4 space-time dimensions no Euclidean instantons are possible for only scalar fields. The theorem is based on a scaling argument: consider the Lagrangian (in 1 time + 4 space dimensions)

$$L = \frac{1}{2} \partial_\mu \phi \, \partial_\mu \phi - V(\phi) \tag{8}$$

and the associated energy integrals

$$V = \int d^4x \, V(\phi) \, , \tag{9}$$

$$T = \frac{1}{2} \int d^4x \, (\underline{\nabla}\phi)^2 \geq 0 \, . \tag{10}$$

Suppose that $\phi(x)$ is a time-independent Euclidean instanton solution, and then consider the trial functions $\phi(\lambda x)$. The energy $E = T + V$ should then be stationary at $\lambda = 1$,

$$E(\lambda) = \lambda^{-2}T + \lambda^{-4}V, \tag{11}$$

$$E'(1) = -2T - 4V = 0 \, . \tag{12}$$

In the normal case, $V(\phi) \geq 0$ and the only time-independent solution is the vacuum where ϕ = constant and $T = V = 0$. In the present case of an unstable vacuum, V can be negative and the theorem is invalid. Put differently, the instanton is possible for scalar fields only if the vacuum is not the lowest energy state.

We now give two examples:

(1)   Skewed Goldstone Model
      Consider the potential

$$V(\phi) = \frac{1}{2} m^2 \phi^2 - \frac{\delta}{3} \phi^3 + \frac{\lambda}{4} \phi^4 \qquad (13)$$

with

$$0 < \frac{\delta}{|m|} \ll \lambda \; .$$

For $\delta = 0$ the vacuum expectation value would be $\pm v$ but for $\delta > 0$ this is shifted to

$$\phi^{\pm} = \pm v + \delta/2\lambda \quad , \quad v = \sqrt{-m^2/\lambda} \; . \qquad (14)$$

One then computes immediately

$$\varepsilon = \frac{2}{3} \delta v^3 \quad , \qquad (15)$$

$$S = 2\sqrt{2} \int_0^v d\phi \sqrt{\frac{1}{2} m^2 \phi^2 + \frac{1}{4} \lambda(\phi^4 + v^4)} \qquad (16)$$

$$= \frac{2}{3} \sqrt{2\lambda} \; v^3 \quad , \qquad (17)$$

$$R_m = 3\sqrt{2\lambda}/\delta \quad , \qquad (18)$$

and hence

$$N = \frac{V \delta^4}{324 \lambda^2} \exp(-36\pi^2 \lambda^2 \, v^3/\delta^3) \; . \qquad (19)$$

Now when $\delta = 0$ the two vacua $\phi = \pm v$ are degenerate and either may be chosen with identical physical predictions. When $\delta > 0$, then so long as $V < \infty$ we may always choose δ such that N is arbitrary small (hence the ϕ^- vacuum becomes effectively stable). Thus there is no abrupt change between $\delta = 0$ and $\delta \neq 0$. When δ

is sufficiently small we may still choose either
vacuum as the physical one. Of course, the physical
predictions will now differ at order δ.

(2) Weinberg-Salam Model of Weak Interactions

    In a weak interaction theory, instanton effects,
being ~$\exp(-1/g^2)$, are exceedingly small. For example
't Hooft[3] considered instanton-induced baryon-number
nonconservation, and found a probability so small that
such a process is unlikely to have occurred even once
in the universe. Thus, as far as violation of
conservation laws is concerned, weak-interaction
instantons are completely negligible in practice.

    In strong interaction Yang-Mills theories,
instantons may be significant and suggestions have
been made to relate them to the quark confinement problem. This is so far only speculative; as a contrary
viewpoint, Salam[9] has recently suggested that spin-two
fundamental fields may be essential to solve this
problem.

    From the point of view of vacuum instability, the
exceeding smallness of instanton processes in weak
interactions is interesting for it implies that a
metastable vacuum can easily have an almost infinite
lifetime.

    In the W-S model, the effective potential with
one-loop corrections reads:

$$V[\phi] = \frac{1}{2} \mu^2 \phi^2 - \lambda \phi^4 + B\phi^4 \ln(\phi^2/v^2) , \quad (20)$$

with $v \simeq 248$ GeV and $B \simeq 10^{-4}$. The Higgs mass M_H^2 and
the volume energy density ε are the important
parameters to study here; they are given respectively
by

$$m_H^2 = V''(v) = 4Bv^2 - 8v^2(\lambda-B) \ , \qquad (21)$$

$$\varepsilon = V(v) - V(0) = (\lambda-B) \ v^4 \ . \qquad (22)$$

If $m_H^2 > 4Bv^2 \equiv m_0^2$, then $\varepsilon < 0$ and the asymmetric vacuum is stable. If $m_H^2 < m_0^2$ the asymmetric vacuum can decay but its lifetime is vast for an appreciable range of m_H^2 .

Defining the ratio

$$r = \frac{2(\lambda-B)}{B} = \frac{m_0^2 - m_H^2}{m_0^2} \qquad (23)$$

one immediately finds

$$R_m = 24\sqrt{2B} \ v \ I(r)/(m_0^2 - m_H^2) \ , \qquad (24)$$

$$S = v^3 \sqrt{2B} \ I(r) \ ,$$

$$N = \frac{V}{R_m^4} \exp(- \frac{\pi^2}{2} R_m^3 S) \ . \qquad (25)$$

In the formulas the integral is implied

$$I(r) = \frac{1}{2} \int_{\bar{x}}^1 dx \sqrt{(1+r) - (1+\tfrac{1}{2}r) \ x + x \ \ell n x - r/2x} \ ,$$

where $x = \bar{x}$ is the zero of the argument of the square root.

Putting in numbers one finds that for $(m_0 - m_H) \sim 1$ GeV the number of bubbles, N, is absolutely negligible for any conceivable $V < \infty$. Thus the physical vacuum could have this degree of instability, and the correct identification of the vacuum is ambiguous.

Finally some concluding remarks: A--highly
speculative. B--serious.

A. Turning briefly to the concept of a doomsday
machine, one would need a future E. O. Lawrence to
invent colliding beams with energies presumably in
excess of the cosmic rays. A catastrophe would be
instigated by a bubble formed (with presently unknown
cross-section) at radius ~0.1 fermi and expanding to
radius 300,000 Km after one second; within the bubble,
electrons might be confined, etc. etc. Mercifully no
observer will be able to see the approach of this
greatest disaster consistent with physical laws--the
boiling of the vacuum.

B. The following points deserve more study:

(1) Does an exact solution for the instanton
vindicate the thin-surface approximation?

(2) Is the estimate of the length scale
justified by systematic calculation of the quantum
corrections?

(3) Can a ϕ^3 theory, usually dismissed for
its unbounded spectrum provide a viable alternative to
the popular ϕ^4 Higgs-Kibble models for weak interactions?
If so, are there any advantages?

REFERENCES

1. P. H. Frampton, Phys. Rev. Letters $\underline{37}$, 1378 (1976). This calculation has recently been justified by S. Coleman (Harvard University preprint).

2. A. A. Belavin et al., Phys. Letters $\underline{59B}$, 85 (1975).

3. G. 't Hooft, Phys. Rev. Letters $\underline{37}$, 8 (1976).

4. R. Jackiw and C. Rebbi, Phys. Rev. Letters $\underline{37}$, 172 (1976).

5. Since the bubble is created with $P_\mu = 0$, its momentum is the same in all Lorentz frames; to put the same statement differently, the bubble expanding at the speed of light appears spherical to all observers. Thus no momentum can be associated with the bubble and no (divergent) phase space integral is necessary here. This point was first made clear to me in the discussion following my seminar at SLAC, November 1976.

6. It is here assumed that non-spherical instantons, necessarily having larger classical action, may be safely neglected.

7. C. Callan and S. Coleman, preprint in preparation.

8. D. Gross, these proceedings.

9. A. Salam, International Symposium "Five Decades of Weak Interactions," City College N.Y. January 1977. Salam suggests that because of the difficulty in establishing confinement for quarks exchanging spin-one gluons unification with gravity, in particular spin-two gluons, might be necessary to ensure confinement through a black-hole-type mechanism.

NONLINEAR DEEP WATER WAVES: A PHYSICAL TESTING GROUND FOR SOLITONS AND RECURRENCE

Henry C. Yuen and Bruce M. Lake

(Presented by Henry C. Yuen)

Fluid Mechanics Department
TRW Defense and Space Systems Group
Redondo Beach, California 90278

INTRODUCTION

The physical problem that we are concerned with is the evolution of nonlinear deep water waves. We shall assume that the waves under consideration are long enough so that the effect of surface tension is small compared to gravity (hence, they are sometimes referred to as "gravity waves"). This requires that the waves be much longer than 10 cm, at which wavelength surface tension and gravitational effects are comparable to each other. We shall assume that the water motion is incompressible, irrotational and inviscid. These are the usual approximations that go with the study of water waves, and are expected to be satisfied under normal circumstances. We shall also assume that the depth of the water is much larger than a typical wavelength. This deep water assumption is the opposite of the shallow water limit which leads to the well known Korteweg-de Vries equation-the first equation found to exhibit permanent, localized waveform solutions now known as solitons.

It is interesting to note that much of the foundation for the study of nonlinear wave phenomena has been laid down from the earliest days of the study of the motion of water. Scott Russell (1844) first reported the observation of a solitary wave in a canal, and the Korteweg-de Vries equation was derived as early as 1895 (korteweg and de Vries, 1895). In the area of deep water waves, the form of the nonlinearity was first obtained by Stokes (1847); in the course of derivation, he introduced for the first time the idea of frequency perturbation caused by nonlinearity, now known as the Stokes' expansion, which contained the central idea of modern nonlinear wave theory.

In this paper, we shall present some of the recent experimental and theoretical findings on the properties of nonlinear deep water waves, taking advantage of the concept of solitons which proves to be of such interest in many other brances of physics.

THE GOVERNING EQUATION

Given a wave system with a Lagrangian, the equations governing the time evolution of the wave-number and the amplitude can be obtained through the averaged variational principle proposed by Whitham (1965, 1967, 1970, 1974).
They are

$$\frac{\partial k}{\partial t} + \frac{\partial \omega}{\partial x} = 0 \ , \qquad\qquad (1)$$

$$\frac{\partial a^2}{\partial t} + \frac{\partial}{\partial x} \left(\frac{\partial \omega}{\partial k} a^2 \right) = 0 \ , \qquad\qquad (2)$$

where k is the wavenumber, ω is the frequency and a is
the amplitude. What remains to be specified is the
dispersion relation, which expresses the frequency ω
as a function of k, or, in the nonlinear case, of k
and a. The linear dispersion relation is

$$\omega = \sqrt{gk \tanh kh} \; , \tag{3}$$

where h is the depth of the water. For deep water,
kh is taken to approach infinity, the tanh term is
taken to be one, and the dispersion relation reduces to

$$\omega = \sqrt{gk} \; . \tag{4}$$

The effect of weak nonlinearity on the dispersion has
been calculated by Stokes, and the resulting dispersion
relation is

$$\omega = \sqrt{gk} \; (1 + \tfrac{1}{2} k^2 a^2) \; . \tag{5}$$

We shall further limit our attention to the
evolution of a wavetrain. By a wavetrain we mean that
there exists a carrier wavenumber k_o which remains
constant throughout the evolution. Thus we can expand
the dispersion relation given by (5) as a series around
$k = k_o$. Letting

$$k = k_o + \tilde{k} \; , \tag{6}$$

we get

$$\omega = \omega(k_o) + \frac{\partial \omega}{\partial k} (k_o)\tilde{k} + \frac{1}{2} \frac{\partial^2 \omega}{\partial k^2}(k_o)\tilde{k}^2 + o(\tilde{k}^2, a^2) \tag{7}$$

and on substitution of (5), we have

$$\omega = \omega_o + \frac{\omega_o}{2k_o}\tilde{k} - \frac{\omega_o}{8k_o^2}\tilde{k}^2 + \frac{1}{2}\omega_o k_o^2 a^2 + o(\tilde{k}^2, a^2) \, , \quad (8)$$

where

$$\omega_o = \sqrt{gk_o} \, . \qquad\qquad (9)$$

The system of equations (1), (2) and (8) governs the evolution of the weakly nonlinear wavetrain in time. By introducing the complex envelope

$$A = a \, e^{i\theta} \, , \qquad\qquad (10)$$

where

$$\frac{\partial\theta}{\partial x} = \tilde{k} \, , \quad \frac{\partial\theta}{\partial t} = -(\omega - (\omega - \omega_o)) \, , \qquad (11)$$

the system can be reduced to one single equation for A:

$$i\left(\frac{\partial A}{\partial t} + \frac{\omega_o}{2k_o}\frac{\partial A}{\partial x}\right) - \frac{\omega_o}{8k_o^2}\frac{\partial^2 A}{\partial x^2} - \frac{1}{2}\omega_o k_o^2 |A|^2 A = 0 \, . \quad (12)$$

This equation is the now familiar nonlinear Schrödinger equation, found to be useful for describing a wide variety of physical situations (see, for example, Scott, Chu and McLaughlin, 1973). It was first derived by Zakharov (1968) using a multiscale method, and by Yuen and Lake (1975) using Whitham's theory. In the next section, we shall examine the properties of its solutions as applied to deep water waves.

ENVELOPE SOLITONS

The initial value problem for equation (12), for initial conditions which decay sufficiently rapidly to zero as $|x| \to \infty$, can be solved exactly by the Inverse

Scattering Method. The spirit of the method, which is applicable in general to a nonlinear evolution equation of the type

$$\frac{\partial u}{\partial t} = N(u),\qquad\qquad(13)$$

where N is a nonlinear operator, can be summarized as follows:

(a) Find two operators L_u and M_u with the property that as u evolves in t according to (13), L_u and M_u satisfy

$$\frac{\partial L_u}{\partial t} = [L_u, M_u],\qquad\qquad(14)$$

where [f,g] denotes the commutator.

(b) Consider the scattering problem

$$L_u \psi = \lambda \psi.\qquad\qquad(15)$$

It can be shown that the spectrum λ is invariant in time if the eigenfunctions ψ satisfy

$$\frac{\partial \psi}{\partial t} + M_u \psi = K_o \psi\qquad\qquad(16)$$

where K_o is a normalization constant. To see this we differentiate (15) with respect to t:

$$\frac{\partial}{\partial t}(L_u) \cdot \psi + L_u \cdot \frac{\partial \psi}{\partial t} = \frac{\partial \lambda}{\partial t} \psi + \lambda \frac{\partial \psi}{\partial t}.\qquad(17)$$

Use of (14) and (15) leads to

$$L_u(\frac{\partial \psi}{\partial t} + M_u \psi) - \lambda(\frac{\partial \psi}{\partial t} + M_u \psi) = \frac{\partial \lambda}{\partial t} \psi. \qquad (18)$$

Thus, if $\partial \psi/\partial t + M_u \psi$ is an eigenfunction of (15) [which is precisely the condition given by (16)], $\partial \lambda/\partial t$ must be zero.

(c) Given an initial condition u_o, calculate the spectrum λ of (15) for $u = u_o$. Since u decays rapidly as $|x| \to \infty$, we can linearize (16) by setting $u = 0$ to obtain the asymptotic behavior of ψ at $|x| = \infty$ for all times. Knowing $\psi(\infty,t)$ and λ, we can attempt to recover u for all times through equation (15). This inverse scattering problem can be reduced to a linear integral equation.

The three steps summarized above demonstrate how the solution of a nonlinear evolution equation (13) can be reduced to the solution of a series of linear equations. The main difficulty is the identification of L_u and M_u. As yet, for a given nonlinear equation, there has not been an established algorithm for determining L_u and M_u, although significant progress has recently been made toward generalizing the Inverse Scattering Method (see Ablowitz, et al, 1973). The forms of L_u and M_u for (12) were given by Zakharov and Shabat (1972). By invoking the mere fact that the spectrum of the associated scattering problem is time invariant, a great deal of information concerning the long time behavior of a localized initial condition can be obtained. It turns out that the members of the discrete spectrum correspond to "solitons" and the continuous spectrum to a "tail," so that the asymptotic properties of an initial condition can be determined by merely examining its spectral contents

in light of the associated scattering problem (15).
This was done in detail by Zakharov and Shabat (1972)
and we shall only summarize the results which are
relevant to deep water waves:

I.   An initial wave-envelope pulse of arbitrary
     shape will eventually disintegrate into a
     number of solitons and an oscillatory tail.
     The number and structure of these solitons
     and the structure of the tail are completely
     determined by the initial conditions.

II. The tail is relatively small and unimportant
     for a pulse initial condition. It disperses
     linearly resulting in a $1/\sqrt{t}$ decay of the
     amplitude.

III. Each soliton is a permanent progressive wave
     solution of (12) in the form

$$S_n = a_n \text{ sech } \sqrt{2} \, k_o^2 a_n [(x-\chi_n) - (\frac{\omega_o}{2k_o} + v_n)t] +$$

$$+ \exp\{- \frac{i}{4} k_o^2 a_n^2 \omega_o t - \frac{4ik_o^2}{\omega_o} v_n [(x-\chi_n) - (\frac{\omega_o}{2k_o} + v_n)t + \Theta_n]\},$$
$$(20)$$

where a_n and v_n characterize the amplitude and
speed [relative to the group velocity $\omega_o/2k_o$]
of the nth soliton, and χ_n and Θ_n represent
its position and phase. Note that the
amplitude is given by a sech profile, the
height of which is inversely proportional to
the effective wavelength. Unlike the solitary
wave solutions of the Korteweg-de Vries equation
for shallow water [see Segur (1973), Hammack
and Segur (1974)], the amplitude a_n and speed
v_n bear no simple relation to each other except

through the fact that they are the imaginary
and real parts of the associated scattering
problem (15).

IV. The solitons are stable in the sense that
they can survive interactions with each other
with no permanent change except a possible
shift in position and phase (corresponding to
a shift in the values of the constants
X_n and Θ_n).

V. The time scale of formation of these solitons
(i.e., the transition time from the initial
to the asymptotic state) is in direct
proportion to the length of the pulse, and
in inverse proportion to the amplitude of the
pulse.

For detailed theoretical discussion of these
properties, the reader is referred to the original work
of Zakharov and Shabat (1972).

These predictions have been tested by performing
carefully controlled experiments, and the agreement
between experiment and theory is found to be very good.
The experiments were performed in a water wave tank 40
feet long, 3 feet wide and 3 feet deep. Waves are
generated at one end by a hinged paddle and absorbed
at the other end by a shallow beach. The paddle motion
can be made to conform to computer prescribed waveforms.
The waves generated are typically in the frequency range
of 2 Hz to 4 Hz. The profiles of the generated waves
are recorded by capacitance wave gauges located at
stations 5 ft, 10 ft, 15 ft, 20 ft, 25 ft and 30 ft
downstream from the wavemaker and displayed on an
oscillograph. Although the output of every amplitude
gauge was linearly proportional to wave amplitude, each

gauge had a slightly different sensitivity. In
recording the measurements shown in Figures 1 through
3, the effects of these sensitivity differences,
together with the first order effects of viscous
dissipation, have been removed by adjusting the
amplifier gain on each oscillograph channel so that the
magnitudes of the amplitude channel outputs were equal
when measuring a linear sinusoidal wavetrain as it
propagated down the tank.

The measurements shown in Figures 1 through 3
illustrate the qualitative comparison between our
experimental observations and the phenomena predicted
by the exact solution. The evolution of three pulses,
which have approximately the same effective duration
but different initial profiles, is shown in Figure 1.
Each pulse has the same carrier frequency ω_o = 2 Hz.
Case A (the left-hand traces) is an envelope soliton
profile which obeys the height-to-width relationship
(20). Case B (the center traces, shown on an amplitude
scale reduced by a factor of 2.5 compared with Cases A
and C) is a sech envelope with amplitude twice that of
the soliton profile. Case C is a sine envelope with the
same amplitude as the soliton profile. In Cases B and
C the initial profile breaks into solitons, while in
Case A, where the initial profile is already that of a
soliton, relatively little happens. This series of
experiments confirms the theoretical prediction that an
initial pulse profile disintegrates into solitons and a
tail,.except when it is itself a perfect soliton as in
Case A. It should be noted in passing that, for deep
water waves, the phase velocity (which is the speed of
the individual wave crests) is roughly twice that of the
group velocity (which, to first order, is the speed of

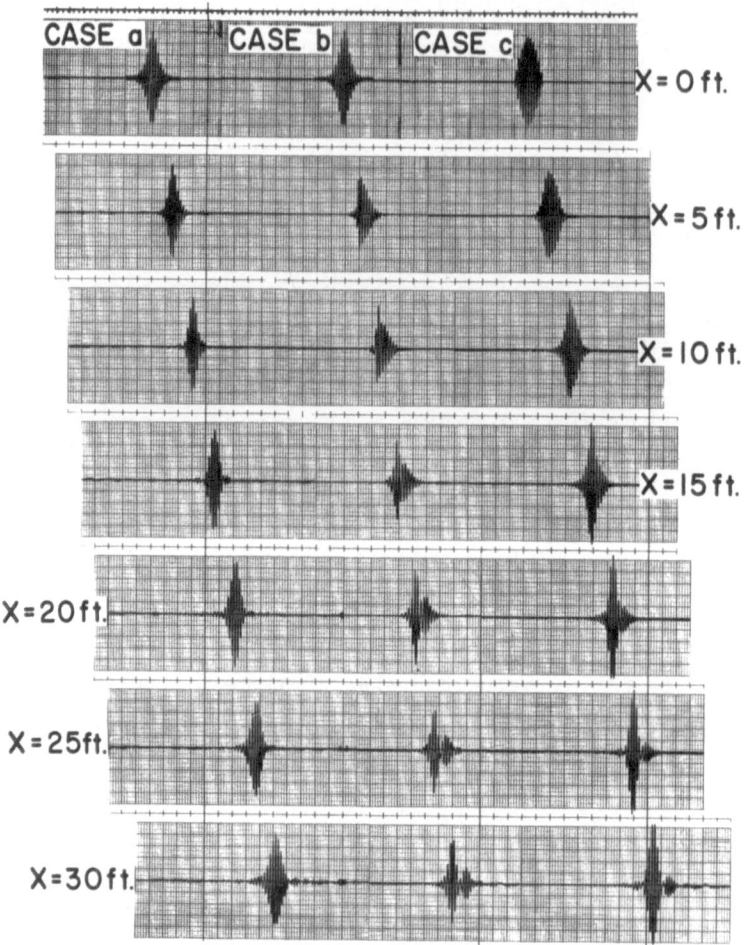

Fig. 1. Evolution of wave pulses. Case A, initial pulse with soliton profile, ω_0 = 2 Hz, initial $(ka)_{max} \simeq 0.14$. Case B, initial pulse with sech profile and amplitude twice that for soliton profile, ω_0 = 2 Hz, amplitude scale of traces reduced by factor of 2.5 compared with Cases A anc C. Case C, initial pulse with sine profile and amplitude equal to that for soliton profile, ω_0 = 2 Hz, initial $(ka)_{max} \simeq 0.14$. In all figures, time increases from left-to-right and the x = 0 ft traces show initial wave pulse inputs to the wavemaker.

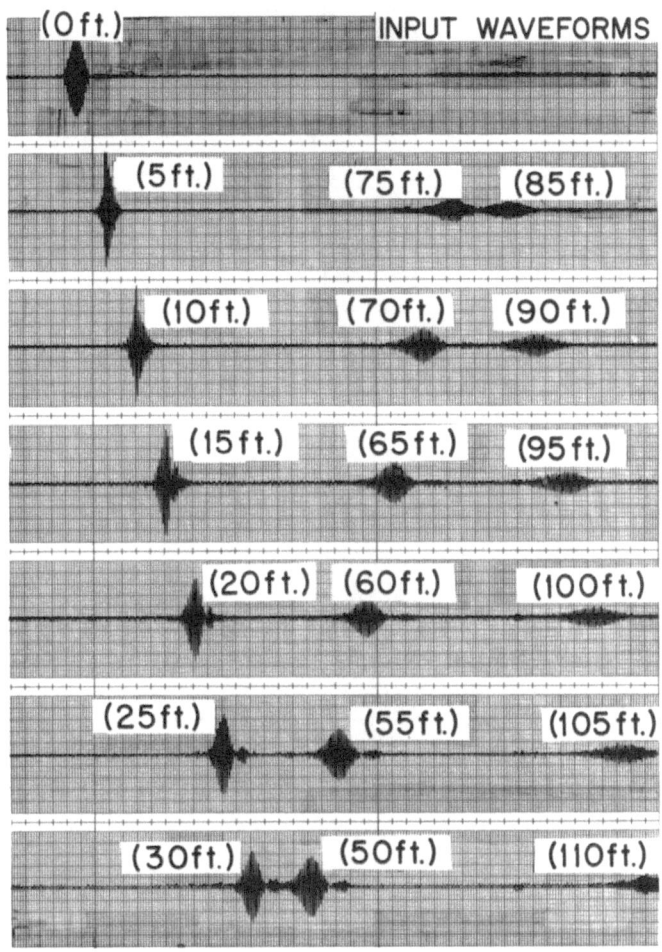

Fig. 2a. Example of wave pulse propagating back and forth between end-walls of wave tank. The initial conditions are the same as for the high frequency pulse in Figure 3, ω_o = 3 Hz, initial $(ka)_{max} \simeq 0.2$. This figure shows the single pulse case used to generate each pulse in the wave pulse collision experiment shown in Figure 2b.

the envelope); therefore, the individual waves outrun
the envelope even though the envelope may be steady.

The head-on collision of two pulses of the same
carrier frequency, ω_0 = 3 Hz, is shown in Figure 2b. The
interaction is accomplished by reflecting the first
pulse off a vertical end-wall (the wave-absorbing beach
was removed) before sending in the second. (The
propagation of a single pulse between the tank end-walls
is shown for reference in Figure 2a.) The pulses were
generated with the same amplitude, frequency and pulse
shape at x = 0. After generating the pulses, the
wavepaddle was left in a vertical position so that the
pulses would reflect from a vertical end-wall at x = 0 ft
as well as at x = 40 ft. The numbers shown on the wave
records in the figure identify the propagation distances
for wave pulses A and B during the process. The solitons
and their formation processes show little effect of
interaction, as can be seen by comparison of pairs of
profiles, such as A (25 ft) (no collision) and B (25 ft)
(one collision) or A (70 ft) (one collision) and B (70 ft)
(two collisions). This verifies the interaction properties
predicted by the theory. No attempts have been made to
measure the phase shifts (if any) quantitatively, but
judging from the data such shifts are very small for all
the cases we have observed.

The interaction of two pulses of different carrier
frequencies is shown in Figure 3. The first is a 3.0-Hz
pulse which propagates more slowly than the second, a
1.5-Hz pulse. The left-hand traces in the figure show
the evolution of the 1.5-Hz pulse alone. The center
traces show the evolution of the 3.0-Hz pulse alone. By
the time it reaches x = 30 ft, the 3.0-Hz pulse has
evolved into two solitons. The right-hand traces show

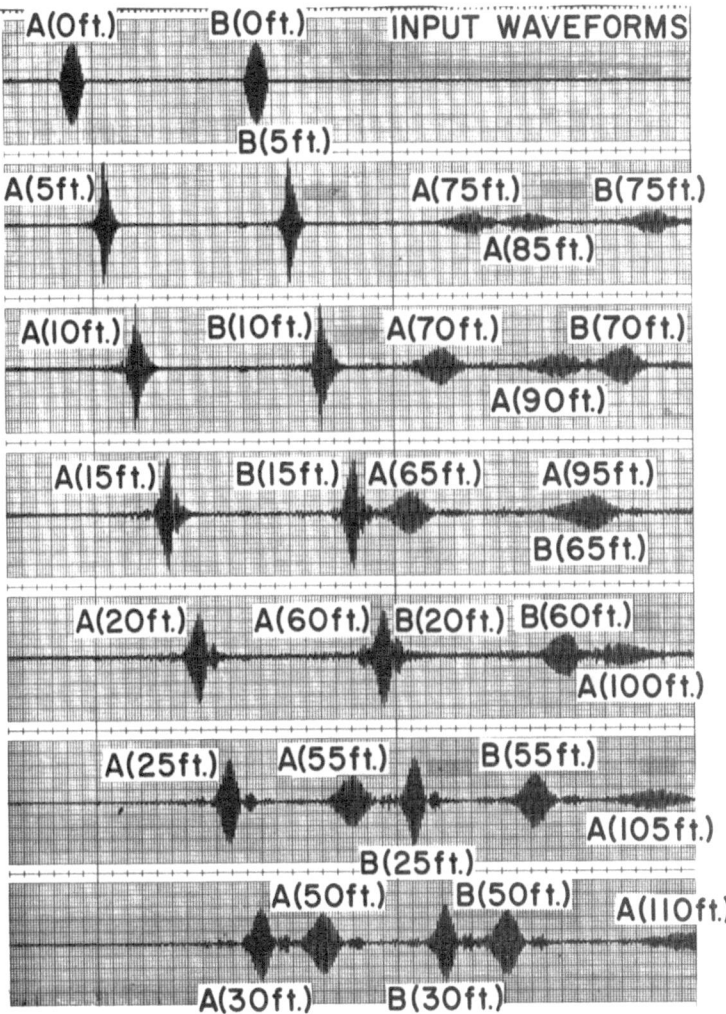

Fig. 2b. Head-on collision of two wave pulses. The initial conditions for each pulse are identical to those for the pulse in Figure 2a. Noted on this figure are the propagation distances for wave pulses A and B during the process.

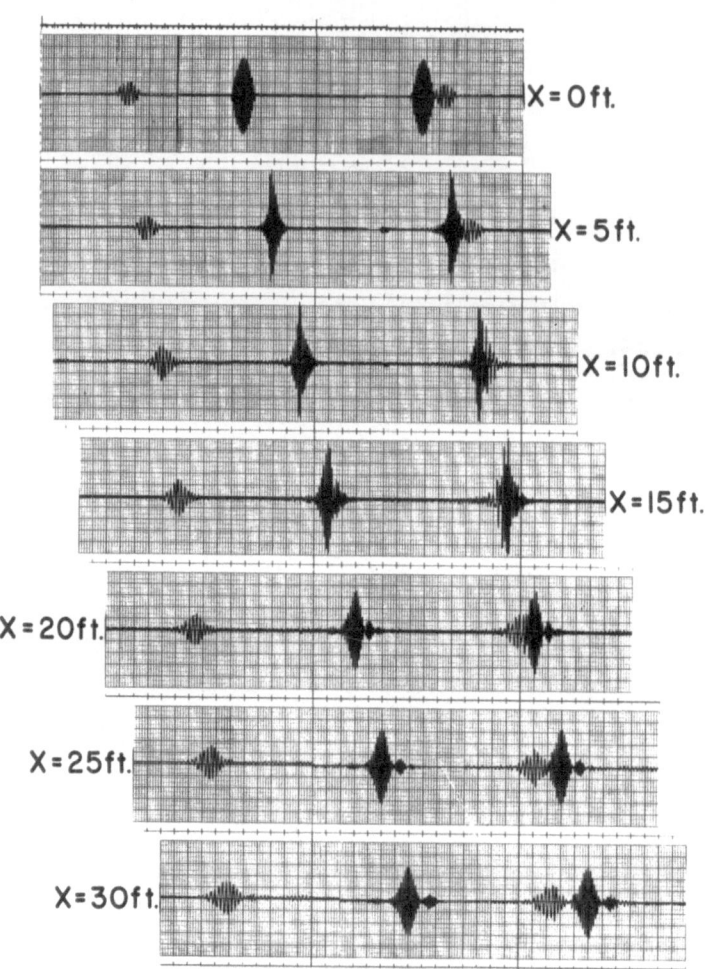

Fig. 3. One wave pulse overtaking and passing through
another wave pulse. Left-hand trace: first pulse alone,
ω_o = 1.5 Hz, initial $(ka)_{max} \simeq 0.10$, six-cycle pulse.
Center trace: second pulse alone, ω_o = 3 Hz, initial
$(ka)_{max} \simeq 0.2$, 12-cycle pulse which disintegrates into
two solitons. Right-hand traces: interaction of the
two pulses.

the evolution and interaction of the two pulses when
the 3-Hz pulse is generated ahead of the 1.5-Hz pulse
at x = 0 ft. As they propagate, the two pulses pass
through one another and emerge in reversed order by
x = 30 ft. Note that despite the long and rather
violent interaction they have undergone, the two pulses
emerge practically unaffected, as can be seen by comparison with the traces obtained when each evolves in
the absence of the other. Strictly speaking, this type
of interaction between pulses of different carrier wave
frequencies cannot be described by a single equation.
However, it has been shown by Oikawa and Yajima (1974)
that solitons with different carrier frequencies also
survive interactions with no permanent change other
than position and phase shifts.

We have also solved equation (12) numerically and
compared the computed results to experiments. The
case we exhibit here corresponds to Case C of Figure 1.
The initial condition for the program is obtained by
curve-fitting the experimentally recorded profile at
x = 5 ft. We then compute the subsequent profiles at
the 5 downstream stations. Since the leading order
effect of dissipation has been scaled out by the gauge
calibrations for the profiles in Figure 1, we have
compared the computed profiles directly with the recorded
profiles without additional adjustments in the amplitudes.
The results are shown in Figure 4.

We see from the comparison that the theoretical
prediction agrees very closely with the experimental
data. In particular, the shape and amplitude of the
largest soliton is extremely well predicted. Some
details of the evolution, including the amplitude of
the leading small soliton, are not exactly predicted.

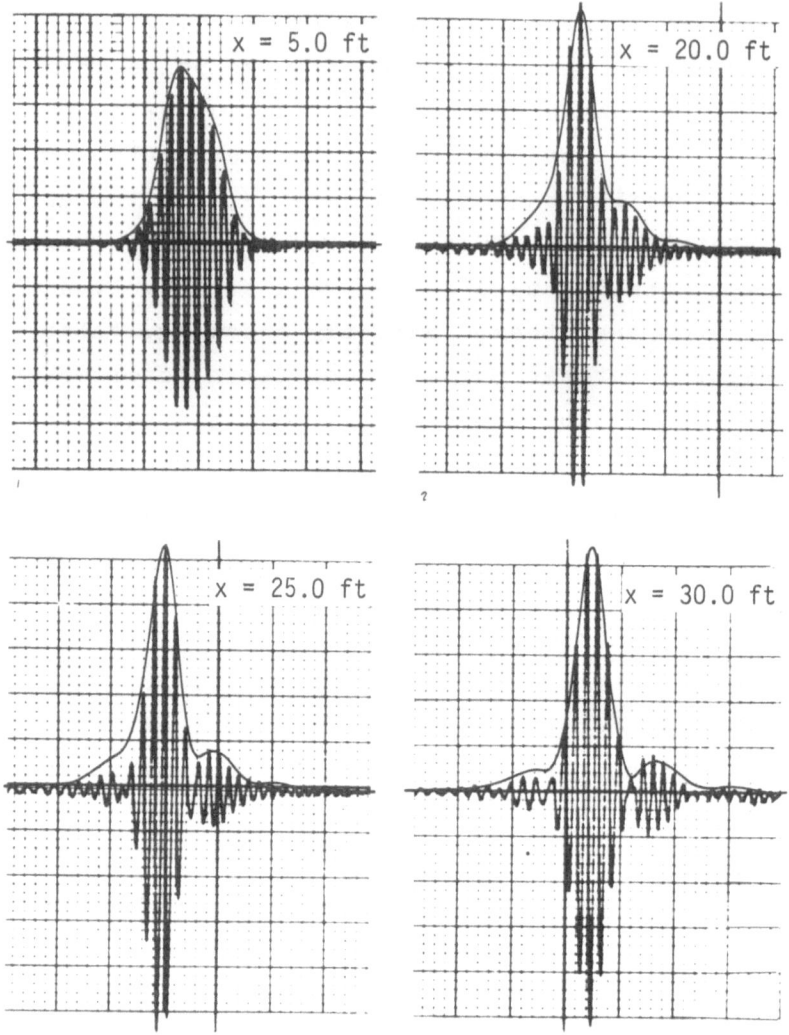

Fig. 4. Comparison of numberical computation with experimental data. The case considered corresponds to Case C in Figure 1. The computed results (shown as envelope curves) are superposed on the oscillograph traces from the experiment. The curve at x = 5.0 ft is the input to the numerical computation, and the other curves show comparison with data at three downstream conditions.

These discrepancies, however, are well within the
range of error we expect to introduce by:

  1. imprecise curve fitting of the recorded
      profile at x = 5 ft for representation
      of the initial condition (especially the
      wavenumber modulation),

  2. neglect of second order dissipative effects
      due to nonlinearity of the equation, and

  3. neglect of higher order terms in ε and ka
      in the derivation of (12).

In fact, in view of these approximations, the
agreement shown in Figure 4 may actually be much
better than one might justifiably expect.

FERMI-PASTA-ULAM RECURRENCE

      We now examine the evolution of a continuous
wavetrain using the nonlinear Schrödinger equation
(12). It was shown by Benjamin and Feir (1967), using
a perturbation analysis on the water wave equations,
that a weakly nonlinear uniform wavetrain is unstable
to modulational perturbation. Their results can be
recovered by a linear stability analysis on the uniform
wavetrain solution of (12). Consider the perturbed
wavetrain

$$A(x,t) = \left[a_o + \varepsilon_+ e^{i\Omega t + iK\left(x - \frac{\omega_o}{2k_o}t\right)} + \varepsilon_- e^{i\Omega t - iK\left(x - \frac{\omega_o}{2k_o}t\right)} \right] \times$$

$$\times \; e^{\frac{i}{2} k_o^2 a_o^2 \omega_o t} \; . \tag{21}$$

The dispersion relation is found to be

$$\Omega^2 = \frac{\omega_o^2 K^2}{8k_o^2} \left(\frac{K^2}{8k_o^2} - k_o^2 a_o^2 \right) , \tag{22}$$

which shows that the wavetrain is unstable (i.e.,
$\Omega^2 < 0$) to perturbations having wavenumber K in the
range

$$0 < K < 2\sqrt{2}\ k_o^2 a_o\ . \tag{23}$$

The maximum instability occurs at

$$K = 2k_o^2 a_o \tag{24}$$

with a growth rate given by

$$Im\Omega = \frac{1}{2}\ \omega_o k_o^2 a_o^2\ . \tag{25}$$

This long wave instability is also known as the
"side-band" instability because it exhibits itself as
a pair of sideband components around the carrier
component in a power spectrum (see Figure 7a). These
instability results have been shown to be in good
quantitative agreement with experimental data (Lake,
et al, 1977).

An interesting question to ask is: what eventually
becomes of the unstable wavetrain? One speculation was
that the wavetrain would rapidly disintegrate, spreading
the energy (initially confined to a few modes around the
carrier) to a broad spectrum of modes, and would
eventually tend to a random end state (see, for example,
Hasselmann, 1967). This speculation was effectively
abandoned with the discovery of soliton solutions, whence
it was proposed that the end-state would be a stable
train of solitons (see, for example, Hasimoto and Ono,
1972). The truth, however, is even stranger.

The evolution of an initially modulated wavetrain

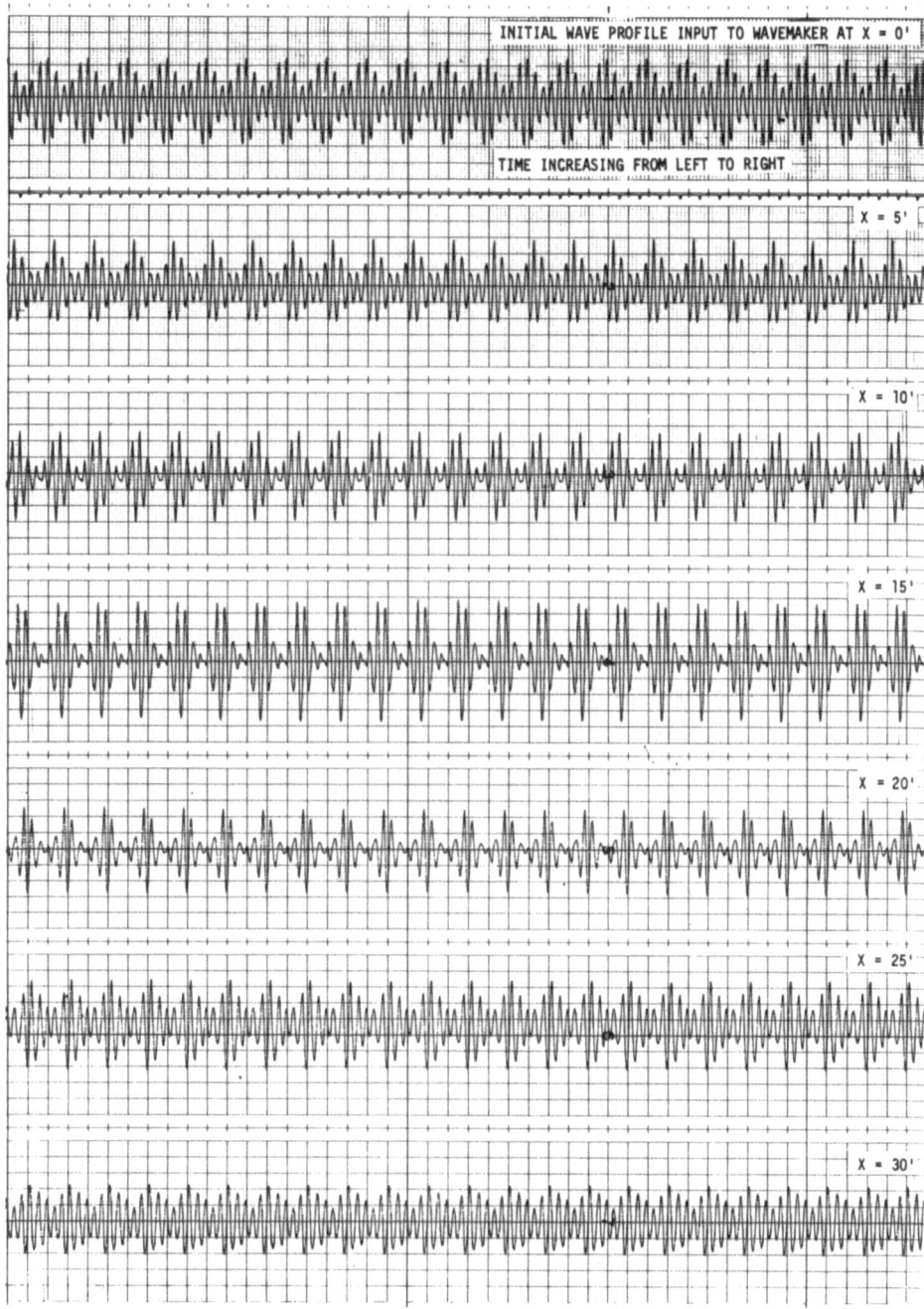

INITIAL WAVE PROFILE INPUT TO WAVEMAKER AT X = 0'

TIME INCREASING FROM LEFT TO RIGHT

X = 5'

X = 10'

X = 15'

X = 20'

X = 25'

X = 30'

Fig. 5. Evolution of a nonlinear wavetrain with strong initial modulation. Average $(ka)_o \simeq 0.2$, $f_o = 2.5$ Hz.

is shown in Figure 5. The magnitude of initial
modulation corresponds to $\varepsilon_{\pm} \doteq 0.25$. This large
modulation is needed so that the phenomena we are
interested in can be observed within the tank length.
It can be seen that the wavetrain achieved a state
of maximum modulation at the x = 15 ft station, at
which stage it resembles a train of solitons. However,
instead of further disintegrating (and tending to a
random end-state), or remaining as a train of solitons,
it demodulates and, at x = 25 ft, the waveform at
x = 5 ft is almost reconstructed. A second example is
shown in Figure 6. The initial wavetrain, which is
relatively unmodulated, achieves a maximum modulation
at x = 20 ft, and returns to an unmodulated state at
x = 30 ft. A third example is shown in Figure 7,
where the power spectra of the wavetrain at three
stages of evolution are shown. It can be seen that
the energy, which is initially confined to a few modes
around the carrier frequency and its harmonics
(at x = 5 ft), is spread to many modes (at x = 10 ft),
but eventually returns to the few original modes
(x = 25 ft). It should be pointed out that we are
measuring the waveforms at intervals of 5 feet; therefore, even though a perfect return to initial conditions
may have occurred somewhere, it is unlikely to be
recorded by any of our wave gauges. In reality, the
effects of dissipation cannot be neglected, and at some
highly modulated states, individual waves in fact may
break; so that a truly perfect recurrence is itself
impossible.

The time periodic return to initial conditions of
the solution of a nonlinear system has been given the
name of Fermi-Pasta-Ulam Recurrence (FPU recurrence),

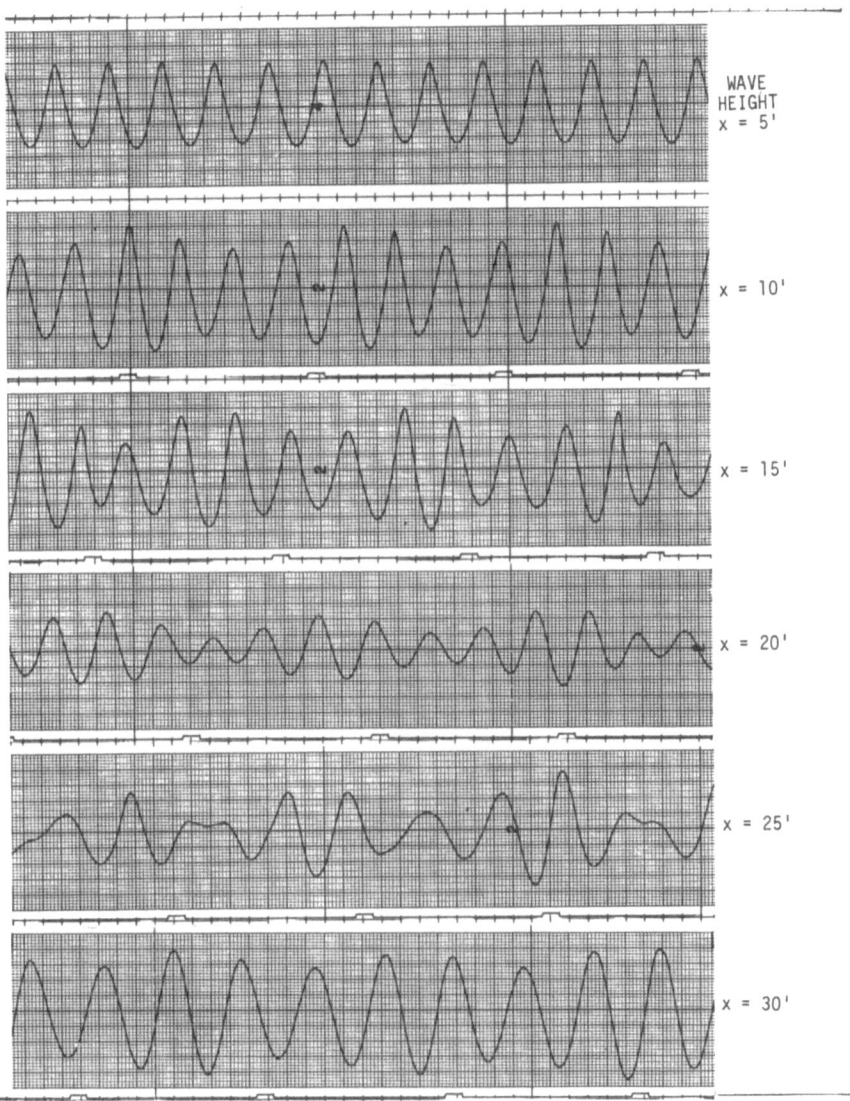

Fig. 6. Evolution of a nonlinear wavetrain with nearly uniform initial amplitude. Average $(ka)_o \simeq 0.23$, $f_o = 3.6$ Hz. Oscillograph records shown on expanded time scale to display individual wave shapes; wave shapes are not exact repetitions each modulation period because modulation period does not contain integral number of waves.

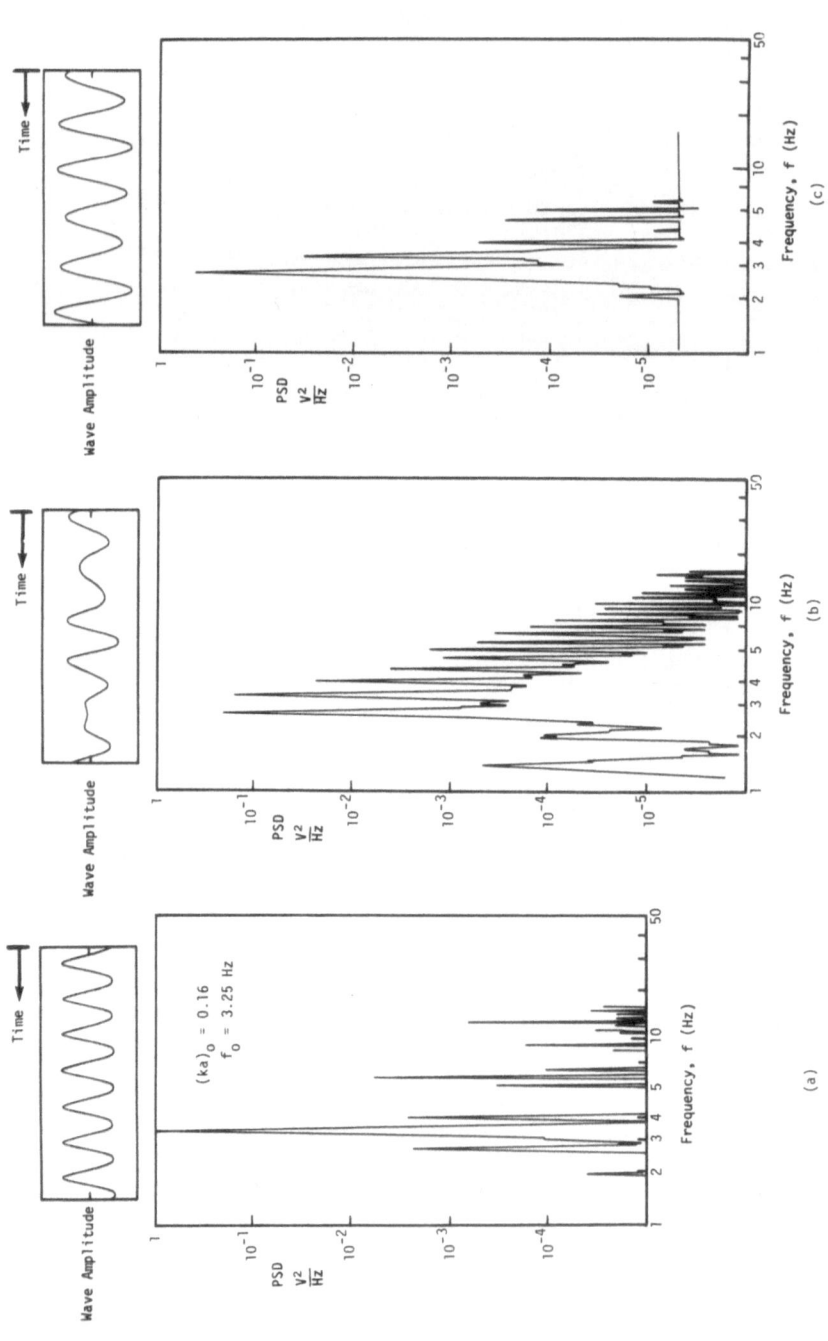

Fig. 7. Evolution of a nonlinear wavetrain with nearly uniform initial amplitude. Power spectra and waveforms at three tank locations: (a) x = 5.0 ft, (b) x = 10.0 ft, (c) x = 25.0 ft.

as the phenomenon was first reported by Fermi, et al
(1955) in relation to some numerical studies performed
on anharmonic lattice vibrations. FPU recurrence for
periodic solutions of the Korteweg-de Vries equation
was found by Zabusky and Kruskal (1965).

Numerical computation of the nonlinear Schrödinger
equation with periodic boundary conditions shows that
the solution does exhibit the FPU phenomenon. An
example of the computation is shown in Figure 8.
Quantitative comparison with experimental data has
also been made; it was found necessary to include a
model for dissipation in the equation to achieve long
time agreement.

DISCUSSION

Admittedly, the concepts of solitons and FPU
recurrence are thus far products of an idealized
equation and idealized experiments. We have assumed
weak nonlinearity, one-space dimensionality, near
monochromaticity, absence of randomness (other than
the noise introduced in the experiments and numerics)
among others. To what extent can the results obtained
under these idealized conditions aid us in our understanding of the complex physical processes taking place
in, say, an ocean environment?

One approach is to systematically relax the
idealized assumptions. Zabarov (1968) derived the
two-space dimensional version of the nonlinear Schrödinger
equation:

$$i\left(\frac{\partial A}{\partial t} + \frac{\omega_o}{2k_o}\frac{\partial A}{\partial x}\right) - \frac{\omega_o}{8k_o^2}\frac{\partial^2 A}{\partial x^2} + \frac{\omega_o}{4k_o^2}\frac{\partial^2 A}{\partial y^2} - \frac{1}{2}\omega_o k_o^2 |A|^2 A = 0,$$

$$(26)$$

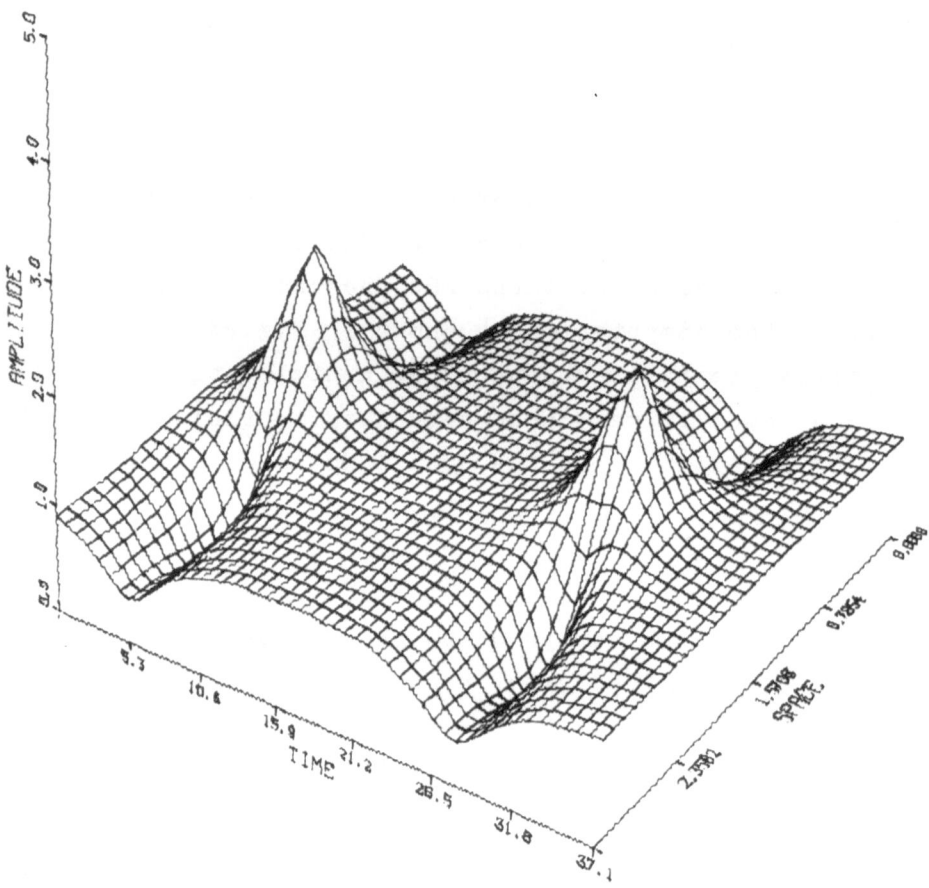

Fig. 8. Fermi-Pasta-Ulam recurrence in nonlinear
Schrödinger equation. The amplitude of the solution
of the nonlinear Schrödinger equation is plotted
against nondimensionalized space ($k\zeta/2$) and time
($1/2\ \omega_o k_o^2 a_o^2 t$). Initial condition contains a 10%
sinusoidal modulation which corresponds to the most
unstable Benjamin-Feir perturbation. It can be seen
that the maximum modulation is attained at TIME = 5.3;
the time required for a complete recurrence cycle is
TIME = 21.2.

where $A = a(x,y,t)e^{i\theta(x,y,t)}$. This equation describes
the modulation in x, y and t of a wavetrain with
carrier wavenumber vector

$$\underset{\sim}{k}_o = (k_o,0) \, . \tag{27}$$

The envelope is given by $a(x,y,t)$, and the perturbations
to the wavenumber vector and frequency are given

$$\underset{\sim}{\tilde{k}} = (\frac{\partial\theta}{\partial x} , \frac{\partial\theta}{\partial y}) \, , \underset{\sim}{\tilde{\omega}} = - \frac{\partial\theta}{\partial t} \, . \tag{28}$$

It can be seen that the soliton solutions of (12) are
also y-independent solutions of (26). The stability
of these one-space dimensional solitons to infinitesimal
cross wave disturbances has been studied by Zakharov
and Rubenchik (1973). They found that the solitons are
unstable to long wave disturbances which are odd in
x with respect to the center of the soliton. A
numerical program has been developed at TRW. Preliminary
computations verify the instability found by Zakharov
and Rubenchik (1973). Computations using periodic
boundary conditions in both x and y indicate that FPU
recurrence is still present in two-space dimensions.
Further study is now under way. The interaction between
two or more wavetrains traveling in different directions
has been studied by Roskes (1976), who calculated the
stability of two interacting uniform wavetrains under
special conditions.

A second approach toward applying the findings on
solitons and FPU to complex physical situations is to
look for underlying principles of these phenomena which
may be generalizable. One such would be the concept of

coherence. The most striking feature of a soliton is
its property of maintaining its identity for a long
time. The same striking feature for a continuous
wavetrain is evidenced by the FPU phenomenon, which
demonstrates the remarkable ability of the nonlinear
wavetrain to "remember" its initial state. This is
in one sense very fortunate, for if it happened that it
is the property of a wavetrain to thermalize and lose
coherence, then it would mean that the band-limited and
deterministic assumptions in the derivation of (12)
would fail at a finite time, and that the wavetrain
concept would be completely useless in studying long
time behavior. The existence of the FPU phenomenon
actually suggests that coherence is maintained (e.g.,
the carrier remains unchanged and the group velocity
remains a constant) even though the spectrum is no
longer narrow-band in a strict sense. Taking this
suggestion a step further, Lake and Yuen (1977) have
proposed that even the complicated spectra of wind-
generated ocean waves may be attributed simply to the
fact that the surface is highly modulated, so that all
the waves may still be coherent in the sense discussed
above. This is almost directly opposite to the
conventional view that the broadband appearance of
ocean wave spectra necessarily implies incoherence and
randomness. Preliminary tests of this proposal using
laboratory and experimental data have produced evidence
in support of such an interpretation of the role of
nonlinearity in determining the properties of wind-
driven ocean waves (Lake and Yuen, 1977).

   Not very long ago the existence of phenomena such
as solitons and recurrence would have been considered
quite unbelievable. When considering the significance

of nonlinearity in the natural sciences, it is perhaps
appropriate that we remember the Queen's advice to
Alice:

> "I ca'n't believe <u>that</u>!" said Alice.
> "Ca'n't you?" the Queen said in pitying
> tone. "Try again: draw a long breath,
> and shut your eyes."
>
> Alice laughed. "There's no use trying,"
> she said: "one <u>ca'n't</u> believe impossible
> things."
> "I daresay you haven't had much practice,"
> said the Queen.
>
> Lewis Carroll, <u>Through the Looking-Glass</u>

REFERENCES

1. Ablowitz, M.J., Kaup, D.J., Newell, A.C., and
    Segur, H., Nonlinear Evolution Equations of
    Physical Significance, Physical Review Letters,
    Vol. 31, No. 2, 9 July, pp. 125-127 (1973).

2. Benjamin, T.B., and Feir, J.E., The Disintegration
    of Wave Trains in Deep Water, Part 1, Theory,
    Journal of Fluid Mechanics, Vol. 27, Part 3, 24
    February, pp. 417-430 (1967).

3. Fermi, E., Pasta, J., and Ulam, S. (1940; First
    Published 1955), Studies of Nonlinear Problems,
    Collected Papers of Enrico Fermi, The University
    of Chicago Press, Vol. 2, No. 266, pp. 978-988 (1962).

4. Hammack, J.L., and Segur, H., The Korteweg-de Vries
    Equation and Water Waves, Part 2. Comparison with
    Experiments, Journal of Fluid Mechanics, Vol. 65,
    Part 2, 28 August, pp. 289-314 (1974).

5. Hasimoto, H., and Ono, H., Nonlinear Modulation of
    Gravity Waves, Journal of the Physical Society of
    Japan, Vol. 33, No. 3, September, pp. 805-811 (1972).

6. Hasselman, K., Discussion [following the paper by
    Benjamin (1967)], Proceedings of the Royal Society
    of London, Series A, Vol. 299, No. 1456, 13 June,
    pp. 76 (1967).

7. Lake, B.M., and Yuen, H.C., A New Model for Nonlinear
    Wind Waves, submitted to Dynamics of Atmospheres
    and Oceans (1977).

8. Lake, B.M., Yuen, H.C., Rungaldier, H. and Ferguson,
    W. E., Nonlinear Deep Water Waves: Theory and
    Experiment, II. Evolution of a Continuous Wavetrain,
    submitted to Journal of Fluid Mechanics (1977).

9. Oikawa, M., and Yajima, N., A Perturbation Approach to Nonlinear Systems. II. Interaction of Nonlinear Modulated Waves, <u>Journal of the Physical Society of Japan</u>, Vol. 37, No. 2, August, pp. 486-496 (1974).

10. Roskes, G.J., Nonlinear Multiphase Deep-Water Wave-trains, <u>The Physics of Fluids</u>, Vol. 19, No. 8, pp. 1253-1254 (1976).

11. Russell, J. Scott, Report on Waves, <u>British Association Reports, 1844</u>, p. 369 (1844).

12. Scott, A.C., Chu, F.Y.F., and McLaughlin, D.W., The Soliton: A New Concept in Applied Science, <u>Proceedings of the IEEE</u>, Vol. 61, No. 10, pp. 1443-1491 (1973).

13. Segur, H., The Korteweg-de Vries Equation and Water Waves. Solutions of the Equation. Part 1, <u>Journal of Fluid Mechanics</u>, Vol. 59, Part 4, 7 August, pp. 721-736 (1973).

14. Stokes, G.G., On the Theory of Oscillatory Waves, <u>Cambridge Transactions</u>, Vol. 8, pp. 441-473 (<u>Papers</u> Vol. 1, pp. 197-229)(1847).

15. Whitham, G.B., Nonlinear Dispersive Waves, <u>Proceedings of the Royal Society of London</u>, Series A, Vol. 283, No. 1393, 6 January, pp. 238-261 (1965a).

16. Whitham, G.B., A General Approach to Linear and Nonlinear Dispersive Waves Using a Lagrangian, <u>Journal of Fluid Mechanics</u>, Vol. 22, Part 2, June, pp. 273-283 (1965b).

17. Whitham, G.B., Variational Methods and Applications to Water Waves, A Contribution to "A Discussion on Nonlinear Theory of Wave Propagation in Dispersive Systems, Organized by B.J. Lighthill and Held on 19-20 May 1966," <u>Proceedings of the Royal Society of London</u>, Series A, Vol. 229, No. 1456, 13 June,

pp. 6-25 (1967a).

18.   Whitman, G.B., Nonlinear Dispersion of Water Waves, Journal of Fluid Mechanics, Vol. 27, Part 2, 2 February, pp. 339-412 (1967b).

19.   Whitham, G.B., Two-Timing, Variational Principles and Waves, Journal of Fluid Mechanics, Vol. 44, Part 2, 11 November, pp. 373-395 (1970).

20.   Whitham, G.B., Linear and Nonlinear Waves (Book), John Wiley and Sons, New York (1974).

21.   Yuen, H.C., and Lake, B.M., Nonlinear Deep Water Waters: Theory and Experiment, The Physics of Fluids, Vol. 18, No. 8, pp. 956-960 (1975).

22.   Zabusky, N.J. and Kruskal, M.D., Interaction of "Solitons" in a Collisionless Plasma and the Recurrence of Initial States, Physical Review Letters, Vol. 15, No. 6, pp. 240-243 (1967).

23.   Zakharov, V.E., Stability of Periodic Waves of Finite Amplitude on the Surface of a Deep Fluid, Journal of Applied Mechanics and Technical Physics, Vol. 2, pp. 190-194 (Zhurmal Prikladnoi Mekhaniki Teknicheskoi Fiziki, Vol. 9, No. 2, pp. 86-94 (1968).

24.   Zakharov, V.E. and Rubenchik, A.M., Instability of Waveguides and Solitons in Nonlinear Media, Soviet Physics JETP, Vol. 38, No. 3, pp. 494-499 (Zh. Eksp. Teor. Fiz., Vol. 65, pp. 997-1011 (1974).

25.   Zakharov, V.E. and Shabat, A.B., Exact Theory of Two-Dimensional Self-Focusing and One-Dimensional Self-Modulating Waves in Nonlinear Media, Soviet Journal of Experimental and Theoretical Physics, Vol. 34, No. 1, January, pp. 62-69 (1972).

SOLITONS AS PARTICLES, AND THE EFFECTS OF PERTURBATIONS

D. J. Kaup

Clarkson College of Technology

Potsdam, New York   13676

A.   INTRODUCTION

In my presentation, I will illustrate how one can use a perturbation theory to determine how various perturbations can effect soliton propagation. These examples which I will present are results obtained recently by Professor Newell and myself, the full details of which are to be published elsewhere.[1] Here, I shall briefly describe the examples and simply discuss some of the important features of the results.

Of course, to avoid any possible future misunderstanding, I want to make it perfectly clear from the start, that in this presentation, when I use the word "soliton," I shall be using it as given by its original definition.[2] Thus, when I use the word "soliton," I do indeed mean that very special solitary wave which preserves its character under collisions, and I certainly do not mean simply localized solitary waves which some have called "solitons." This latter usage of the word has generated a confusion of sorts, which invariably follows a misdefinition.

Naturally, the reason why I choose to remain with
the original definition[2] is i) to keep confusion to a
minimum, and ii) so that I may use the very powerful and
exact theory which is associated with true solitons.
Briefly stated,[3,4,5] the method of the "inverse scattering transform" (IST), which is used for solving those
equations which have soliton solutions, is very analogous
to the method of the Fourier transform. As indicated
below,

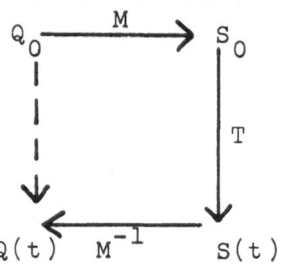

we take the initial data, Q_0, and map it into a set of
"scattering data," S_0, by solving a direct scattering
problem, treating Q_0 as a potential. Under this map,
the initial data has been decomposed (nonlinearly) into
action-angle variables, whose time evolution is almost
trivial. Thus, we can easily determine the scattering
data at any other later time, $S(t)$, and from the inverse
scattering problem, one can at least in principle re-
construct the final solution, $Q(t)$, from the scattering
data.

Before describing how we calculate the effect of
perturbations on solitons, we shall give a brief description of the scattering data and these special nonlinear
evolution equations. For the most part, we shall be
working with those equations which are exactly solvable

by the "Zakharov-Shabat" (ZS) inverse scattering transform. In this case, we take[3,6]

$$v_{1x} + i\zeta v_1 = q v_2 , \qquad (2a)$$

$$v_{2x} - i\zeta v_2 = -q^* v_1 , \qquad (2b)$$

as the eigenvalue problems were $q(x,t)$ is the field.
This q is mapped into the ZS scattering data by first
defining ψ to be the solution of (2) which satisfies

$$\psi \to \binom{0}{1} e^{+i\zeta x} \text{ as } x \to +\infty . \qquad (3)$$

Then for real ζ, the "reflection coefficient," $\rho(\zeta)$, is
found from

$$\psi \to a(\zeta) \begin{pmatrix} \rho(\zeta) e^{-i\zeta x} \\ \\ e^{i\zeta x} \end{pmatrix} \text{ as } x \to -\infty , \qquad (4)$$

where $a(\zeta)$ is a "transmission coefficient." Similarly,
whenever $a(\zeta) = 0$ for ζ in the upper half ζ-plane (with
$a\rho \neq 0$), we have a bound state, and the corresponding
value of ζ is the bound-state eigenvalue. For a more
complete description of this scattering data, the reader
is referred to the literature.[3,6] Here we shall simply
state what it is, being

$$S = \{\rho(\xi) \ \xi \text{ real}; \quad (\zeta_k, \beta_k)|_{k=1}^N \}, \qquad (5)$$

where ρ is the reflection coefficient, ζ_k is the kth
bound state eigenvalue and β_k is the normalization coefficient of the associated bound state eigenfunction.

The class of nonlinear equations which are solvable by (2) can be compactly described[3] by

$$
\begin{bmatrix} q_t^* \\ q_t \end{bmatrix} = 2\Omega(L^A) \begin{bmatrix} -q^* \\ q \end{bmatrix} , \qquad (6)
$$

where Ω is any ratio of entire functions and the operator L^A is

$$
L^A = \frac{-i}{2} \begin{bmatrix} \partial_x + 2q^* \int_{-\infty}^x dy\, q & 2q^* \int_{-\infty}^x dy\, q^* \\ \\ -2q \int_{-\infty}^x dy\, q & -\partial_x - 2q \int_{-\infty}^x dy\, q^* \end{bmatrix} . \qquad (7)
$$

For example, if we take $\Omega(\zeta) = -2i\zeta^2$, from (6) and (7) we find

$$
iq_t = -q_{xx} - 2(q^*q)q, \qquad (8)
$$

which is the nonlinear Schrödinger[6] (NLS) equation. Of course, by choosing different forms for $\Omega(\zeta)$, one may generate all possible nonlinear equations, solvable by the ZS-IST.

In a straightforward manner, one can show[3] that if q evolves according to (6) then the scattering data evolves according to

$$
\zeta_{k,t} = 0, \qquad (9a)
$$

$$
\beta_{k,t} = 2\Omega(\zeta_K)\beta_k , \qquad (9b)
$$

$$
\rho(\xi)_{,t} = 2\Omega(\xi)\rho(\xi). \qquad (9c)
$$

Of course, from (9) it should be clear that these are essentially the equations of motion for the action-angle variables.

Let's now return to (6) and consider an almost exactly solvable case, where in general

$$
\begin{pmatrix} q_t{}^* \\ q_t \end{pmatrix} = 2\Omega(L^A) \begin{pmatrix} -q^* \\ q \end{pmatrix} + \varepsilon \begin{pmatrix} F_1 \\ F_2 \end{pmatrix} , \tag{10}
$$

with ε as a small parameter and $F_1 = F_2{}^*$. Then if we define the following "squared eigenfunctions,"

$$
\Psi(\zeta) \equiv \begin{pmatrix} \psi_1{}^2(\zeta) \\ \psi_2{}^2(\zeta) \end{pmatrix} , \tag{11a}
$$

$$
\Psi \equiv \begin{pmatrix} \psi_1{}^2(\zeta_k) \\ \psi_2{}^2(\zeta_k) \end{pmatrix} , \tag{11b}
$$

$$
\chi_k \equiv \frac{\partial \Psi(\zeta)}{\partial \zeta} \bigg|_{\zeta = \zeta_k} , \tag{11c}
$$

one can show[1,7,8,9] that

$$
\zeta_{k,t} = \varepsilon <F|\Psi_k> , \tag{12a}
$$

$$
\beta_{k,t} + \frac{a''_k}{a'_k} \beta_k \zeta_{k,t} = 2\Omega(\zeta_k)\beta_k + \varepsilon\beta_k <F|\chi_k> , \tag{12b}
$$

$$
\rho(\xi)_{,t} = 2\Omega(\xi)\rho(\xi) + \varepsilon <F|\Psi> , \tag{12c}
$$

where

$$<F|G> \equiv \int_{-\infty}^{\infty}(F_1G_1+F_2G_2)dx. \qquad (13)$$

For N-soliton solutions, one can explicitly construct Ψ, Ψ_k and χ_k; thus given F, we can evaluate the inner products in (12). How this is done is described completely in Refs.1, 7 and 8.

B.   DAMPING AND FORCING TERMS

Our first example is a NLS soliton under the influence of damping and an oscillating forcing term.[10] Thus we take

$$iq_t = -q_{xx} - 2(q*q)q - i\Gamma q - iEe^{i\omega t}, \qquad (14)$$

where $\Omega(\zeta) = -2i\zeta^2$ and $F_2 = -\Gamma q - Ee^{i\omega t}$. When $F_2 = 0$, the one soliton solution of (14) is given by

$$q = \frac{-2i\eta e^{-2i\sigma_1}}{\cosh2\theta_1} , \qquad (15a)$$

where

$$\sigma_1 = \xi x + \bar{\sigma}_1 , \qquad (15b)$$

$$\theta_1 = -\eta x + \bar{\theta}_1, \qquad (15c)$$

$$\bar{\theta}_{1,t} = -\Omega_r, \qquad (15d)$$

$$\bar{\sigma}_{1,t} = -\Omega_i, \qquad (15e)$$

and Ω_r and Ω_i follow from

$$\Omega(\zeta_1) = \Omega_r(\xi,\eta) + i\Omega_i(\xi,\eta). \qquad (16)$$

In (15) and (16), we have choosen our scattering data to be

$$\rho(\xi) = 0, \qquad (17a)$$

$$N = 1, \qquad (17b)$$

$$\zeta_1 = \xi+i\eta \ (\eta>0), \qquad (17c)$$

$$\beta_1 = -2\eta e^{-2\bar{\theta}_1} e^{-2i\bar{\sigma}_1} \qquad (17d)$$

Now, first note that in the absence of any perturbations, $q \approx e^{2i\Omega_i t}$, whereas the forcing term goes as $e^{i\omega t}$. Clearly, if $\omega = 2\Omega_i$, we can have a resonance which for the NLS occurs when

$$\omega = 4\eta^2 - 4\xi^2. \qquad (18)$$

Next, we note that this resonance depends also on the amplitude of the soliton, η, and thus we may expect the soliton to adjust its amplitude as it attempts to resonate with the forcing term. If we consider a zero velocity soliton (due to damping, one expects the velocity to be zero), then $\xi = 0$ in (17c) and the first order perturbation result is

$$\eta_t = -2\Gamma\eta + \frac{\pi}{2}E\sin\chi, \qquad (19a)$$

$$\chi_t = \omega - 2\Omega_i, \qquad (19b)$$

where

$$\chi \equiv \omega t + 2\bar{\sigma}_1 \ . \tag{20}$$

Note in (19a) that depending on the phase of the soliton, $\bar{\sigma}_1$, the forcing term can act either with or against the damping, and as η changes, Ω_i will also ($\Omega_i = 4\eta^2$ for NLS) change, causing χ to change too.   The simplest solution of (19) is the steady state solution where, for NLS,

$$\eta = \frac{1}{2}\omega^{1/2} \ , \tag{21b}$$

$$\sin\chi = \frac{2\Gamma\omega^{1/2}}{\pi E} \ . \tag{21b}$$

The solution of (21b) for $0<\chi<\frac{\pi}{2}$ is a stable solution, while the solution for $\frac{\pi}{2}<\chi<\pi$ is unstable.   Typical paths in the η-χ plane are described in Ref. 1.

Note that due to (21b), the soliton can only phase-lock to the forcing term if ω is in the frequency range

$$0<\omega< \frac{\pi^2 E^2}{4\Gamma^2} \ , \tag{22}$$

which depends on the ratio of the forcing term to the damping.   Furthermore, at fixed ω, there is a threshold value for E if phase-locking is to occur, which is

$$E > \frac{2}{\pi} \Gamma\omega^{1/2} \ . \tag{23}$$

For a further discussion of this system, the reader is referred to References 1 and 10.

C. THE PERTURBED SINE-GORDON EQUATION

Since the sine-Gordon equation is relativistic, it can be used as a one-dimensional relativistic particle model, and as we shall see here, the soliton solutions (kinks, antikinks, and breathers) of the sine-Gordon equation do behave as relativistic particles under the influence of perturbations. This equation was first solved exactly[11] by using light-cone coordinates and the ZS - IST, where the equation of motion is

$$\phi_{xt} = \sin\phi .\qquad\qquad (24)$$

It has also been solved[12,13] in "laboratory coordinates," where the equation of motion is then

$$\phi_{TT} - \phi_{XX} + \sin\phi = 0,\qquad\qquad (25)$$

and where these coordinates are related by

$$x = \frac{1}{2}(X+T) ,\qquad\qquad (26a)$$

$$t = \frac{1}{2}(X-T).\qquad\qquad (26b)$$

Either of these methods of solution may be used to study effects of perturbations, and to first order, they both are equivalent.[1]

The simplest soliton solution is the kink or anti-kink for which the ZS scattering data is

$$\rho(\xi) = 0, \qquad (27a)$$

$$N = 1, \qquad (27b)$$

$$\zeta_1 = i\eta, \qquad (27c)$$

$$\beta_1 = \pm\, 2\eta e^{-2\bar{\theta}_1}; \qquad (27d)$$

then the solution is

$$\phi = 4\tan^{-1}[e^{\pm 2\theta_1}], \qquad (28)$$

where

$$\theta_1 = -\eta x + \bar{\theta}_1, \qquad (29a)$$

$$\bar{\theta}_{1,t} = \frac{-1}{4\eta} . \qquad (29b)$$

If we transform to X and T variables, (29) becomes

$$\theta_1 = -\frac{1}{2}\gamma_1(X - V_1 T) + \theta_{10}, \qquad (30)$$

where

$$V_1 = \frac{1 - 4\eta^2}{1 + 4\eta^2} , \qquad (31a)$$

$$\gamma_1 = \eta + \frac{1}{4\eta} = [1 - V_1^2]^{-1/2}. \qquad (31b)$$

As an example, if we take

$$\phi_{TT} - \phi_{XX} + \sin\phi = -\Gamma\phi_T + \varepsilon g(X), \qquad (32)$$

then one finds[1]

$$\partial_T(\gamma_1 V_1) = -\Gamma \ \gamma_1 V_1 - \frac{\partial U}{\partial X_0}, \qquad (33a)$$

$$\partial_T X_0 = V_1, \qquad (33b)$$

where $\bar{X}_0$ is the value of X at which θ_1 in (28) is zero (and thus, is the center of the kink or antikink) and

$$U(X_0,T) = \frac{\varepsilon}{8} \int_{-\infty}^{\infty} g(X)\phi_{,X} dX. \qquad (34)$$

In (32), Γ is a damping and $g(X)$ represents an impurity which can either slow down or speed up a kink. Equations (33) show that, even with perturbations, a kink acts like a relativistic particle where $\gamma_1 V_1$ is the relativistic momentum and U is a scalar potential. The strength of the scalar potential is given as the average of g times $\phi_{,X}$, and the force arises from the change in U as the central position of the kink is changed.

A much more interesting case is when a kink and an antikink collide in the presence of damping,[1] or damping and a forcing term.[14] In the first case, the kink-anti-kink always decays into a breather, which is like the process

$$\text{fermion} + \text{antifermion} \to \text{meson} \qquad (35)$$

In the second case, decay into a breather need not always occur, and McLaughlin and Scott are currently determining the threshold conditions.

D.   DENSITY GRADIENTS

In many cases, it is possible for the coefficients in an equation to vary as the position changes, and these variations can speed up, slow down, and possibly even trap solitons. Using the NLS again as an example, we take

$$iq_t = -q_{xx} - 2(q^*q)q + 2qf(x), \qquad (36)$$

and will consider two cases: $f(x) = \alpha x$ and $f(x) = \alpha x^2$. As pointed out by Liu and Chen,[15] the first case is exactly solvable and although not pointed out by them, can also be solved by a combination of a phase and coordinate transformations.[16] To show this, simply take

$$X = x + 2\alpha t^2, \qquad (37a)$$

$$T = t, \qquad (37b)$$

$$q = \phi e^{-2i\alpha t(x + \frac{2}{3}\alpha t^2)}; \qquad (37c)$$

then (36) becomes

$$i\phi_T = -\phi_{XX} - 2\phi(\phi^*\phi), \qquad (38)$$

which is again the unperturbed NLS. Thus, if we have a soliton solution to (38), it must move at constant velocity in (X,T) space, or $X = VT$. Then by (37)

$$x = Vt - 2\alpha t^2, \qquad (39)$$

and the soliton moves as though it was being acted on by

a constant force. For $\alpha > 0$, a soliton moving initially
to the right will eventually be reflected. Note that
its amplitude never changes.

Let's now consider the second case where $f(x) = \alpha x^2$.
Here, we cannot solve it exactly so we resort to first
order perturbations. Using Eqs. (15), (16), and (17),
we find the first order equations of motion[1] for the
scattering data to be

$$\xi_t = \alpha \bar{\theta},$$
(40a)

$$\eta_t = 0,$$
(40b)

$$\bar{\theta}_{1t} = -4\xi\eta,$$
(40c)

$$\bar{\sigma}_{1t} = 2(\xi^2 - \eta^2) - \alpha \left(\frac{\pi^2}{32\eta^2} + \frac{\bar{\theta}_1}{4\eta} + \frac{3\bar{\theta}_1^2}{2\eta^2} \right).$$
(40d)

Note again that (to first order) the soliton's amplitude,
2η, does not change. By (15a) and (15c), the central
position of the soliton ($\theta_1 = 0$) is given by $x_0 = \bar{\theta}_1/\eta$,
so from (40a) and (40c) we find

$$\ddot{x}_0 + (4\alpha\eta)x_0 = 0.$$
(42)

Again, the soliton is acting like a particle (harmonic
oscillator for $\alpha > 0$). For $\alpha > 0$, which corresponds to a
minimum in $f(x)$, the particle (soliton) is trapped and
simply oscillates back and forth with a frequency $2(\alpha\eta)^{1/2}$
which is amplitude dependent. On the other hand, for
$\alpha < 0$, which corresponds to a maximum in $f(x)$, the particle
motion is hyperbolic with an acceleration away from $x = 0$.

E.   THE DAMPED KdV

As our last example, we will discuss the damped KdV and use it as an illustration of what happens when we perturb a system solvable by the Schrödinger IST. In this case, instead of (2), we have

$$v_{xx} + [\zeta^2 + q(x,t)]v = 0, \qquad (43)$$

and we define the scattering data in the following manner. Let ψ be a solution of (43) which satisfies

$$\psi \to e^{i\zeta x} \text{ as } x \to + \infty . \qquad (44)$$

Then as $x \to - \infty$,

$$\psi \to a[e^{i\zeta x} - \rho e^{-i\zeta x}], \qquad (45)$$

which defines ρ, the reflection coefficient. Again, whenever $a = 0$ for ζ in the upper half ζ-plane, we have a bound state. We take the scattering data to be

$$S_- = \{\rho(\xi); \quad (\zeta_k, \beta_k)|_{k=1}^N\}, \qquad (46)$$

where β_k is again the associated normalization coefficient.

Those nonlinear equations which are exactly solvable by the Schrödinger IST can be given in the general form[3] of

$$q_t + \partial_x[P_0(M)q] = 0, \qquad (47)$$

where P_0 is any ratio of entire functions, and the operator M is given by

$$M = -\frac{1}{4} \partial_x^2 - q - \frac{1}{2} \int_x^\infty q,_y \, dy \; , \qquad (48)$$

which satisfies

$$M(\psi^2) = \zeta^2 \psi^2 . \qquad (49)$$

For example, if $P_0(M) = -4M$, then (47) and (48) give the KdV equation

$$q_t + q_{xxx} + 6q_x q = 0 . \qquad (50)$$

When q evolves according to (47), one can show that the scattering data then evolve according to[3]

$$\zeta_{k,t} = 0, \qquad (51a)$$

$$\beta_{k,t} = 2i\zeta_k \beta_k P_0(\zeta_k^2) \; , \qquad (51b)$$

$$\rho_t = 2i\zeta\rho P_0(\zeta^2) . \qquad (51c)$$

Let us now consider the effects of perturbations on (47). If we rewrite (47) as

$$q_t + \partial_x [P_0(M)q] = \varepsilon F, \qquad (52)$$

then (51) becomes

$$\zeta_{k,t} = \frac{\epsilon}{2i\zeta_k a_k'} \int_{-\infty}^{\infty} F\psi_k^2 \, dx, \qquad (53a)$$

$$\beta_{k,t} + (\frac{a_k''}{a_k'} + \frac{1}{\zeta_k})\beta_k\zeta_{k,t} = 2i\zeta_k\beta_k P_0(\zeta_k^2)$$

$$+ \frac{\epsilon}{2i\zeta_k(a_k')^2} \int_{-\infty}^{\infty} F(\frac{\partial\psi^2}{\partial\zeta})\Big|_{\zeta=\zeta_k} dx, \qquad (53b)$$

$$\rho_{,t} = 2i\zeta\rho P_0(\zeta^2) + \frac{\epsilon}{2i\zeta a^2} \int_{-\infty}^{\infty} F\psi^2 \, dx. \qquad (53c)$$

Note that although (53) is basically the same as (12), it differs in that it has singular terms such as ζ_k^{-1} and ζ^{-1}. This is due in part to the solutions of (43) being singular in general as $\zeta \to 0$, and will show up later in the effects of the perturbations.

Let's now consider the effects of damping on the KdV equation where

$$q_t + q_{xxx} + 6q_x q = -\Gamma q . \qquad (54)$$

In plasmas, (54) would model a damped ion-acoustic wave, and in water waves it would model a shallow water wave propagating in a variable depth channel, where Γ is then proportional to the depth gradient.[1] Considering only a single soliton, we take the scattering data initially to be

$$\rho(\xi) = 0, \tag{55a}$$

$$N = 1, \tag{55b}$$

$$\zeta_1 = i\eta, \tag{55c}$$

$$\beta_1 = 2i\eta e^{2\bar{\theta}_1}, \tag{55d}$$

and the one soliton solution is

$$q = 2\eta^2 \text{sech}^2\theta, \tag{56}$$

where

$$\theta = \bar{\theta}_1 - \eta x . \tag{57}$$

From (56) and (57), the central position of the soliton is $\bar{x} = \bar{\theta}_1/\eta$, and from (51), (53) and (55), the equations of motion for η and $\bar{x}$ are

$$\eta_t = -\frac{2}{3} \Gamma\eta, \tag{58a}$$

$$\bar{x}_t = 4\eta^2 + \frac{1}{3}\Gamma/\eta . \tag{58b}$$

Equation (58a) shows that a shallow water wave entering a deeper depth will have its amplitude decreased exponentially. Equation (58b) shows that the effect of this perturbation is to always increase the soliton's velocity above that of the unperturbed case. This is reasonable since we can expect a decaying soliton to emit radiation, which for the KdV can only travel in the negative x direction. Thus conservation of linear momentum requires

the soliton to receive an impulse in the positive x
direction.

In the previous cases which we have considered,
one could always show that the radiation part of the
spectrum would remain unexcited and of order ε.[1] However, due to the ζ^{-1} terms in (53c), that is no longer
the case here, and a significant amount of the continuous spectrum is excited in this case. This becomes
most clear when we consider the infinity of conservation laws associated with (52), the first two of which
are the integrals of q and q^2. When (54) is satisfied,
the conserved quantities are no longer exactly conserved,
but rather

$$\partial_t \int_{-\infty}^{\infty} q\,dx = -\Gamma \int_{-\infty}^{\infty} q\,dx \ , \tag{59a}$$

$$\partial_t \int_{-\infty}^{\infty} q^2 dx = -2\Gamma \int_{-\infty}^{\infty} q^2 dx, \tag{59b}$$

and if we assume that the solution is predominantly a
soliton, from (56) and (58a), we have

$$\partial_t \int_{-\infty}^{\infty} q_s\,dx = -\frac{2}{3}\Gamma \int_{-\infty}^{\infty} q_s\,dx, \tag{60a}$$

$$\partial_t \int_{-\infty}^{\infty} q_s^2 dx = -2\Gamma \int_{-\infty}^{\infty} q_s^2 dx \ . \tag{60b}$$

Comparison of (59) and (60) shows that the soliton is
changing its amplitude so as to satisfy (59b) for the
total energy, but does not satisfy (59a) for the area or
"mass flow." Of course since the first integral goes as
η and the second as η^3, it is impossible for a soliton
solution to satisfy both of these simultaneously. To

explain the difference between (59a) and (60a), we only
have to determine how the continuous spectrum evolves
due to (54). From (53c), one finds

$$\rho_t = -8i\xi^3\rho - \frac{2i\pi\Gamma e^{2i\xi\bar{x}}}{3a\sinh(\pi\xi/\eta)} . \qquad (61)$$

Clearly, the major contribution to $\dot{\rho}$ will come from
$\xi \underset{\sim}{\,} 0$, and upon assuming that the soliton follows the unperturbed path, $\bar{x} = x_0 + 4\eta^2 t$, one finds that the contribution of the continuous spectrum to q is approximately[1]

$$q_c \underset{\sim}{\,} - \frac{\Gamma}{6\pi\eta} \int_{-\infty}^{\infty} \frac{d\xi}{\xi} \begin{pmatrix} \sin 2\xi(x-x_0-4\eta^2 t) \\ -\sin 2\xi(x-x_0+4\xi^2 t) \end{pmatrix} . \qquad (62)$$

The first integral is exactly $\pi \text{sgn}(x-x_0-4\eta^2 t)$ while the
second one is essentially $-\pi \text{sgn}(x-x_0+4\bar{\xi}^2 t)$, for some $\bar{\xi}$.
Thus q_c is essentially nonzero only if $x_0 - 4\bar{\xi}^2 t<x<x_0 +$
$4\eta^2 t$, which is the region between the soliton and radiation, and here q_c is approximately the constant value of

$$q_c \underset{\sim}{\,} - \frac{\Gamma}{3\eta} . \qquad (63)$$

In terms of water waves, this implies that a KdV
soliton moving into deeper water will leave a _depression_
of the water level behind it, which extends back to where
the radiation solution starts, at about $x \underset{\sim}{\,} x_0$. As the
soliton moves forward, the length of this depression increases, and this is just the required amount to explain
the difference between (59a) and (60a). We have

$$\partial_t \int_{-\infty}^{\infty} q_c dx \underset{\sim}{} \partial_t \int_{x_0}^{\bar{x}} q_c dx \underset{\sim}{} \bar{x}_t q_c$$

$$\underset{\sim}{} - \frac{1}{3} \Gamma(4\eta) , \qquad\qquad (64)$$

which follows from (58b) and (63). Since the area under
the soliton is 4η and the area under the continuous spectrum is of order Γ, we have

$$\partial_t \int (q_c + q_s) dx \underset{\sim}{} -\Gamma \int q dx, \qquad\qquad (65)$$

showing that the formation of the depression does account for this discrepancy.

F.   SUMMARY

From these examples, one can see that true solitons
are indeed dynamic entities and do behave like particles.
They can resonate with applied forces, relativistic
(sine-Gordon) solitons behave like relativistic particles,
collisions can be inelastic (kink + antikink → breather),
density gradients can trap solitons, etc. Although there
are many examples still to be looked at (such as kink-
antikink collision in the ϕ^4 theory), it is only due to
the special nature of true solitons that we can obtain
such comprehensive results. First, since the IST is a
canonical transformation, we know what the complete set
of normal modes for the unperturbed system is. Due to
this, the fully nonlinear states (true solitons) will
evolve independent of the linear-like states (radiation),
regardless of their configuration. Thus perturbations
can only weakly couple these modes together, allowing
one to actually follow the perturbed system out to a

time of order ε^{-1}.

In conclusion, although unperturbed soliton systems do appear to be "dull" in a way, when we take the next step and consider other effects on these systems, we find that they are indeed dynamical objects and do exhibit a rich variety of interactions.

ACKNOWLEDGMENTS

Reaearch supported in part by the National Sceince Foundation and the Office of Naval Research.

REFERENCES

1. D. J. Kaup and A. C. Newell, "Solitons as Particles
    and Oscillators," preprint (submitted to Proc. Royal
    Soc.)

2. Zabusky, N. J. and Kruskal, M.D., Phys. Rev. Lett.,
    15, 240 (1965).

3. Ablowitz, M. J., Kaup, D. J., Newell, A. C., and
    Segur, H., Studies in Applied Mathematics, 53, 249
    (1974).

4. Flaschka, H., and Newell, A. C., Lecture Notes in
    Physics, 38, 355, (1975). Dynamical Systems, Theory
    and Applications. Edited by J. Moser, Springer
    Verlag.

5. M. J. Ablowitz, D. J. Kaup and A. C. Newell,
    Journal of Math. Phys., Vol. 15, No. 11, 1852 (1974).

6. Zakharov, V. E., and Shabat, A.B., JETP, 34, 62 (1972).

7. Kaup, D. J., SIAM J. Appl. Math. 31, 121 (1976).

8. Newell, A.C., 1976, The Inverse scattering transform.
    A review article which will appear in a Springer-
    Verlag volume on Solitons. Editors: R. Bullough
    and P. Caudrey.

9. J.P. Keener and D.W. McLaughlin, "A Green's Function
    for a Linear Equation Associated with Solitons,"
    (preprint).

10. Kaup, D.J. and Newell, A.C., Prediction of a Nonlinear,
    Oscillating Dipolar Excitation in One-Dimensional
    Condensates, submitted to Phys. Rev. Letts.

11. Ablowitz, M.J., Kaup, D.J., Newell, A.C. and Segur,
    H., Phys. Rev. Lett., 30, 1262 (1973).

12. Takhtajan, L. and Faddeev, L.,Teor. Mat. Fiz., 21,
    160 (1974).

13. Kaup, D.J., Studies in Appl. Math., 54, 165 (1975).

14. D.W. McLaughlin and A.C. Scott, "Fluxon Interactions,"
    (preprint).

15. H.H. Chen and C.S. Liu, "Solitons in Nonuniform
    Medium," (preprint).

16. This was apparently noted by several different in-
    dividuals at about the same time.

THE FORMULATION OF VARIATIONAL PRINCIPLES BY MEANS OF CLEBSCH POTENTIALS

Hanno Rund*

Department of Mathematics

The University of Arizona, Tucson, Arizona 85721

I. INTRODUCTION

Although Clebsch representations have appeared sporadically in the literature on classical hydro-dynamics,[1] the applicability of such representations to physics in general, and variational principles in particular, seems to have been overlooked until fairly recently. This is probably due to the fact that the standard formulation of such representations, originally due to A. Clebsch,[2] is subject to severe restricitions, as is immediately evident from the following enunciation: Given any differentiable vector field $\vec{X}$ on a 3-dimensional Euclidean space E_3, there exist scalar functions ψ, P, Q of the coordinates of E_3, such that $\vec{X}$ can be represented in the form

$$\vec{X} = \nabla\psi + Q\nabla P. \qquad (1.1)$$

---

*This research was sponsored in part by NSF GP 40370.

The aforementioned restricitions refer to the dimension
of the space, as well as to its Euclidean character.
Thus, when applications of a general nature are envisaged,
one is compelled to seek extensions (if such are
feasible) which are not subject to the limitations
inherent in (1.1).

However, before we proceed to discuss such possible
generalizations, the relevance of representations such
as (1.1) should be briefly clarified. Consider, for
example, the integral of the inner product $\vec{X} \cdot \vec{X}$ over
a fixed region G of E_3, and suppose that one wishes to
obtain the conditions which must be satisfied in order
that this functional assume an extreme value for a class
of vector fields $\vec{X}$ with prescribed boundary conditions
on ∂G. Unfortunately, formulated thus, this is not a
well-posed problem in the calculus of variations. This
is due to the fact that the given Lagrangian $\vec{X} \cdot \vec{X}$ does
not contain any derivatives, which implies that the
usual criteria, such as the Euler-Lagrange equations,
and the conditions of Legendre and Weierstrass, are
inapplicable. However, when $\vec{X}$ is replaced in the
Lagrangian by the representation (1.1), an equivalent
variational problem is obtained for the new field
functions ψ, P, Q (the so-called Clebsch potentials), and,
since derivatives of ψ and P have thus been injected
into the Lagrangian, the problem is now well-posed.

Obviously one would expect generalizations of (1.1)
to give rise to similar techniques which are applicable
to more general situations, and it is the object of the
present lecture to describe some consequences of this
scheme. As will be indicated below, there are at
least two distinct types of generalized Clebsch
representations. The first of these, to be described in

section 2, is concerned with the representation of
arbitrary skew-symmetric tensor fields, which is
obviously of some relevance to electrodynamics. It is
assumed that an electromagnetic field on a 4-dimensional
Riemannian space-time manifold is represented by a
skew-symmetric tensor field F_{hj}, whose behavior is
governed by an invariant variational principle. In the
Lagrangian of the latter the role of the field functions
is played by the Clebsch potentials of F_{hj}, and it is
found that the Euler-Lagrange equations give rise to the
complete set of Maxwell's equations. Since this
technique does not depend on the assumption that F_{hj}
is equivalent to the curl of a 4-potential, the
possibility of the existence of magnetic charge
distributions is not excluded a priori; in fact, a
modified variational principle yields the extended
Maxwell equations of the kind previously postulated by
Schwinger[3] and others for field theories admitting
magnetic as well as electric charge distributions.
Moreover, when solutions of the corresponding Einstein-
Maxwell equations in vacuo are considered for the case
of a static electromagnetic field and a spherically
symmetric metric, it is found that the electric and
magnetic charges of the central gravitating mass which
gives rise to these fields occur on precisely the same
footing, namely as constants of integration.

The second of the aforementioned generalizations
of (1.1) is given in section 2, in which applications
to nonrelativistic Hamiltonian systems are discussed.
A generalized Hamilton-Jacobi theory is thus obtained,
which gives rise in a natural manner to an associated
multiple integral variational problem. The Euler-Lagrange
equations of the latter imply not only the field equations,

but also the canonical equations of motion; this is in sharp contrast to the standard approach to classical field theories, in which the equations of motion result from an entirely separate single integral variational principle. In this sense, therefore, the method of Clebsch potentials can be regarded as a unifying influence.

In the final section 4 similar techniques are applied to relativistic Hamiltonian systems. It is found, in particular, that the generally accepted equations of motion of a perfect fluid in general relativity are thus obtained, although, again in contrast to the standard approach, no assumption what-soever with regard to the structure of the energy-momentum tensor is made. Moreover, a single variational principle in terms of Clebsch potentials gives rise not only to the Einstein field equations in the presence of matter, but also implies (in all rigor) that the motion of free particles must take place along the geodesics of the underlying space-time manifold.

II. APPLICATIONS TO ELECTRODYNAMICS

The first of the aforementioned generalizations[4] of the Clebsch representation (1.1) can be formulated as follows: Let V_n denote a differentiable manifold[5], referred to local coordinates x^h, on which there is given an $(n-2)$-index geometric object field $X_{h_1 \ldots h_{n-2}}$, which is differentiable and skew-symmetric in all subscripts. Then there exist $(n-3)$-index fields $Y_{j_1 \ldots j_{n-3}}$, and $(n-1)$ functions $F^{(1)}, \ldots, F^{(n-1)}$, such that the given field may be represented (at least locally) in the form

$$X_{h_1 \cdots h_{n-2}} = \delta^{j_1 \cdots j_{n-2}}_{h_1 \cdots h_{n-2}} \frac{\partial Y_{j_1 \cdots j_{n-3}}}{\partial x^{j_{n-2}}} +$$

$$F^{(n-1)} \frac{\partial(F^{(1)}_{h_1}, \ldots, F^{(n-2)}_{h_{n-2}})}{\partial(x^{h_1}, \ldots, x^{h_{n-2}})} , \qquad (2.1)$$

where $\delta^{\cdots}_{\cdots}$ denotes the generalized Kronecker delta.
Clearly (2.1) reduces to (1.1) when n=3, and accordingly
(2.1) is called a _generalized Clebsch representation_,
the fields which appear on the right-hand side being
referred to as _Clebsch potentials_. Obviously the latter
are not unique; in fact, they are subject to a fairly
wide class of gauge transformations.

Because of the skew-symmetric character of the
given field, the representation (2.1) is of immediate
relevance to electrodynamics when n = 4, with x^4 = ict.
Indeed, let us suppose that our V_4 is endowed with a
Riemannian metric tensor field g_{hj}, it being assumed
that $g = |\det(g_{hj})| \neq 0$. Let us also suppose that an
electromagnetic field on V_4 is represented as usual
by a skew-symmetric tensor field F_{hj}. As a direct
consequence of (2.1) we may always express F_{hj} in the
form

$$F_{hj} = f_{hj} + ib_{hj} , \qquad (2.2)$$

where[6]

$$f_{hj} = \psi_{j|h} - \psi_{h|j} + R \frac{\partial(P,Q)}{\partial(x^j, x^h)} , \qquad (2.3)$$

and

$$b^{hj} = g^{-1/2} \varepsilon^{hj\ell k} (\phi_{k|\ell} + W \frac{\partial U}{\partial x^\ell} \frac{\partial W}{\partial x^k}) . \qquad (2.4)$$

Clearly the Lagrange density defined by

$$L_o = \frac{1}{4} \sqrt{g} \, F^{hj} \, F_{hj},\qquad\qquad(2.5)$$

gives rise to an invariant integral of a well-posed problem in the calculus of variations, provided that the Clebsch representations (2.2)-(2.4) are substituted in the integrand (2.5). In fact, it may be shown[7] that the corresponding Euler-Lagrange equations for the Clebsch potentials $\psi_h, P, Q, R, \phi_h, U, V, W$ imply that the representations (2.3) and (2.4) are reduced to the simple form

$$f_{hj} = \psi_{j|h} - \psi_{h|j} \,, \quad ib^*_{hj} = \phi_{h|j} - \phi_{j|h} \,;\quad(2.6)$$

moreover, these Euler-Lagrange equations also give rise to the Maxwell equations in their classical form:

$$(\sqrt{g} \, F^{hj})_{|j} = 0, \; (\sqrt{g} \, F^{*hj})_{|j} = 0.\qquad(2.7)$$

At this stage it should be emphasized that this derivation of Maxwell's equations depends solely on two assumptions: namely, the skew-symmetry of F_{hj}, and the variational principle embodied in (2.5). This is in sharp contrast to the standard treatment of these equations, in which it is assumed a priori that F_{hj} is the curl of a given 4-potential, it being recalled that such an assumption implies the validity of the second member of (2.7) as an identity, while only the first member emerges as the Euler-Lagrange equation associated with the Lagrangian (2.5). Within the present context, however, the complete set of Maxwell's equations is obtained as a system of Euler-Lagrange equations, no prior assumption as to the existence of a 4-potential having been made. The signigicance of this observation is not merely of a mathematical nature, as is evident from the fact that the existence of a 4-potential $(\vec{A}, \phi)$, for which $\vec{H} = \nabla \times \vec{A}$, precludes the possibility of the existence of magnetic charge distributions.

This suggest the introduction of a Lagrangian which is more general than (2.5), namely

$$L = L_o - \psi_h J^h + \phi_h S^h , \qquad (2.8)$$

in which ψ_h, ϕ_h once more refer to the Clebsch representation (2.3), (2.4), and where J^h, S^h respectively denote electric and magnetic current densities. The resulting Euler-Lagrange equations again imply the representation (2.6), together with <u>the extended</u> <u>Maxwell equations</u>

$$(\sqrt{g}\ F^{hj})_{|j} = J^h , \quad (\sqrt{g}\ F^{hj})|_j = - S^h , \qquad (2.9)$$

which, at least in flat space-time, are identical with those previously postulated by Schwinger[8] and others. In passing we remark that in flat space-time the representation (2.6) may be expressed in the form

$$\vec{E} = -\nabla V - \frac{1}{c}\frac{\partial\vec{\psi}}{\partial t} - \nabla \times \vec{\phi}, \quad \vec{H} = -\nabla U - \frac{1}{c}\frac{\partial\vec{\phi}}{\partial t} + \nabla \times \vec{\psi},$$
$$(2.10)$$

where U and V are given by $\psi_4 = iV$, $\phi_4 = iU$. When $S^h = 0$, the second member of (2.9) implies that the field ϕ_h can be removed by means of a gauge transformation: in this manner classical electromagentic theory (without magnetic charge distributions) may be retrieved.

Thus far it has been tacitly assumed that the under-lying metric g_{hj} has been prescribed, regardless of the electromagnetic field. However, it is now obvious how a single variational principle for the determination of

the behavior of both fields may be constructed. Let
R_{hj} denote the Ricci tensor of V_4, so that $R = g^{hj}R_{hj}$
is the curvature scalar. The Euler-Lagrange equations
associated with the density

$$L = a\sqrt{g}\ R + L_o \ , \qquad\qquad (2.11)$$

where $a = (16\pi\kappa)^{-1}c^{-4}$, again give rise to the
representation (2.6), together with the extended
Einstein-Maxwell equations in vacuo, the latter consisting of (2.7) and

$$R^{hj} = - \frac{8\pi\kappa}{c^2}\ T^{hj} \ , \qquad\qquad (2.12)$$

where the electromagnetic energy-momentum tensor density
must necessarily[9] occur in the form

$$T^{hj} = -2c^{-2}\ \frac{\partial L_o}{\partial g_{hj}} \ . \qquad\qquad (2.13)$$

In order to determine the effect which the presence
of magnetic charge distributions may have on the
gravitational field, it is advisable to consider the
case of a spherically symmetric line element of the form

$$ds^2 = c^2 e^\nu dt^2 - (e^\lambda dr^2 + r^2 d\theta^2 + r^2\sin^2\theta d\phi^2), \quad (2.14)$$

where λ,ν are functions solely of the radial coordinate
r, together with a static field defined by

$$\psi_h = (0,0,0,iV(r)), \quad \phi_h = (0,0,0,iU(r)) \ . \qquad (2.15)$$

The Einstein-Maxwell equations then imply that the real
functions U, V must satisfy the differential equations

$$\frac{dU}{dr} = \frac{\varepsilon}{r^2}\ e^{(\lambda+\nu)/2} \ , \quad \frac{dV}{dr} = \frac{\gamma}{r^2}\ e^{(\lambda+\nu)/2} \ , \quad (2.16)$$

where ε and γ are (real) constants of integration,
which are here interpreted respectively as the electric
and magnetic charges of the central gravitating mass
which generates the gravitational field corresponding
to the metric (2.14). Moreover, the field equations
also dictate that the functions λ and ν which appear
in (2.14) are given by

$$e^{-\lambda} = e^{\nu} = 1 - \frac{2m}{r} + \frac{4\pi\kappa}{c^4 r^2} (\varepsilon^2 - \gamma^2) . \qquad (2.17)$$

When both ε and γ vanish, this reduces to the
Schwarzschild solution, while the case $\gamma = 0$ gives rise
to the Reissner-Nordström metric. However, it should
be stressed that the Lagrangian (2.11) refers to the
vacuum by virtue of the structure of its component
(2.5) and therefore does not a priori contain any
built-in elements with regard to electric or magnetic
charges; indeed, these charges occur in (2.16) as a
consequence of the field equations, namely as constants
of integration, and, as such, arise on an entirely
equal footing.

The phenomenon that $-\gamma^2$ (rather than $+\gamma^2$) appears
in (2.17) is somewhat surprising. This is due to the
fact that the energy-momentum tensor, as defined by
(2.13), assumes the following explicit form when
evaluated for the fields defined by (2.14) and (2.15):

$$(T_h^j) = \frac{1}{2}c^{-2} (\varepsilon^2 - \gamma^2) \sqrt{g} \, r^{-4} \, \mathrm{diag}[-1,1,1,-1]. \quad (2.18)$$

Thus a thorough appraisal of the tensor (2.13) appears to
be warranted. The definition (2.13) automatically
ensures[9] that the following conditions are satisfied
by the energy-momentum tensor density: (i) it is
symmetric; (ii) it is trace-free; (iii) its divergence

vanishes as a consequence of Maxwell's equations (2.7);
(iv) it is invariant under arbitrary guage transformations of the Clebsch potentials; (v) it reduces to the energy-momentum tensor of classical electrodynamics in the
absence of magnetic charges. However, in addition to
these desirable features, there are also two possible
drawbacks: (vi) when the rate at which work is being
done on a test particle carrying both electric and
magnetic charges is evaluated in terms of the
4-divergence of T_4^j, the resulting expression for the
modified Lorentz force does not reduce to that postulated
by Schwinger[10] for the case of flat space-time; (vii)
the value of $-T_4^4$ is not necessarily nonnegative when
magnetic charges are present.

It is possible to construct[11] an alternative
energy-momentum tensor density θ^{hj} which displays the
properties (i) - (v), but which does not suffer from
the drawbacks (vi) and (vii). Moreover, the solutions
of the alternative Einstein-Maxwell equations, obtained
from (2.12) by the replacement of T^{hj} by θ^{hj} on the
right-hand side, give rise to a metric as specified by
(2.17) in which $+\gamma^2$ occurs instead of $-\gamma^2$. Unfortunately,
there does not in general exist a Lagrangian such that
θ^{hj} can be expressed in the form (2.13), and consequently
the alternative Einstein-Maxwell equations cannot be
derived from a variational principle. Clearly the
ultimate choice between these energy-momentum tensor
densities must await the outcome of future developments.

III. HAMILTONIAN SYSTEMS (NON-RELATIVISTIC THEORY)

The second generalization of the Clebsch
representation (1.1) referred to in the introduction is
an immediate consequence of a known theorem[12] on Pfaffian

forms. Let $p_j(x^h)$ denote the components of a type
(0.1) tensor field on a differentiable manifold V_n.
This field defines an invariant 1-form $\omega = p_j dx^j$. An
application of the aforementioned theorem to ω leads to
the following Clebsch representation:

$$p_j = \frac{\partial \psi}{\partial x^j} + Q^\alpha \frac{\partial P_\alpha}{\partial x^j} \quad \text{(summation over } \alpha\text{)}, \quad (3.1)$$

where

$$\alpha = 1,\ldots,(n-1)/2 \text{ when n is odd}, \quad (3.2)$$

and

$$\alpha = 1,\ldots,n/2 \ \underline{\text{and}} \ \psi = 0 \text{ when n is even}. \ (3.3)$$

Let us now consider a classical holonomic dynamical
system of n degrees of freedom, specified by some given
Hamiltonian function $H(t,x^j,p_j)$, where t denotes the
time, and x^j,p_j are the generalized coordinates and
momenta respectively, the former being regarded as the
local coordinates of the underlying manifold V_n. It is
assumed that H is of class C^2 and invariant under
arbitrary transformations of the coordinates; accordingly
the momenta p_j are the components of a type (0,1) tensor
field. They are therefore susceptible to a representation
of the form (3.1), in which, however the Clebsch
potentials[13] may be functions of t also as a consequence
of the possible dependence of p_j on t.

It is now supposed that we are given a field
$p_j(t,x^h)$, which satisfies the canonical equations

$$\frac{dx^j}{dt} = \frac{\partial H}{\partial p_j} , \quad \frac{dp_j}{dt} = - \frac{\partial H}{\partial x^j} . \quad (3.4)$$

This has the following implications[14] in terms of the representation (3.1): there exists a 'superpotential' $\Phi = \Phi(t,Q^\alpha,P_\alpha)$ such that the Clebsch potentials satisfy the differential equations

$$\frac{dQ^\alpha}{dt} = \frac{\partial\Phi}{\partial P_\alpha} \, , \quad \frac{dP_\alpha}{dt} = -\frac{\partial\Phi}{\partial Q^\alpha} \, , \qquad (3.5)$$

where Φ is related to the given Hamiltonian according to the relation

$$\frac{\partial\psi}{\partial t} + Q^\alpha \frac{\partial P_\alpha}{\partial t} + \Phi(t,Q^\beta,P_\beta) + H(t,x^h, \frac{\partial\psi}{\partial x^h} + Q^\alpha \frac{\partial P_\alpha}{\partial x^h}) = 0,$$
$$(3.6)$$

and coversely.

It is readily seen that this construction entails a substantial generalization of the concept of the complete figure of the calculus of variations in the sense of Carathéodory.[15] It will be recalled that this figure is characterized by a representation of the canonical momentum as the gradient of a characteristic function $S(t,x^h)$:

$$p_j = \frac{\partial S}{\partial x^j} \, , \qquad (3.7)$$

where S is required to be a solution of the Hamilton-Jacobi equation

$$\frac{\partial S}{\partial t} + H(t,x^j, \frac{\partial S}{\partial x^j}) = 0. \qquad (3.8)$$

When a representation such as (3.7) is indeed admissible, we may choose the potentials in (3.1) as follows:

$$\psi = 0, \quad Q^\beta = \delta_1^\beta, \quad P_\beta = \delta_\beta^1 S, \quad \Phi = 0.$$

Under these circumstances, the relation (3.6) reduces

to (3.8), which, together with (3.7), is known to imply
the second member of the canonical equations (3.4).
Thus <u>the partial differential equation</u> (3.6) <u>for the
Clebsch potentials is to be regarded as a generalized
Hamilton-Jacobi equation</u>.

The essence of this generalization may be illustrated as follows: Whenever (3.7) holds, the 'generalized vorticity' tensor

$$\omega_{hj} = \frac{\partial p_j}{\partial x^h} - \frac{\partial p_h}{\partial x^j} \qquad\qquad (3.9)$$

vanishes identically. However, it is found in the
general case that the equations (3.5) merely imply that
<u>the Lie derivative of</u> ω_{hj} <u>with respect to the vector
field</u> $\partial H/\partial p_j$ <u>vanishes</u>; from this it may be inferred that
the tensor (3.9) may be represented in the form

$$\omega_{hj}(t,x^p) = \omega^o_{\ell k} \, A^\ell_h(t,x^p) A^k_j(t,x^p) \, , \qquad (3.10)$$

where $\omega^o_{\ell k}$ is the value of $\omega_{\ell k}$ at some initial time
$t = t_o$. Thus, in particular, if $\overset{\cdot}{\omega}_{\ell k} = 0$ at $t = t_o$,
this property will persist for subsequent values of t,
in which case the representation (3.7) is valid, as is
immediately evident from (3.9).

The principal advantage of the method of Clebsch
potentials derives from the fact that one may associate
a multiple integral variational problem with the system
(3.5) - and hence with the canonical equations (3.4).
Indeed, let us introduce an auxiliary function $\nu(t,x^h)$,
in terms of which the following Lagrange density may be
constructed:

$$L_o = \nu[\frac{\partial \psi}{\partial t} + Q^\alpha \, \frac{\partial P_\alpha}{\partial t} + H(t,x^h,p_h) + \Phi(t,Q^\beta,P_\beta)] . \quad (3.11)$$

This Lagrangian defines an (n+1)-fold integral problem
in the calculus of variations for the functions
$(\psi, Q^{\alpha}, P_{\alpha}, \nu)$, it being understood that the arguments p_j
which appear in H are to be expressed in terms of (3.1).
It is found that <u>the system of Euler-Lagrange equations
associated with</u> (3.11) <u>implies the equation of continuity</u>

$$\frac{\partial \nu}{\partial t} + \frac{\partial}{\partial x^j} \left(\nu \frac{\partial H}{\partial p_j}\right) = 0 , \qquad (3.12)$$

<u>together with the canonical equations</u> (3.5), <u>and con-
versely</u>. Moreover, when the Hamiltonian involves some external potential field $V(t, x^h)$, whose behavior is governed
by a variational principle in terms of a separate Lagrangian, the density (3.11) may be augmented by the latter.
<u>The resulting single variational principle then gives rise
not only to the equations of motion, that is, to the</u>
canonical equations (3.4), <u>but also to the field equations</u>

The corresponding conservation laws can now be determined in accordance with standard techniques based on
Noether's theorem. More directly, the Hamiltonian complex[16]
may be derived from (3.11). This complex is not invariant
under general gauge transformations of the Clebsch
potentials, but it may be rendered gauge-invariant by the
addition of suitable terms which do not affect the value
of the divergence of the complex. The following energy-
momentum tensor is thus obtained:

$$\overset{o}{T}{}^A_B = \begin{pmatrix} -L\,\delta^j_h + \nu p_h \dfrac{\partial H}{\partial p_j} \,, & \nu(p_{n+1} + \Phi)\dfrac{\partial H}{\partial p_j} \\[4mm] \nu p_h \,, & -\,L_o + \nu(p_{n+1} + \Phi) \end{pmatrix} ,$$

$$(3.13)$$

where $A, B = 1, \ldots, n+1$, and

$$p_{n+1} = \frac{\partial \psi}{\partial t} + Q^\alpha \frac{\partial P_\alpha}{\partial t} . \tag{3.14}$$

According to the general theory, the tensor (3.13) satisfies $n + 1$ divergence relations, the most important of which is of the form

$$\frac{\partial \overset{o}{T}{}_{n+1}^{\;\;j}}{\partial x^j} + \frac{\partial \overset{o}{T}{}_{n+1}^{\;n+1}}{\partial t} = \nu \frac{\partial H}{\partial t} , \tag{3.15}$$

which gives rise directly to a conservation law whenever $\partial H / \partial t = 0$.

A special case of particular importance may be discussed within this context when H is the (non-relativistic) Hamiltonian of a particle in an electromagnetic field $(\vec{E}, \vec{H})$, while, at the same time, the Lagrangian (3.11) is augmented by the expression $\frac{1}{2}(E^2 - H^2)$, together with suitably chosen terms which represent a pressure field.[17] The Euler-Lagrange equations of the resulting variational principle give rise directly to Maxwell's equations, together with the generally accepted equations of motion of a magnetofluid, provided that the auxiliary function $\nu(t, x^h)$ is interpreted as the particle number density [which is entirely reasonable in view of the continuity equation (3.12)]. Moreover, the term $- T_4^{\;4}$, evaluated according to (3.13), coincides precisely with an expression for the energy density of a magnetofluid as derived by Bernstein et. al.[18] on the basis of a completely different approach.

IV. HAMILTONIAN SYSTEMS (RELATIVISTIC THEORY)

It is possible to establish a theory for relativistic Hamiltonian systems along lines somewhat analogous to

those described in the previous section. However,
fundamental differences result from the fact that the
time t can no longer be singled out as a preferred
independent variable; instead, an arbitrary parameter
τ must be introduced. This implies that the initial
theory has to be based on a parameter-invariant
variational principle, that is, on a scalar Lagrangian
$L = L(x^j, \dot{x}^j)$, with $\dot{x}^j = dx^j/dt$, which is homogeneous
of the first degress in $\dot{x}^j$. Thus the canonical
momenta are now defined by the equations $y_j = L \, \partial L/\partial \dot{x}^j$;
since it is assumed that these relations can be solved
for $\dot{x}^j = \phi^j(x^h, y_h)$, this gives rise to the invariant
Hamiltonian[19]

$$H(x^j, y_j) = L(x^h, \phi^h(x^j, y^j)). \qquad (4.1)$$

This function, which is not directly related to the
energy, is homogeneous of the first degree in y_j by
virtue of this construction, while the canonical
equations associated with the variational problem
defined by L are to be expressed in the form

$$\frac{dx^j}{dt} = H \, \frac{\partial H}{\partial y_j} \;, \quad \frac{d}{dt}\left(\frac{y_j}{H}\right) = - \frac{\partial H}{\partial x^j} \;. \qquad (4.2)$$

These equations assume the formal structure of their
nonrelativistic counterparts (3.4) when the parameter
τ is chosen such that H = 1. [This can always be
achieved by setting $d\tau = L(x^h, dx^h)$, but in general it is
advisable to choose τ in a more convenient manner which
is usually suggested by the specific problem at hand.]
        Restricting ourselves to the case when n is even
(since the most interesting physical situations occur
when n = 4), we choose the following Clebsch
representation for y_j:

$$y_j = P_\alpha \frac{\partial Q^\alpha}{\partial x^j} \ . \tag{4.3}$$

The validity of the canonical equations (4.2) now implies
the existence of a superpotential $\Phi(Q^\alpha, P_\alpha)$, in terms
of which the Clebsch potentials satisfy the equations

$$\frac{dQ^\alpha}{d\tau} = \Phi \frac{\partial \Phi}{\partial P_\alpha}, \ \frac{d}{d\tau} \left(\frac{P_\alpha}{\Phi}\right) = -\frac{\partial \Phi}{\partial Q^\alpha} \ , \tag{4.4}$$

the function Φ being related to H by the partial
differential equation

$$H\left(x^h, P_\alpha \frac{\partial Q^\alpha}{\partial x^h}\right) - \Phi(Q^\alpha, P_\alpha) = 0, \tag{4.5}$$

and conversely. Again, this equation can be regarded
as a generalized Hamilton-Jacobi equation.

The general theory proceeds along the lines
described in the previous section. In particular, if
we introduce a scalar density field $\nu(t, x^h)$, by means
of which the Lagrangian of an n-fold integral variational
problem is defined as

$$L_o(x^h, Q^\alpha, P_\alpha, \nu) = \nu \left[H\left(x^h, P_\alpha \frac{\partial Q^\alpha}{\partial x^h}\right) - \Phi(Q^\alpha, P_\alpha)\right], \tag{4.6}$$

the resulting Euler-Lagrange equations give rise to the
equation of continuity

$$\frac{\partial}{\partial x^j} \left(\nu \frac{\partial H}{\partial y_j}\right) = 0, \tag{4.7}$$

together with the equations (4.4) - and hence also to
the canonical equations (4.2).

Thus far the gravitational field has played no role
in the theory. In order to introduce such fields, let
us assume that n = 4, the underlying manifold being

endowed with a Riemannian metric g_{hj} in the sense of
section 2. It is then possible to construct a
Lagrange density

$$L = \sqrt{g}\ R + 2ikL_o - 2k\sqrt{g}\ V, \qquad (4.8)$$

where k is a constant and V a scalar function representing possible pressure terms whose structure will be
determined presently. The term L_o in (4.8) has the
same form as (4.6), but it must be realized that the
dependence of H on x^h may now explicity involve - inter
alia - the metric tensor also. The resulting Euler-
Lagrange equations which correspond to the variation
of g_{hj} and v give rise to the field equations

$$\sqrt{g}\ (R^{hj} - \frac{1}{2}\ Rg^{hj}) = -\ k\ T^{hj}\ , \qquad (4.9)$$

where the matter energy-momentum tensor is necessarily
defined by

$$T^{hj} = -iv\ \frac{\partial H}{\partial g_{hj}} + \sqrt{g}\ g^{hj}\ V\ . \qquad (4.10)$$

Since the divergence of the left-hand side of (4.10)
vanishes identically, it follows that

$$T^{hj}\big|_j = 0\ , \qquad (4.11)$$

and this conclusion must be consistent with the remaining
Euler-Lagrange equations which correspond to a variation
of the Clebsch potentials Q^α, P_α. In fact, it is found
that this consistency requirement almost completely
specifies the structure of the functions H and V.

The following result is crucial to the consistency
argument: Let $H(x^h, g_{hj}, y_h)$ be (i) a scalar, (ii)
homogeneous of the first degree in y_h, (iii) of class

C^1 in g_{hj}. Then H must possess the form

$$H = U^{-1}(x)(g^{hj}y_h y_j)^{1/2}, \qquad (4.12)$$

where U is an arbitrary scalar function of the positional coordinates only. Since the conditions (i) and (ii) are implied by the construction (4.1), while (iii) shall now be assumed, we shall suppose henceforth that H is given by (4.12) with some as yet unspecified function U. Accordingly we can express the canonical equations (4.2) in a more explicit form, it being recalled that, according to the general theory, these equations are implied by the Euler-Lagrange equations associated with L_o and hence also by those associated with L. Also, at this stage it is convenient to specify the parameter τ, which has been quite arbitrary thus far, by setting $\tau = s$, where $ds^2 = -g_{hj} dx^h dx^j$. Then, putting $u^j = dx^j/ds$, it is found that the equations (4.2) are equivalent to

$$U \frac{Du^j}{Ds} + \frac{dU}{ds} u^j + g^{hj} \frac{\partial U}{\partial x^h} = 0, \qquad (4.13)$$

where D/Ds denotes the absolute derivative. Moreover, with the aid of (4.12) the tensor (4.10) can be expressed as

$$T^{hj} = \frac{1}{2} U\nu u^j u^h + \sqrt{g}\, g^{jh} V. \qquad (4.14)$$

The explicit form of the aforementioned consistency condition can now be derived in terms of (4.13) and (4.14): indeed, this condition is completely equivalent to the requirement that the functions U, V are related according to the partial differential equation

$$\frac{1}{2}\, \mu\, \frac{\partial U}{\partial x^j} = \frac{\partial V}{\partial x^j} \, , \tag{4.15}$$

where μ is the scalar function $v g^{-1/2}$. This suggests the introduction of the function

$$F = \frac{1}{2}\, \mu\, U - V, \tag{4.16}$$

in terms of which (4.13) may be written as

$$(F + V)\, \frac{D u^j}{D s} + \frac{dV}{ds}\, u^j + g^{hj}\, \frac{\partial V}{\partial x^h} = 0, \tag{4.17}$$

while (4.11) and (4.14) then yield

$$\frac{dF}{ds} + (F + V) u^h_{\ |h} = 0 \, . \tag{4.18}$$

These equations display a structure which is strikingly similar to that of the generally accepted equations of motion of a perfect fluid in the general theory of relativity[20]. In fact, these systems may be identified with each other if F and V are interpreted respectively as the energy density ρ and the pressure p as measured by an observer in a co-moving frame. For, under these circumstances, the relation (4.18) gives rise to the equation

$$\frac{du}{ds} + p\, \frac{d}{ds}\, (\frac{1}{\mu}) = 0 \, , \tag{4.19}$$

where $u = \rho/\mu$ is the internal energy, while the energy-momentum tensor (4.14) now assumes the generally accepted form

$$T^{hj} = \sqrt{g}\, [(p + \rho) u^h u^j + p g^{hj}]. \tag{4.20}$$

It should be emphasized that <u>this form of</u> T^{hj} <u>is a direct consequence of the theory</u>; it has not, as is usually the case, been assumed <u>a priori</u>. Moreover, the scalar U which appears in the expression (4.12) for the Hamiltonian turns out to be identical with 2w, where w is the specific enthalpy.

Finally, we remark that the special case of an incoherent matter field, whose particles do not interact, is obtained from the above analysis by putting V = p = 0. It then follows from (4.15) that U = const., and (4.13) reduces to

$$\frac{Du^j}{Ds} = 0, \tag{4.21}$$

which, by virtue of the special choice of the parameter, is simply the differential equation satisfied by the geodesics on V_4. It is therefore concluded that <u>the single variational principle defined by the Lagrange density</u> (4.8), <u>with</u> V = 0, <u>not only yields the gravitational field equations</u> (4.9), <u>but also implies that the motion of free particles takes place along the geodesics of the space-time manifold</u>.

REFERENCES

1. See, for instance, J. Serrin, Handbuch der Physik VIII/1, 125(1959); C. Eckart, Phys. Fluids 3, 421 (1960).

2. A. Clebsch, J. reine angew. Math. 54, 293(1857); ibid. 56, 1(1859).

3. J. Schwinger, Phys. Rev. D12, 3105(1975); Science 165, 757(1969).

4. H. Rund, Topics in Differential Geometry (edited by H. Rund and W. F. Forbes, Academic Press, New York, 1976), p. 111.

5. Unless otherwise specified, lower case Latin indices range from 1 to n, where $n \geq 3$; the summation convention is operative throughout.

6. Vertical bars followed by a subscript denote covariant differentiation in terms of the Christoffel symbols defined by the metric tensor.

7. For the proofs of these statements the reader is referred to H. Rund, J. Math. Phys. 18, 84(1977), and a subsequent article to appear in the same journal.

8. Reference, footnote 3.

9. D. Lovelock and H. Rund, Tensors, Differential Forms, and Variational Principles (Wiley-Interscience, New York, 1975), pp. 302-304.

10. Reference, footnote 3.

11. Second reference, footnote 7.

12. E. Goursat, Lecons sur le Probléme de Pfaff (Hermann, Paris, 1927), pp. 32-42.

13. For the purposes of the present discussion it is assumed that n is odd; only minor modifications are required for the case when n is even.

14. For the proofs of these and subsequent statements, the reader is referred to a forthcoming article in the "Archive for Rational Mechanics and Analysis".

15. C. Carathéodory, _Variationsrechnung_ _und_ _partielle_ _Differentialgleichungen_ _erster_ _Ordnung_ (Teubner, Leipzig und Berlin, 1935).

16. H. Rund, The Hamilton-Jacobi Theory in the Calculus of Variations (Van Nostrand, London and New York, 1966; augmented and revised edition, Krieger Publishing Company, Huntington, New York, 1973), p. 240.

17. In the absence of pressure terms the Lagrangian thus constructed reduces to a Lagrangian suggested by Seliger and Whitham for the study of plasma waves [Proc. Roy. Soc. Ser. A 305, 1(1968)].

18. I. B. Bernstein, E. A. Frieman, M. D. Kruskal, and R. M. Kulsrud, Proc. Roy. Soc. Ser. A. 244, 17(1958), p. 19.

19. This construction is motivated and discussed in all detail in Chapter 3 of reference 16.

20. J. Ehlers, Akad. Wiss. Mainz, Math.-naturw. Kl. 1961, No. 11(1961), p. 32.

COHERENT STRUCTURES IN FLUID DYNAMICS

N. J. Zabusky

University of Pittsburgh

Pittsburgh, PA 15260

In the last decade we have experienced a conceptual shift in our view of turbulence. For flows with strong velocity shears, near boundaries, density gradients, magnetic fields or other organizing characteristics, many now feel that the spectral or wave-number space description has inhibited fundamental progress.

The next "El Dorado" lies in the mathematical understanding of coherent structures in weakly dissipative fluids: the formation, evolution and interaction of metastable, vortex-like solutions of nonlinear, partial differential equations. By juxtaposing recent experimental results with computer simulations, I hope to focus on essential analytical directions.

The impact of modern flow visualization methods and of conditional sampling methods has been reviewed recently by Davies and Yule (1975), and Roshko (1976) for classical fluids. Fetter (1974) and P.H. Roberts and Donnelly (1974) also have reviewed the unusual properties of liquid helium, a quantum fluid. Below 1°K inviscid classical fluid theory applies to vortex dynamics. I

will concentrate on the simulations of unstable shear
flows, in particular the two-dimensional street of
<u>finite-area</u> <u>vortical</u> <u>regions</u> (FAVRs) or "vortex cores",
as studied by Zabusky and Deem (1971) and Christiansen
and Zabusky (1973); and cusped wave formation in stratified media with a simple velocity shear (Deem 1977).

    I will omit a discussion of the work that deals with
the statistical aspects of apparently chaotic hydrodynamical turbulence (e.g. Leslie 1973, Orszag 1976).
The approach taken, working with correlation functions
and Fourier spectra, has deemphasized the importance of
studying coherent structures and their interactions in
physical space as argued by Saffman (1968, 1971).

    I will also omit a discussion of the recent work
that attempts to draw insight on continuum turbulence
from the solution of a few ordinary differential equations, like that proposed by Lorenz. The first transition to instability, and perhaps the second, may be
"fitted" within the solution set of a few nonlinear
ordinary differential equations. However, if the excitation parameter exceeds a threshhold, these systems
exhibit a "chaotic" behavior. I am convinced that these
trajectories have little to do with the turbulence we
observe in the richer nonlinear continuum system.

    In the following sections I will:

    (i)   Present the equations for two-dimensional
           hydrodynamics (formally equivalent to a
           collisonless plasma with a strong magnetic
           field, the so-called guiding center plasma).

    (ii)   Discuss the contemporary experimental two-
           dimensional fluid dynamical situation.

    (iii)   Discuss the interrelationship of laboratory
           experiments and properties of numerical

simulations obtained with finite-difference and particle- field codes.

(iv) Discuss a not-so-familar situation: the evolution of perturbations to incompressible sheared-and-stratified fluids described by the Boussinesq equations show standing "cat-eyes" and propagating cusped waves.

(v) Discuss numerical solutions of new algorithms that are being developed to gain insight into the dynamics of two-and three-dimensional coherent structures.

(vi) Discuss recent experiments of three-dimensional wakes and analytical stationary states in three dimensions.

TWO-DIMENSIONAL HYDRODYNAMICS

Equations of Motion and Algorithms

The time-dependent incompressible Navier-Stokes equations are

$$\nabla \cdot \underset{\sim}{u} = 0, \tag{1}$$

$$\partial_t \underset{\sim}{u} + \underset{\sim}{u} \cdot \nabla \underset{\sim}{u} = -\rho^{-1} \nabla p + \nu \nabla^2 \underset{\sim}{u}, \tag{2}$$

where ρ is the constant density and ν^{-1} is proportional to the Reynolds number. In two dimensions the vorticity $\underset{\sim}{\zeta} = (0,0,\zeta)$ has one component in the z-direction (orthogonal to the plane of motion) and one can introduce a stream function, $\underset{\sim}{\psi} = (0,0,\psi)$, such that

$$\underset{\sim}{u} = \nabla \times \underset{\sim}{\psi} = u(x,y)\ \hat{\underset{\sim}{x}} + v(x,y)\ \hat{\underset{\sim}{y}}, \tag{3}$$

$$\zeta = \nabla \times \underset{\sim}{u} = -\nabla^2 \psi , \qquad (4)$$

or

$$\zeta = -\nabla^2 \psi , \qquad (5)$$

where $\underset{\sim}{\nabla} = \hat{x}\partial_x + \hat{y}\partial_y$ and $\zeta = \zeta(x,y)$ and $\psi = \psi(x,y)$.

If we take the curl of (2) and use the definitions, we obtain

$$\partial_t \zeta + [\zeta,\psi] = \nu\nabla^2 \zeta , \qquad (6)$$

where

$$[A,B] = A_x B_y - B_x A_y , \qquad (7)$$

and subscripts denote partial differentiation ($A_x \equiv \partial_x A$). At $\nu = 0$ ("infinite" Reynolds number), (2) reduces to the Euler equations and (6) becomes

$$\partial_t \zeta + \underset{\sim}{u} \cdot \underset{\sim}{\nabla} \zeta = \partial_t \zeta + \psi_y \zeta_x - \psi_x \zeta_y = 0. \qquad (8)$$

Eq. (8) is the Liouville equation of a Hamiltonian system whose particles at $\underset{\sim}{x}_i = (x_i, y_i)$ have characteristic velocities

$$\dot{x}_i = \psi_y(x_i) \equiv u(x_i) \text{ and } \dot{y}_i = -\psi_x(x_i) \equiv v(x_i).$$
$$(9)$$

If, for the moment, we imagine boundaries at infinity or periodic boundary conditions, then any state of the system is described by the vorticity distribution $\zeta(x,y)$ and can evolve into all other states subject to the constraints imposed by the conservation laws

$$\underset{\sim}{P} = \rho \iint \underset{\sim}{u} \, dxdy \qquad (\text{linear momentum}), \qquad (10)$$

$$\underset{\sim}{L} = \rho \iint \underset{\sim}{r} x \underset{\sim}{u} \, dxdy \qquad (\text{angular momentum}), \qquad (11)$$

$$E = \frac{1}{2} \rho \iint |\underset{\sim}{u}|^2 dxdy = \frac{1}{2} \rho \iint \zeta \psi \, dxdy \quad (\text{kinetic energy}),$$
$$(12)$$

$$\int A(\zeta) d\zeta = \text{constant} \qquad (\text{vorticity areas}), \ (13)$$

where integrals are taken over the infinite region or the periodic box. $A(\zeta)d\zeta$ is the area between two vorticity contours ζ and $\zeta + d\zeta$ and Helmholtz's theorem tells us that these areas are convected with the fluid. Hence, it is convenient to study systems where the vorticity density ζ is constant and takes on the values $+\zeta_0$, 0 and $-\zeta_0$. The area conservation law is simply the conservation of area within the contours surrounding these regions.

The Zabusky-Deem wake solutions were obtained by solving (1) and (2) on a compact staggered mesh. In their second-order finite-difference algorithm the pressure is obtained by inverting Poisson's equation such that incompressibility (1) is enforced at each time step. In the limit $\nu = 0$, the algorithm "semi-conserves" energy

$$E = \langle \underset{\sim}{u} \cdot \underset{\sim}{u} \rangle = \frac{1}{N} \sum_{m,n} (u^2_{m+\frac{1}{2},n} + v^2_{m,n+\frac{1}{2}}),$$

but does not conserve enstrophy

$$\langle \zeta^2 \rangle = \frac{1}{N} \sum_{m,n} \zeta^2_{m+\frac{1}{2},n+\frac{1}{2}} \qquad .$$

By "semiconserve" we mean that temporal derivatives and
the time variable are assumed continuous. Note, ζ is
not a part of the marching calculation, but is obtained
outside the evolutionary loop from a discrete compact
version of (4), and hence is a sensitive diagnostic for
the quality of the solution.

The Christiansen-Zabusky, $\nu = 0$, wake simulations
were made with the fundamentally different VORTEX parti-
cle/field code, based on (5) and (9). The particles are
positively and negatively signed vortex filaments of
constant magnitude, and the velocity fields are obtained
from the stream function by inverting (5) using a second-
order finite-difference algorithm*. J. P. Christiansen
(1973) developed the VORTEX code and studied some of its
properties for a single circular vortex of constant
vorticity ζ_0. He and his colleagues applied it to numerous hydrodynamical and plasma guiding-center configurations, as reviewed in Roberts and Christiansen (1972).

In VORTEX no attempt is made to enforce conservation
or 'semiconservation' of mass (local incompressibility
of the flow), momentum or energy. These quantities are
monitored to allow one to assess the quality of the run.
Although the velocity field calculated at the mesh points
is solenoidal, the interpolations used to obtain the

---

*That is, $\zeta(x,y)$ is computed at lattice locations by a
 spatial filtering process (area-weighting) applied to
 the point filaments. Eq. (5) is inverted using the
 usual second-order five-point approximation and u and v
 are evaluated at lattice points by taking first central
 differences of ψ (Christiansen and Hockney 1971). u
 and v are obtained at filament locations by linear in-
 terpolation using the four nearest lattice points. The
 velocity fields convect the filaments and the ordinary
 differential equations (9) are solved by a second-order
 leap-from scheme.

velocity of a vortex filament introduce a local violation of incompressibility which is manifest in the
figures below as a fine wavelike structure on the large-
area vortices, e.g. Figs. 8 and 9 (t = 8·75) and the
dispersion of long thin "arms" of vorticity. These
shortwave-length errors do not apparently contribute
to large-scale motions for the duration of our runs.

HISTORICAL REMARKS
Theory, Analysis and Numerical Computation
Equations (1) and (2) or (5) and (9) were known to
19th century mathematicians, and in 1888, Basset noted,
"The mathematical difficulties of solving this problem
[evolution of interacting FAVRs] when the initial distribution of the vortices and the initial forms of their
cross-section are given are very great; and it seems
impossible in the present state of analysis to do more
than obtain approximate solutions in certain cases."
For nonlinear time-dependent wake problems, nothing of
penetrating significance has been done analytically.
There are numerous papers dealing with the linearized
stability of point vortex systems beginning with von
Kármán and Bénard. There is Onsager's (1949) stimulating
concept of "negative temperature". This property is
associated with the commonly occurring coalescence or
"fusion" of like-signed FAVRs in two-dimensional hydro-
dynamics. This situation prompted von Neumann (1946)
(see Goldstine and von Neumann (1963)) to write,

"Our present analytical methods seem unsuitable for
the important problems arising in connection with nonlinear partial differential equations and in fact with
virtually all types of nonlinear problems... we are
up against an important conceptual difficulty which

tends to obscure the great physical and mathematical
regularities that do exist."

"...really efficient high-speed computing devices
may in the field of nonlinear partial differential
equations as well as in many other fields which are
now difficult or entirely denied of access, provide
us with those heuristic hints which are needed in
all parts of mathematics for genuine progress...[and]
should ultimately lead to important analytical advances."

The discovery of the soliton and the remarkable
analytical developments that followed provide a cogent
example of the synergetic mode of working advocated by
von Neumann, Ulam (1960, Chap. 8) and Zabusky (1967,
1973). With modern high-speed, digital parallel pro-
cessors and sophiticated electronic graphical output
systems, we can continue to solve judiciously-posed
problems and obtain insights as to rewarding analytical
directions. We do this below for wake models and point
to key impasse problems.

TWO-DIMENSIONAL WAKES

Past Experiments (before 1970)

Experimentalists have had their share of problems
interpreting pictures and probe measurements. Strouhal
(1878) observed the periodic shedding of vortex ag-
gregations behind a bluff body moving in a fluid and the
subsequent downstream formation of a vortex street.
Lamb (6th ed., p. 680) states, "...it is to the formation of a double train of vortices that we must look
for the explanation of many acoustical phenomena" and
refers to early visualization experiments by Ahlborn
(1902), Bernard (1908); and Prandtl (1927). Yet, the
pictures of Taneda (1959 and 1965), Hama (1957, 1962,

and 1963), Zdravkovich (1969), and the hot-wire data of
Sato and Kuriki (1961) show many features of moderate-
Reynolds-number wakes at moderate distances downstream
from their source that are not well understood.

Our slowly evolving comprehension is recognized in
the reviews of Rosenhead (1953), Townsend (1956), Willie
(1960), Morkovin (1964), Betchov and Criminale (1967),
Stuart (1971), Berger and Willie (1972). Morkovin's
paper is a first attempt at an analytical/computational/
empirical review and evaluation of this problem area.
In the abstract and elsewhere he notes, "As the Reynolds
number increases, the resulting unsteady three-dimensional, interacting, vortical patterns appear to be the
key to some of the perplexing experimental observations."
The specific examples below demonstrate that the essence
does lie in understanding the dynamical interaction of
FAVRs.

Recent Experiments

As experimental resolution improved, the importance
of coherent interactions in the evolution of vortex
streets was appreciated. For example:

- The generation of FAVRs by fixed and oscillating
  circular cylinders (Griffin and Votax 1972, and
  Griffin and Ramberg 1974). For example, P.
  Bearman (1969) has clarified the sharp transition in Strouhal number from 0.18 to 0.45 as the
  Reynolds number (based on body diameter) passes
  through 4.2×10^5 and increases to 1.6×10^6.
  For Re $> 6 \times 10^5$, the slightest imperfection
  introduces three dimensional fluctuations and a
  broad-band signal slightly downstream of the
  cylinder.

- The growth of small amplitude perturbations into
  FAVRs beyond flat plates. See, for example, the
  photographs of Taneda (1959 and 1965) where
  $Re < 1.64 \times 10^5$ (1965, Fig. 3, Re = 8150, where
  Re is based on the length of the plate).

- The nutation (slow pitching) or rotation of FAVRs.
  For example, the hot-wire measurements of Sato
  and Kuriki (1961) implicitly reveal this rotation as pointed out by Zabusky and Deem (1971).
  A similar rotation was also observed in 2D jet
  experiments by Beavers and Wilson (1970).

- The deformation of FAVRs, for example as discussed by Durgin and Karlsson (1971).

- The "breakdown" and "reformation" of streets
  containing FAVRs. See, for example, the photographs by Taneda (1959 and 1965) for cylinders,
  where $Re \leq 304$.

- The "persistence of von Kármán vortices at much
  greater downstream distances and much higher
  Reynolds numbers" was reaffirmed by Papailiou
  and Lykoudis (1974).

A desire to understand "turbulent" mixing layers
has led Winant and Browand (1974), Brown and Roshko (1974),
and Browand and Weidman (1976) to study instabilities in
free-shear layers. These narrow regions of velocity-
change (vorticity regions of one sign) are unstable and
yield persistent FAVRs of one sign that eventually coalesce. For such flows our concept of the effective
Reynolds stress is changing. It is not some random,
chaotic motion that causes it, but rather an average
over the motions induced by the passage of interacting
FAVRs.

Analysis and Computer Simulations

Wille (1960) discusses the inadequacy of the point-vortex street wake model. He remarks "...the application of refined mathematical models leads to the result that all two-dimensional vortex streets are unstable" (p.275). He concludes "... the theory of vortex streets that comes under the heading stability is not yet satisfactory" (p. 277). Although he discusses possible stabilization mechanisms, he omits the idea advocated by Christiansen and Zabusky (1973, Sections 6 and 9) on the basis of their numerical simulations. They find that finite-amplitude surface deformations of FAVRs seem to stabilize those flows that are marginally stable when represented by a point vortex model (namely those with the von Kármán ratio $(b/a) = 0.281$).

Computer simulation studies of the stability of wake-like configurations have been carried out for several years. Calculations by Abernathy & Kronauer (1962) on the evolution of a laminar wake with a rectangular velocity profile arising from two rows of vortex filaments demonstrated the formation of the von Kármán vortex street. They also observed the capture or trapping of vortex filaments in regions of opposite-signed vorticity. This was also seen in calculations by Fromm and Harlow (1963).

The linear stability theory for symmetric and asymmetric configurations of point vortices begun by von Kármán (1911 and von Kármán and Rubach 1912) has been elaborated by many including Kochin (1939) and Kochin, Kiebel & Roze (1964). For two opposite-signed streets of point vortex filaments the symmetric configuration is unconditionally unstable. The asymmetric (staggered) configuration on the other hand is only stable if the

transverse-to-longitudinal separation ratio b/a is
0.281. However, as emphasized by Kochin et al. (1964,
pp. 226-234), this is a necessary condition: "A first-
order perturbation theory shows that the positions of
vortices in a street with b/a = 0.281 will separate by
a finite amount." Wille (1960) discusses these difficulties.

    The unclear experimental/theoretical situation
prompted Zabusky and Deem (1971) to investigate lowest-
mode perturbations to unstable wake-like profiles, and
they found remarkably good agreement with the probe
measurements of the Sato-Kuriki (1961) experiment. R.H.
Hardin and F.D. Tappert joined with Zabusky and Deem at
Bell Laboratories in 1971 in a broad-ranging investigation of two-dimensional hydrodynamic and magnetohydro-
dynamics motions*. The results were collected in a film
record that I will show excerpts of today. I will never
forget the first time that Hardin showed the film - it
was a supernova experience which revealed key links and
exposed new analytical and computational directions.
In particular, a second-mode excitation initially produces
the staggered vortex street observed by Zabusky and Deem.
At a later time, it breaks down and likesigned FAVRS
coalesce. One obtains a vortex street of double the
wavelength - a feature of Taneda's experiment.

    In 1971-1972, Zabusky and Christiansen investigated
the stability of staggered-wakes of FAVRs with VORTEX.

---

*In 1971 we were all members of the Computational Physics
Research Department at Bell Laboratories, Inc. Whippany,
N.J. D. Montgomery, who was visiting us at the time,
drew inspiration for his later work on the meaning of
Onsager's negative temperature from these interactions
(Montgomery, 1977).

In their film one sees wave-like surface deformations
on FAVRs of moderate and large size, coalescence of
like signed FAVRs and fission of a FAVR in selfconsistent
shear field.

Flat-Plate Wake Experimental-Simulation Comparisons

Zabusky and Deem set out to simulate the Sato and
Kuriki (1961) experiment and made the ansatz that time
elapsed during a numerical calculation corresponds to
downstream distance (a Taylor-like hypothesis). They
placed their "flow" in a rectangle with fixed boundary
conditions at $y = \pm \frac{1}{2} L_y$ and periodic boundary condition at $x = (0, L_x)$. Fig. 1 is a schematic of the experiment and computational realization. The first
rectangle shows a Gaussian profile

$$U(y) = U_0 - U_{c0} \exp(-y^2/\Delta^2), \qquad (14)$$

measured by Sato and Kuriki near their plate, and the
second the numerical results obtained. We perturbed
the Gaussian profile with velocities $(u'(x,y), v'(x,y))$
obtained from the eigen-function solutions of the (inviscid) Rayleigh equation. The wavelength of the perturbation was L_x (the first harmonic) and the finite-
difference mesh was (128 × 128). Fig. 2 shows the
evolution of isovorticity contours. In frame (a) one
clearly sees that the derivative of the Gaussian
$2y\Delta^{-2} e^{-(y^2/\Delta^2)}$ changes sign, that is, the upper dotted
lines are negative regions of vorticity and the lower
solid lines are positive. Also apparent is the long
wavelength perturbation. The flow is unstable and at
(c) FAVRs have formed, joined by elongated and structured
appendages or "arms" (Patnaik, et al. call the FAVRs,
"cores" and the arms, "braids".) The Reynolds number is

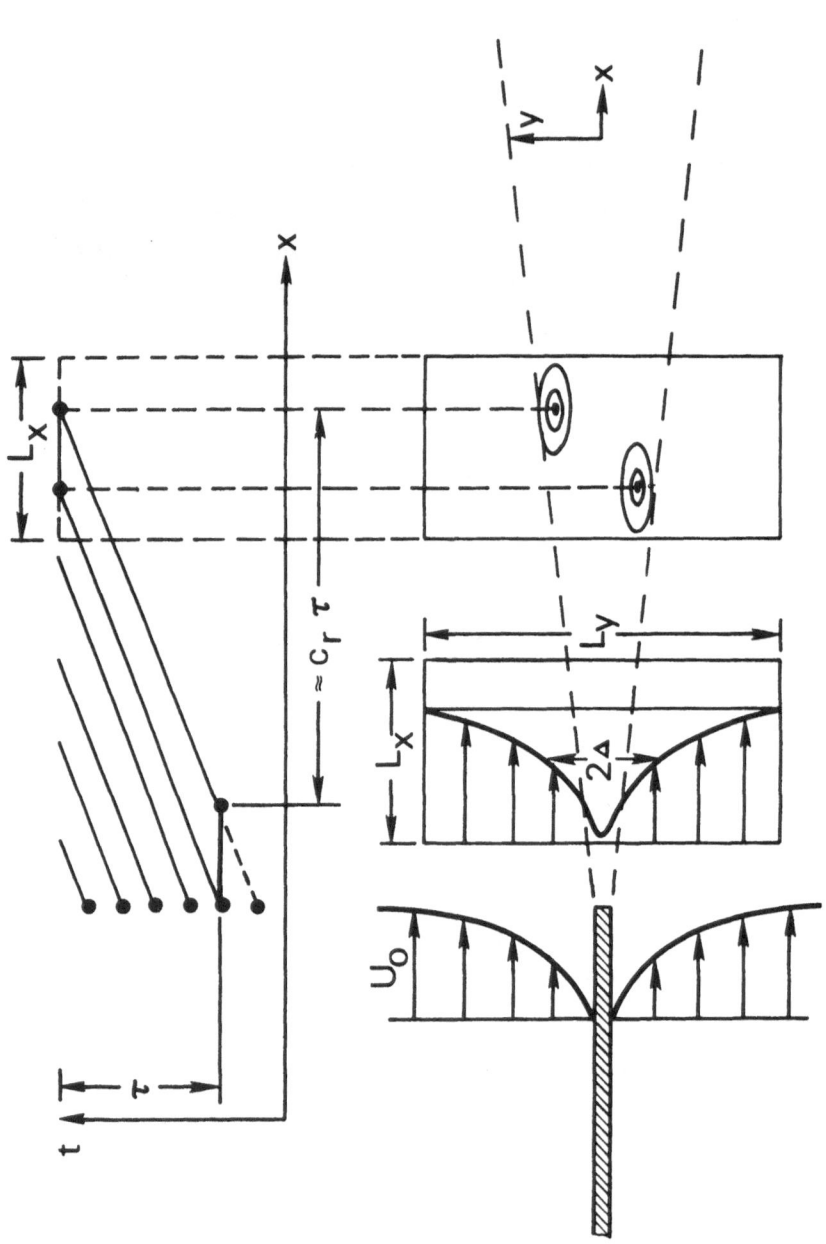

Fig. 1. Schematic of two-dimensional flat-plate wake experiment of Sato and Kuriki and the computer set-up in the rectangular box $L_x \times L_y$. Trajecotries of primary FAVRs in (x,t) space are shown above, Coalescence occurs further downstream.

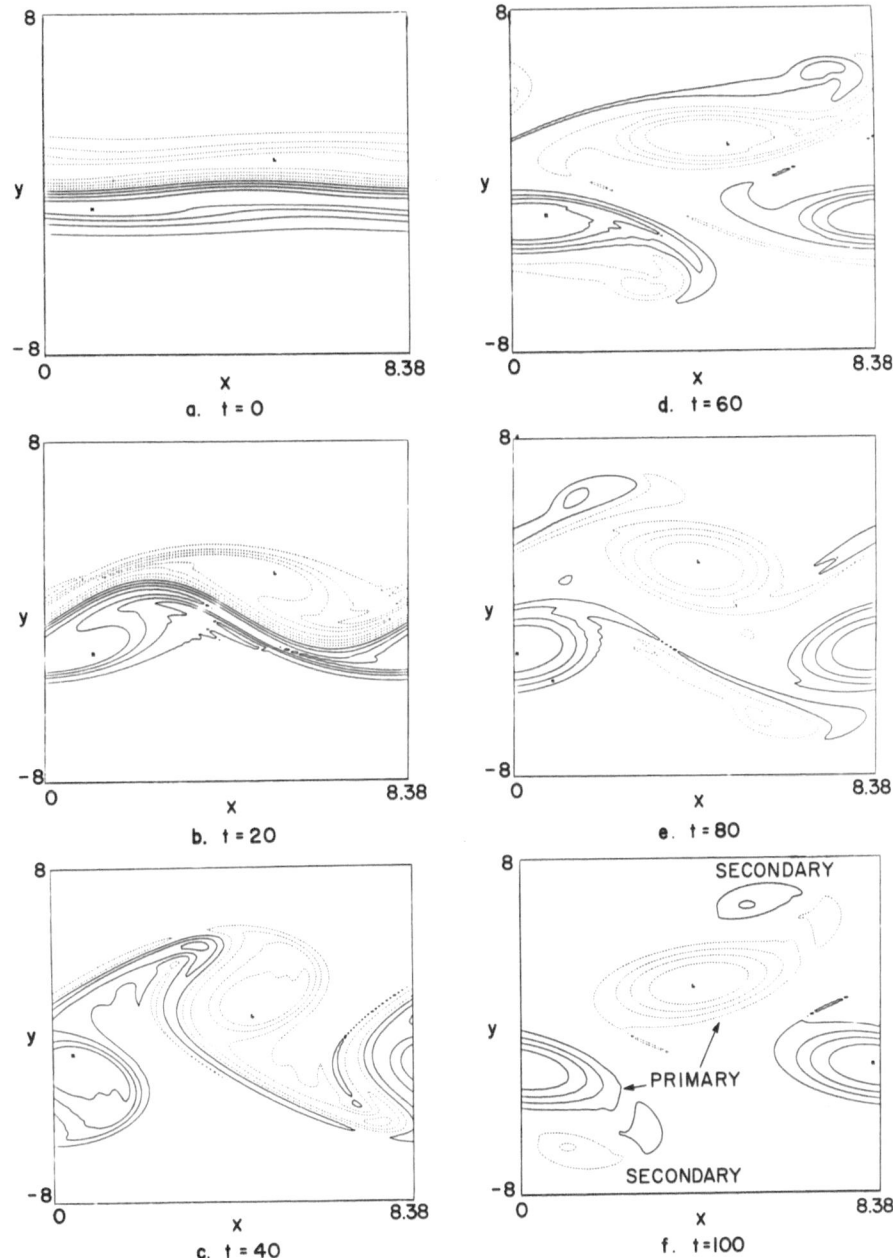

Fig. 2. Constant vorticity contours (9 levels) for a perturbed Gaussian profile (Zabusky and Deem, 1971) at times (a) t = 0, (b) t = 20, (c) t = 40, (d) t = 60, (e) t = 80, (f) t = 100.

Re = $(\Delta U_0/\nu)$ = 750 and weak but finite dissipation smooths the flow, as is evident in subsequent frames. Two new features emerge in (d) and persist. First, regions of opposite signed vorticity have been en-trained by the dominant or "primary" FAVRs. That is, in the upper-half of the frames one sees secondary regions of vorticity moving past primary FAVRs (solid contours past dotted contours). The second feature is the seemingly elliptical vortex region (exaggerated because of different vertical-horizontal figure scales) whose major axis undergoes a pitching or "nutation" with respect to horizontal (x) axis.

The nutation phenomenon is a computational dis-covery which is also seen in the constant pressure con-tours in Fig. 3. The vertical array of ellipses shows the particular contour surrounding the low pressure point in the lower half of the flow. During $0 \leq t \leq 30$ the contour's major axis is tilted upward (evolution of initial state) and then the pitching motion begins - associated with the interaction of primary FAVRs. The nutation frequency is also evident in Fig. 4, which shows the normalized downstream modal energy variations, that is, discrete representations of

$$E_k = \frac{1}{L_y} \int_{-\frac{1}{2}L_y}^{+\frac{1}{2}L_y} [u_k^2(y) + v_k^2(y)]\,dy, \qquad (15)$$

where

$$\begin{pmatrix} u_k \\ v_k \end{pmatrix} = \frac{1}{L_x} \int_0^{L_x} \begin{pmatrix} u(x,y,t) \\ v(x,y,t) \end{pmatrix} \exp(i\,2\pi kx/L_x)\,dx. \qquad (16)$$

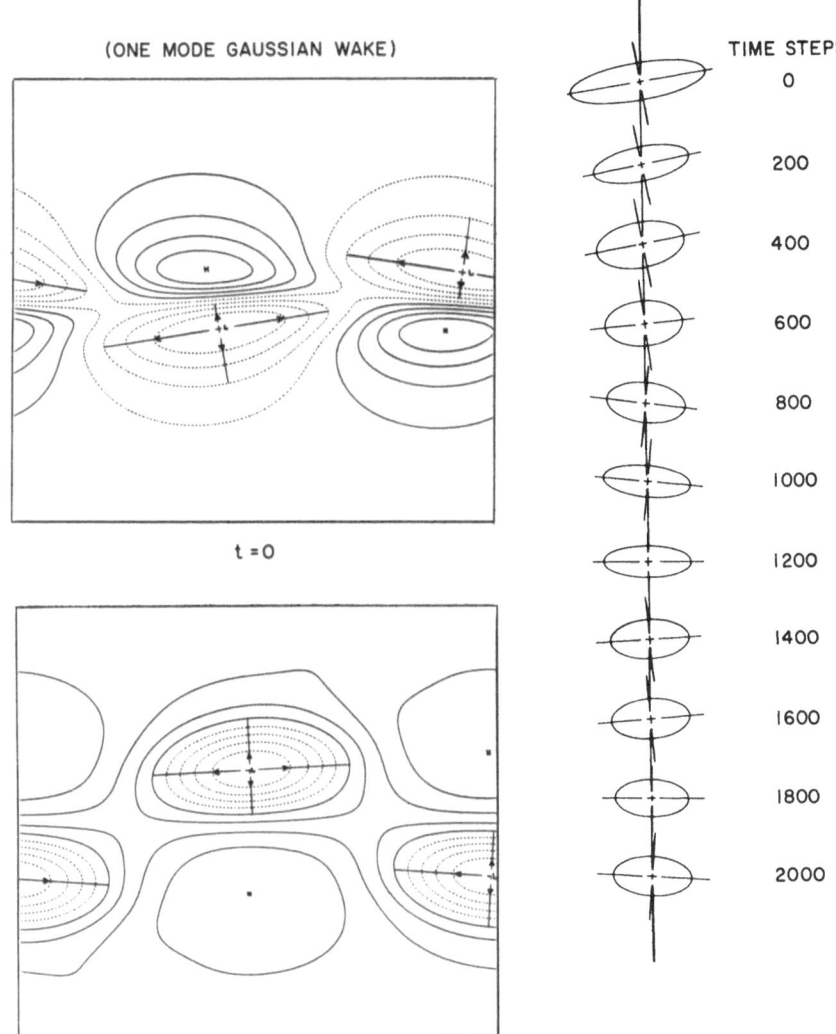

Fig. 3. Constant pressure contours (9 level) for a perturbed Gaussian profile (Zabusky and Deem, 1971) at (a)
t = 0 and (b) t = 100. The vertical array of ellipses
shows the contour pattern surrounding the low pressure
point in the lower half of the flow at intervals of 200
computation steps (or 10 units of time). The horizontal
and vertical scales are the same as in Fig. 2. They differ in such a way that ellipses appear more elongated.

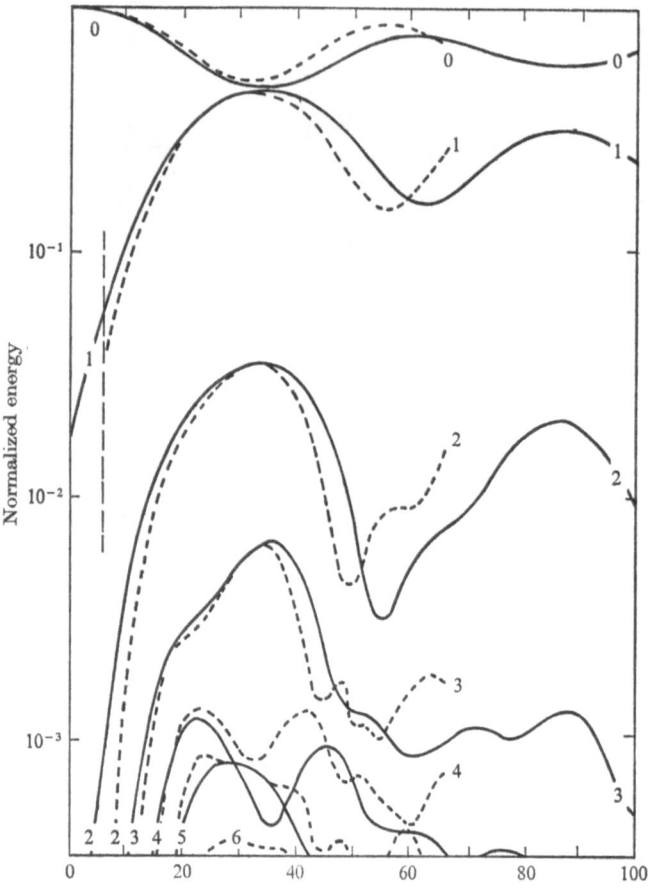

Fig. 4.   Downstream modal energy $E_k(t)/E(t)$ for the Gaussian wake (Zabusky and Deem, 1971). _____, R = 750; _ _ _ _, R = ∞.

The nutation frequency is due mainly to an interaction between the mean flow and first harmonic. Recently, A. Miura and T. Sato (1977) extended J.T. Stuart's (1971) work by allowing amplitude and phase variation. They derived Duffing's equation and seem to predict the nutation frequency. Note that Fig. 4 indicates a slight Reynolds number dependence, for (the dotted curves were obtained for an inviscid run ($\nu = 0$ or Re $= \infty$).

When the results of this calculation were compared to the Sato & Kuriki measurements, the essential features were found to be in agreement. In particular, the "cross-stream phase" of the first harmonic

$$\tan \theta_1 = \frac{\int_0^{L_x} u \sin(2\pi x/L_x)dx}{\int_0^{L_x} u \cos(2\pi x/L_x)dx}$$

was particularly revealing (Zabusky and Deem 1973, Fig. 9, row 4).

Growth of Most Unstable Mode and Coalescence of FAVRs

We also undertook a run to investigate how the most unstable mode grew, that is, the mode which had the largest growth rate, according to linear theory. We performed an inviscid run with a "Bickley" profile

$$u(y) = U_{c0} \operatorname{sech}^2(y/\Delta), \qquad (17)$$

perturbed by five equal-amplitude unstable modes (the k = 3 mode had the largest growth rate). Fig. 5 shows the results and we quote directly from the paper (p. 337) "...The constant pressure contours between Figs. 15(a) [this paper, Fig. 5(a)] and 15(c) reveal an effect

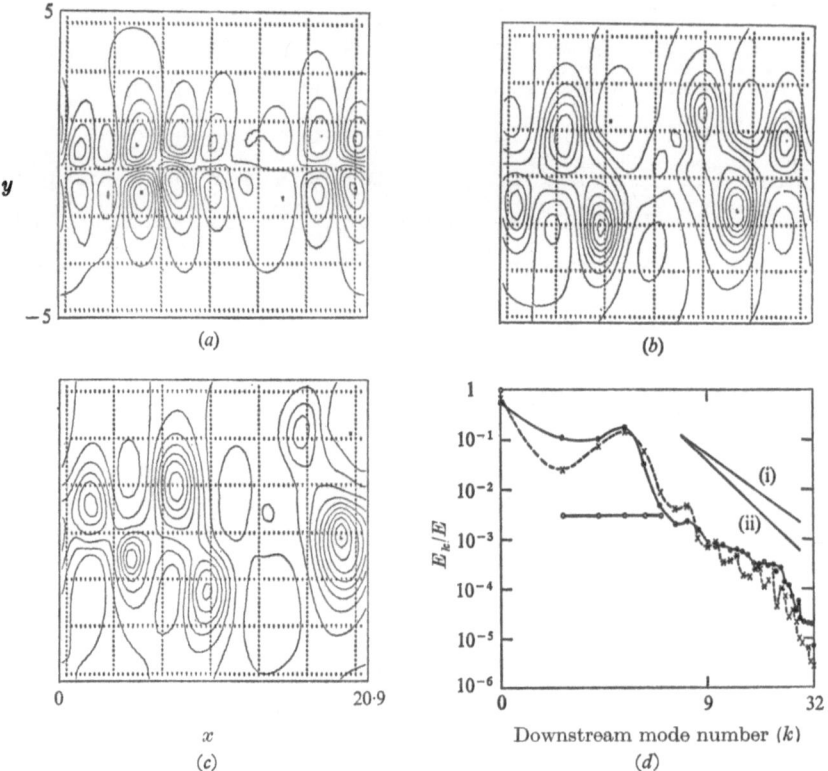

Fig. 5. Five-model perturbed Bickley profile (Zabusky and Deem, 1971). Pressure contours: (a) t = 0; (b) t = 18; (c) t = 30; (d) Downstream modal energy spectra: o, t = 0; ×, t = 18; •, t = 30. Also line (i) is $(k+1)^{-3}$ and line (ii) is $(k+1)^{-4}$. Note, initially mode 3 is most unstable.

absent in previous sections, namely, the gradual <u>merging</u>
<u>of</u> <u>smaller</u> <u>into</u> <u>larger</u> vortices* [italics this paper].
This effect is seen more clearly in Fig. 15(d), which
gives the corresponding development of modal energies.
At t = 0, the five equally excited modes together with
the mean or k = 0 mode are indicated by open circles.
At t = 18, the k = 3 (most unstable) mode dominates the
remaining modes k > 0, as one would expect from linear
theory. However, by t = 30 we note a large increase in
the energy content of the k = 1 subharmonic as a gradual
merging of smaller vortices (Batchelor 1953)."

    <u>Formation</u> <u>and</u> <u>Stability</u> <u>of</u> <u>a</u> <u>Staggered</u> <u>Vortex</u> <u>Array</u>
    The experimental and computational results in con-
junction with von Kármán's linearized analysis highlighted two related questions concerning stability.
First, if one measures the transverse-to-longitudinal
distances (b/a) between adjacent FAVRs in Fig. 2 (transverse, positive-to-negative FAVR centers; longitudinal,
positive-to-positive or negative-to-negative FAVR centers)
one obtains b/a = 0.47. (In Zabusky and Deem this was
reported as 0.24, a misprint). Yet, the von Kármán
criterion for a point vortex array is marginal stability
only for b/a = 0.281 and instability for all other ratios.
Obviously, the restrictive periodic boundary conditions
and a single first-harmonic initial excitation yield a
solution that cannot exhibit coalescence. Would coalescence occur if either of these restrictions are
removed? Second, the breakdown and rearrangement of a

---

* Essentially, we have lost 7 contour "islands" in evolving
from (a) to (c). We define a contour island as at least
<u>two</u> localized, nearly concentric, closed contours.

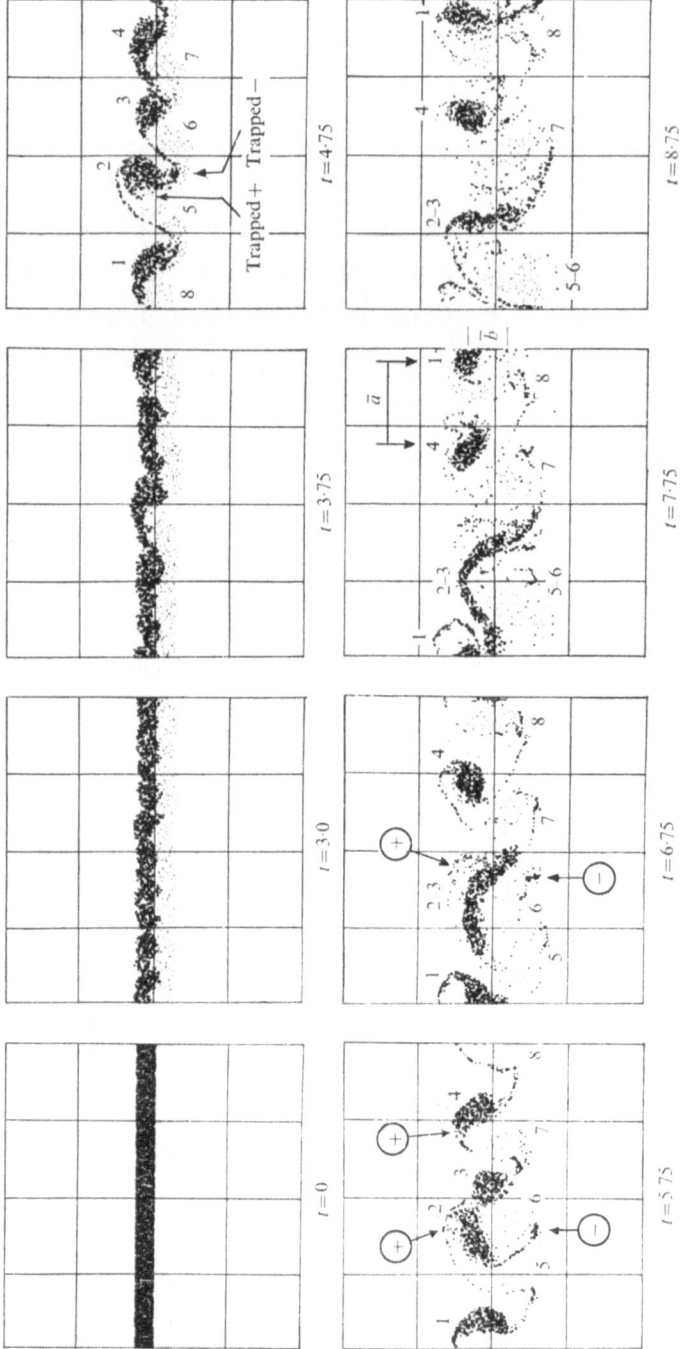

Fig. 6. Instability of a laminar wake. (Christiansen and Zabusky, 1973).
Small random perturbations imposed on regions of constant vorticity: upper
(dark), negative; lower (right) positive. Boundaries: fixed in y, periodic
in x.

vortex street into one of longer wavelength, as observed
by Taneda, associated with this instability?

Hardin et al. (1971) perturbed the Gaussian profile
with a single second harmonic and observed the shear
profile evolve into a staggered vortex array (i.e., with
two positive and two negative FAVRs per period.) After
many vortex nutation periods, the array became unstable,
a sudden rearrangement took place and the like-signed
FAVRs coalesced.

Christiansen and Zabusky (1973) undertook a more
careful study of this phenomenon with the VORTEX code
on a 64 × 64 mesh. First they repeated the Hardin calculation with different initial conditions - a triangular
velocity profile with a random small perturbation as
shown in Fig. 6. The dark upper region (t = 0) is a
region of constant negative vorticity, the light lower
regions are positive. (Dark FAVRs rotate clockwise and
light FAVRs rotate counterclockwise.) At t = 4.75 we
see that the fourth harmonic is dominant and the shear
flow has evolved into 4 positive (1,2,3 and 4) and negative (5,6,7,8) unequal FAVR's. At t = 5.75, No. 2 and
No. 3 are beginning to coalesce. (Beyond t = 6.75,
numerical dispersion errors have begun to accumulate).
At t = 4.75 and 5.75, we see the same entrainment or
"trapping" of small regions of opposite signed vorticity
that gave rise to the secondary FAVRs in Fig. 2. Trapping and rotation have been observed by Zdravkovich
(1969, Sec. 3, Fig. 3) and can also be seen in Fig. 11 of
Abernathy and Kronauer (1962). Thus, we have an enhanced transport of material across a shear flow due to
convection by FAVRs that were formed as a result of a
linear instability of the system.

To clarify the roles of FAVR area, initial shape

and (b/a) ratio, we undertook a series of runs beginning
with the configuration shown in Fig. 7. Figs. 8 and 9
show that two different circular areas at b/a = 0.281
translate, rotate and deform, but with no breakdown of
the flow. The translation and rotation are expected
from point vortex theory. The distortion of the FAVR
surfaces (elliptical, m = 2; triangular, m = 3, etc.)
had already been noted by Christiansen (1974). Fig. 10
shows the unstable case, b/a = 0.6. We see: coalescence
of Nos. 1 and 2 at t = 6; trapping of (+ or light)
vorticity by No. 2; distortion of FAVR surfaces; coalescence of Nos. 3 and 4 between 19.0 < t < 20.5; and
"egection" of vortex arms at the elongated FAVR at t =
20.5 becomes more elliptical at t = 23.5.

Fig. 11 shows the case (b/a) = 0, a standing wave
with zero velocity. Such arrangements ((b/a) < 0.2)
can be generated in the laboratory by oscillating a
cylinder transverse to a flow (Griffin and Votax 1972,
and Griffin and Ramberg 1974). The initial condition is
four alternating signed cricular FAVRs with numbers 1, 3
and 2 at y = 0 and 4 displaced upward by one lattice
interval. We note: rotation; strong deformations; rise
of No. 4; splitting or "fission" of No. 2 into fragments
(2a) and (2b) in the shear field created by Nos. 3 and 4;
ejection of one arm by No. 3 and its entrainment by (2b);
and coalescence of No. 1 and fragment (2a) at t = 12.5.

We conclude that the measured growth rates of finite-
area vortex streets differ from those of corresponding
point vortex systems (von Kármán systems). For small
and moderate areas the difference is weakly dependent
upon the area and shape, but strongly dependent upon (b/a)
as illustrated in Fig. 12. In this figure we conjecture
that there exists a region (α_1, α_2) on the b/a axis,

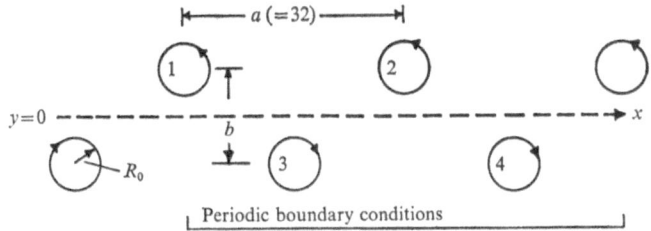

Fig. 7. Schematic of initial FAVR arrangement.

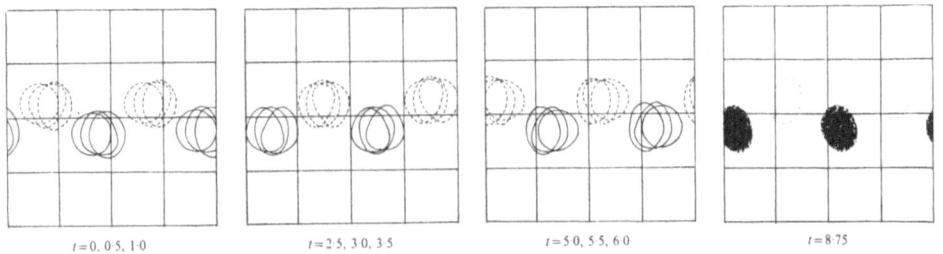

Fig. 8. Small circular FAVRs; b/a = 0.281; fixed-y boundaries.

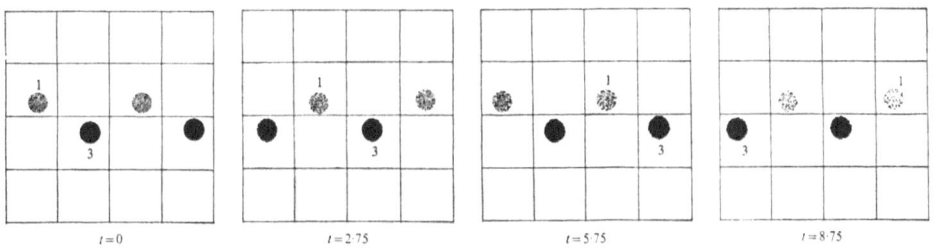

Fig. 9. Large circular FAVRs. b/a = 0.281 periodic-y boundaries. The first three frames show only the vortex boundaries. The smallest time indicated refers to the left-most contour in each frame.

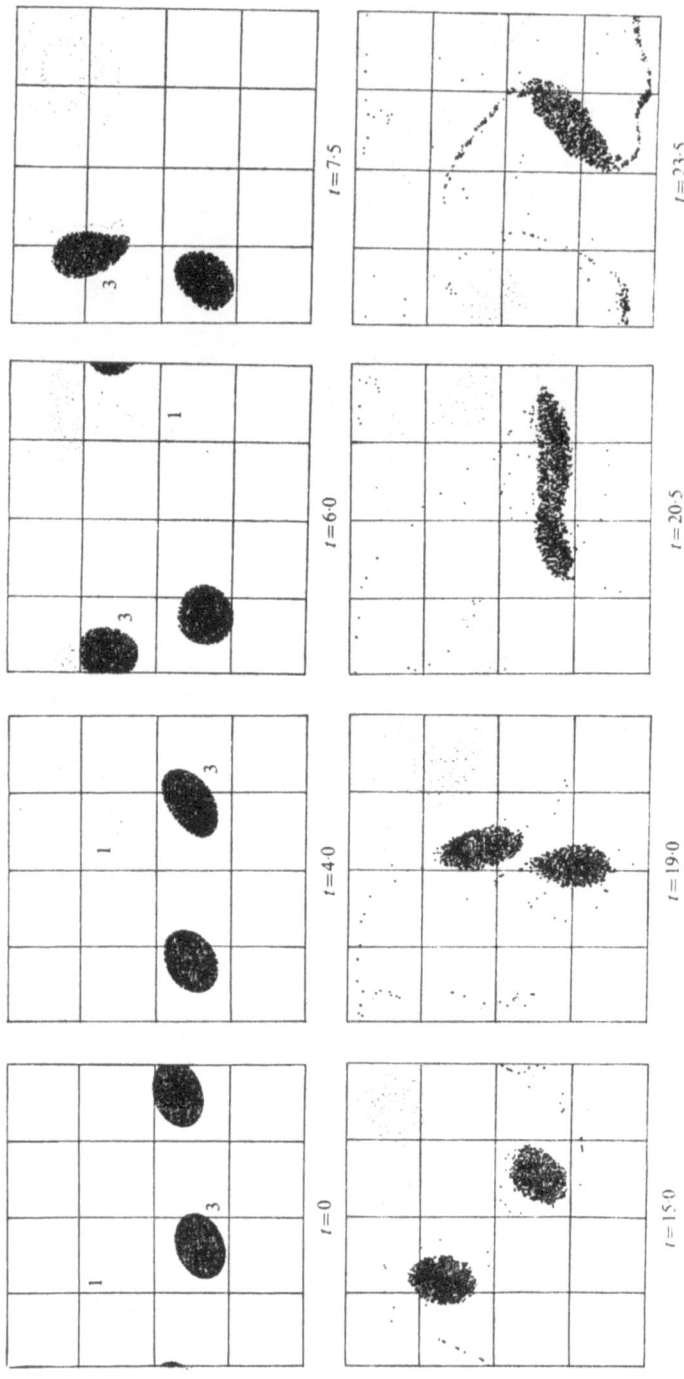

Fig. 10 Large elliptical FAVRs. b/a = 0.6; fixed-y boundaries.

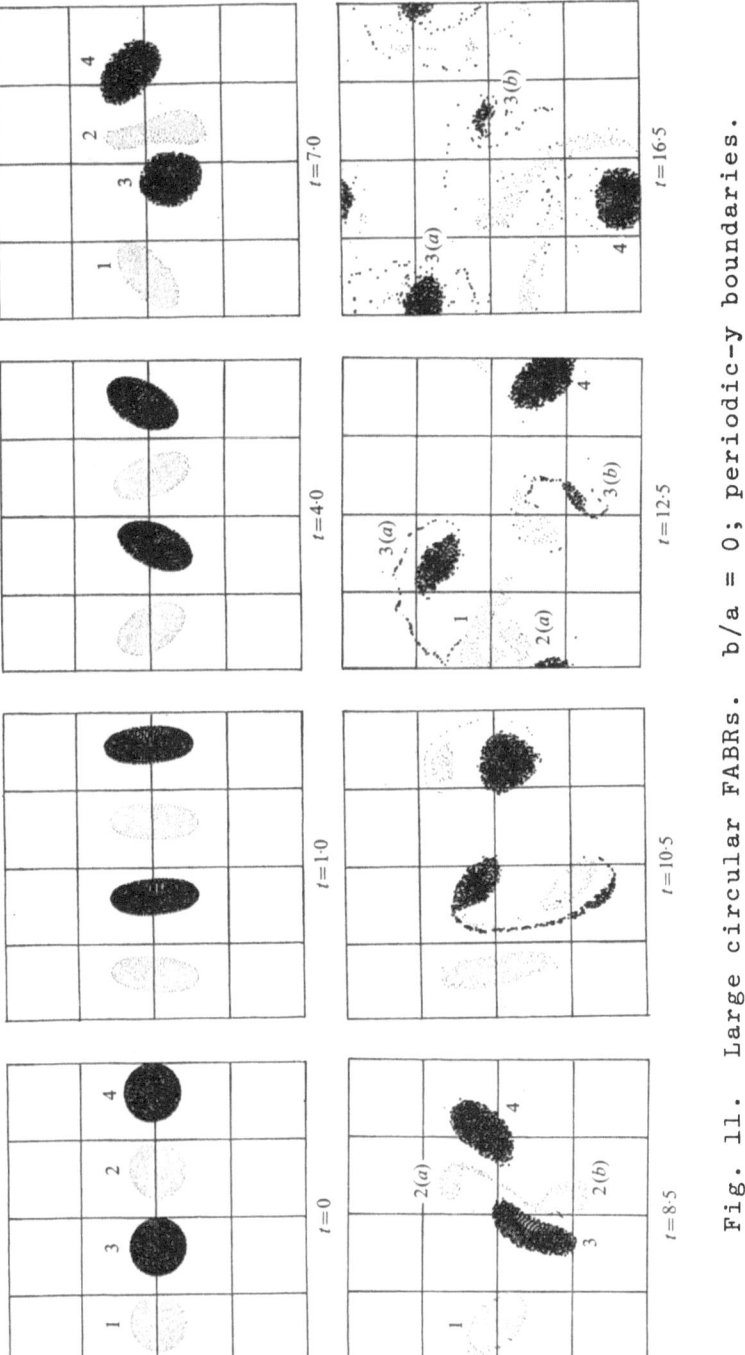

Fig. 11. Large circular FABRs. b/a = 0; periodic-y boundaries.

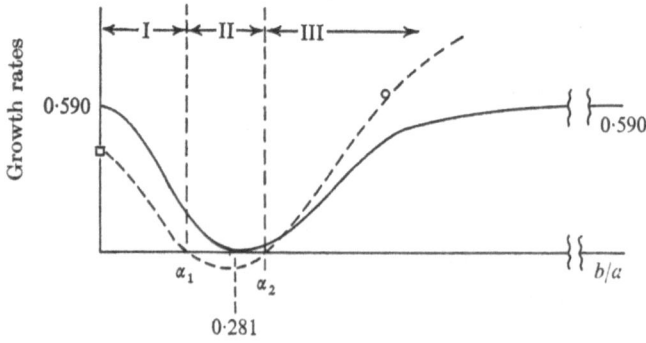

Fig. 12. Growth rates. _____, perturbation theory (von
Kármán dispersion relation for asymmetric point vortex
system). _ _ _, corresponding FAVRs.

$$\alpha_1 < 0.281 < \alpha_2,$$

where the growth rate is negative, that is, the presence
of additional degrees of freedom associated with surface
deformations of finite-area vortex systems is stabilizing.
Furthermore, because the vorticity is distributed, the
growth rate is reduced in regions I and II and increased
in region III.

The mathematization of these consistent features
constitutes the problems we pose. In particular, if we
clarify the role of the wave-like deformations or distortions on FAVRs, we will deepen our insights on stability of arrays of coherent structures.

STRATIFIED FLUIDS

Metastable coherent structures are nearly universal.
The Kelvin-Helmholtz instability and the evolution of
two-dimensional shear flows have a distinguished and long
lineage. Yet the coherent phenomena associated with
the nonlinear evolution of stratified shear flows, while
less familiar, are important for fundamental and applied
reasons. Woods and Wiley (1972) suggested that many of
the observed details of thermocline temperature structure
(affecting vertical heat transport in the ocean) can be
attributed to shear instabilities caused by long internal
waves propagating on sheets.

Equations and key results are presented below.
Comparisons with experiment and more details on the
features of the flow can be obtained by consulting the
references.

"Cats-Eyes" and Cusped Waves in Stratified Fluids
The normalized Boussinesq equations describe the

coupled evolution of normalized velocity normalized u
and advected (and dissipated) normalized density ρ (or
temperature T)

$$\partial_t \underset{\sim}{u} + \underset{\sim}{\nabla} \cdot (\underset{\sim}{uu}) = - \underset{\sim}{\nabla} p - Fr^{-1} \rho \hat{\underset{\sim}{z}} + Re^{-1} \nabla^2 \underset{\sim}{u}, \tag{18}$$

$$\partial_t \rho + \underset{\sim}{\nabla} \cdot (\rho \underset{\sim}{u}) = Pe^{-1} \nabla^2 \rho , \tag{19}$$

$$\underset{\sim}{\nabla} \cdot \underset{\sim}{u} = 0, \tag{20}$$

where

$$\underset{\sim}{\nabla} = \hat{\underset{\sim}{x}} \partial_x + \hat{\underset{\sim}{z}} \partial_z = (\partial_x, \partial_z),$$

$$\underset{\sim}{u} = (u(x,z,t), w(x,z,t)),$$

is the two-dimensional velocity in the vertical x - z
plane, p is the nondimensional pressure, $\hat{\underset{\sim}{z}}$ is the unit
normal in the vertical direction, Fr is the Froude number

$$Fr = g\delta/U_0^2 ,$$

Re is the Reynolds number

$$Re = \delta \ U_0/\nu, \tag{21}$$

ν is the kinematic velocity, Pr is the Prandtl number

$$Pr = \nu/\kappa, \tag{22}$$

(κ is the thermal diffusivity) and Pe is the Peclet number

$$Pe = Re \ Pr = U_0 \ \delta/\kappa. \tag{23}$$

Patnaik, et al. (1976) used values typical of the
atmosphere (Pr = 0.72) while Deem (1977) used values
typical of the ocean, Pr = 10 and Fr = 10^{-3}. Note, in
water at 20°C perturbations in density, $\Delta\rho$, are proportional to perturbations in temperature, ΔT. That is,
$(\Delta\rho/\Delta T) = -0.057 \ (\rho_0/T_0)$ and thus ρ in (19) could be
replaced by T.

The initial state was composed of a small perturbation obtained from a linear eigenvalue problem plus the
unperturbed profiles $\bar{u}(z)$ and $\bar{\rho}(z)$

$$u(z) = U_0 \ erf(\pi^{\frac{1}{2}} \ z/2\delta), \qquad (24)$$

and

$$\rho(z) = \rho_0 \ exp[- \frac{\beta}{R} \ erf(\pi^{\frac{1}{2}} \ zR/2\delta)]. \qquad (25)$$

Here, δ is the velocity half-width and δ/R is the half-
width of the density profile and

$$(\beta/R) = sinh^{-1}(\Delta\rho/2\rho_0), \qquad (26)$$

where $\Delta\rho$ is the total density difference across the
layer.

The instability of these stationary profiles is
governed by the local gradient Richardson number

$$Ri(z) = - g \ (\frac{d\ln \bar{\rho}}{dz}) \ (\frac{1}{(d\bar{u}/dz)^2})$$

$$= J \ exp[- \frac{\pi}{4} \ (R^2 - 2)z^2], \qquad (27)$$

where

$$J = \beta g\delta/U_0^2, \qquad (28)$$

is the <u>overall</u> Richardson number based on the layer
thickness δ. For R < √2, (shear profile narrower than
density profile) the gradient Richardson number has a
minimum at the layer center, and profiles (24) and (25)
are <u>stable</u> for J ≥ 1/4. For R > √2 (shear profile
wider than density profile) Ri(z) approaches zero for
large z. For example, for R = 3.5 (see Deem, 1977, Fig.
2), one obtains an unstable nonpropagating mode for
J < 1/4 and an unstable propagating mode for J > 1/4.
Furthermore, as J is increased above 1/4, the most unstable propagating mode (associated with perturbations
at the flow edges) is of shorter wavelength.

Table I contains parameters and key features of
the flows shown in Fig. 13 through 20.

The most unstable wavelength was used and was the
largest wavelength in the box. Patnaik, <u>et al</u>., in their
study of the cats eye (J < 1/4), also used longer wave-
lengths and observed coalescence.

Fig. 13 shows the isopycnics (constant density
contours) and Fig. 14 the constant vorticity contours
for R = 1, e.e., the <u>nonpropagating</u> mode that evolves
into the cat's eye. This is the regime considered by
Patnaik, <u>et al</u>. Figs. 15 and 16 are the isopycnics and
isovorticity contours, respectively, for R = 3.5. Since
J = 0.1, we obtain the nonpropagating mode over a larger
region of space, and a larger spiral forms (t = 40),
which diffusion smooths into a larger cat's eye.

Figs. 17, 18 and 19 give similar contours for J = 0.4
and the same R = 3.5 and Re = 100. Now we have a <u>propa-
gating</u> mode and the upper density maximum evolves into
an overturned cusp. The vorticity diagram shows the
evolution of a high region (apparent at t = 60) which
travels at nearly $\frac{1}{2}$ U_0 (the "average" speed) and grows

Table I: Parameter Survey for Unstable Stratified Shear Flows

Fig. No. (Below)	Fig. No. (Deem, '77)	Contours of	R	J	Re	Feature
13	4	ρ	1	0.1	100	*nonpropagating-to- "cats-eye" with nutation.
14	5	ω,ψ	1	0.1	100	
15	9	ρ	3.5	0.1	100	nonpropagating-to- "cats-eye" with nutation.
16	10	ω,ψ	3.5	0.1	100	
17	12	ρ	3.5	0.4	100	propagating-to- cusped wave.
18	13	ω,ψ	3.5	0.4	100	
19	14	ρ,ω,ψ	3.5	0.4	100	continues above
20	15	ρ	3.5	0.4	200	propagating-to-breaking cusped wave.

*Results similar to Patnaik et al. (1976)

1. All runs Pr = 10.

2. Vorticity $\omega = -\partial_z u + \partial_x w$.

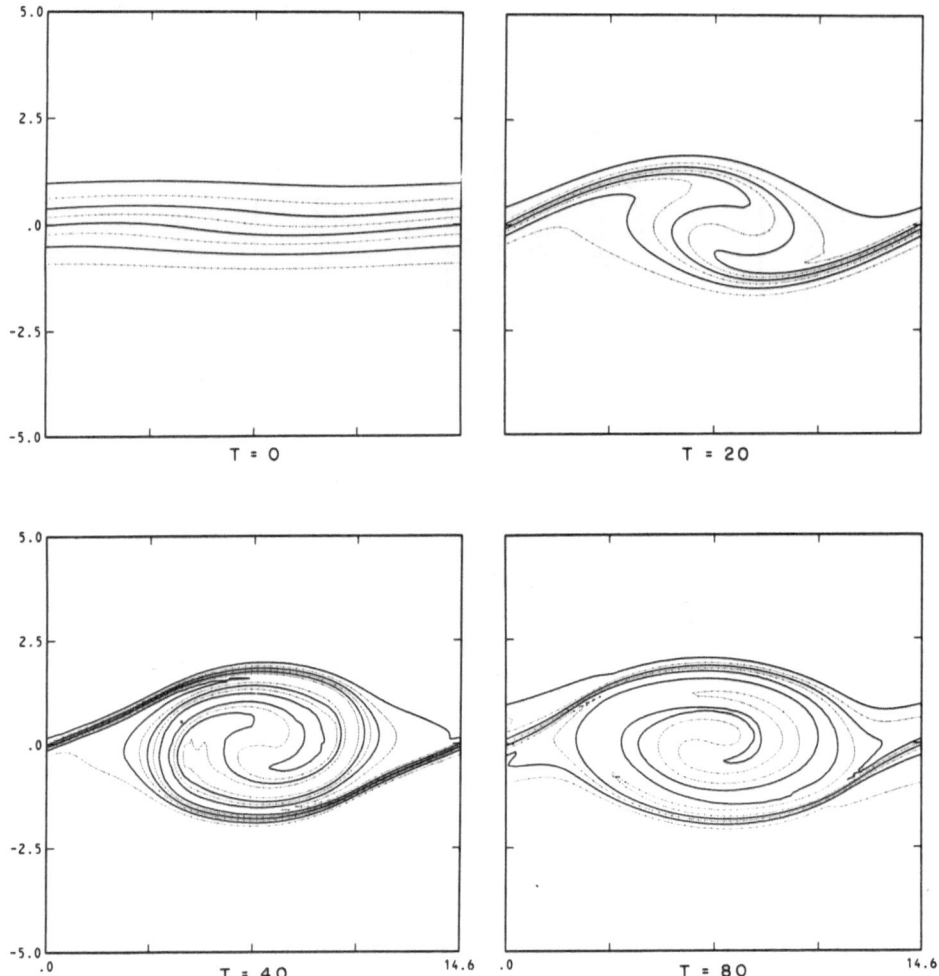

Fig. 13.   Constant density contours for the Boussinesq
equations.   R = 1, J = 0.1, α = 0.43, Re = 100, Pr = 10
(Deem 1977, Run 1).

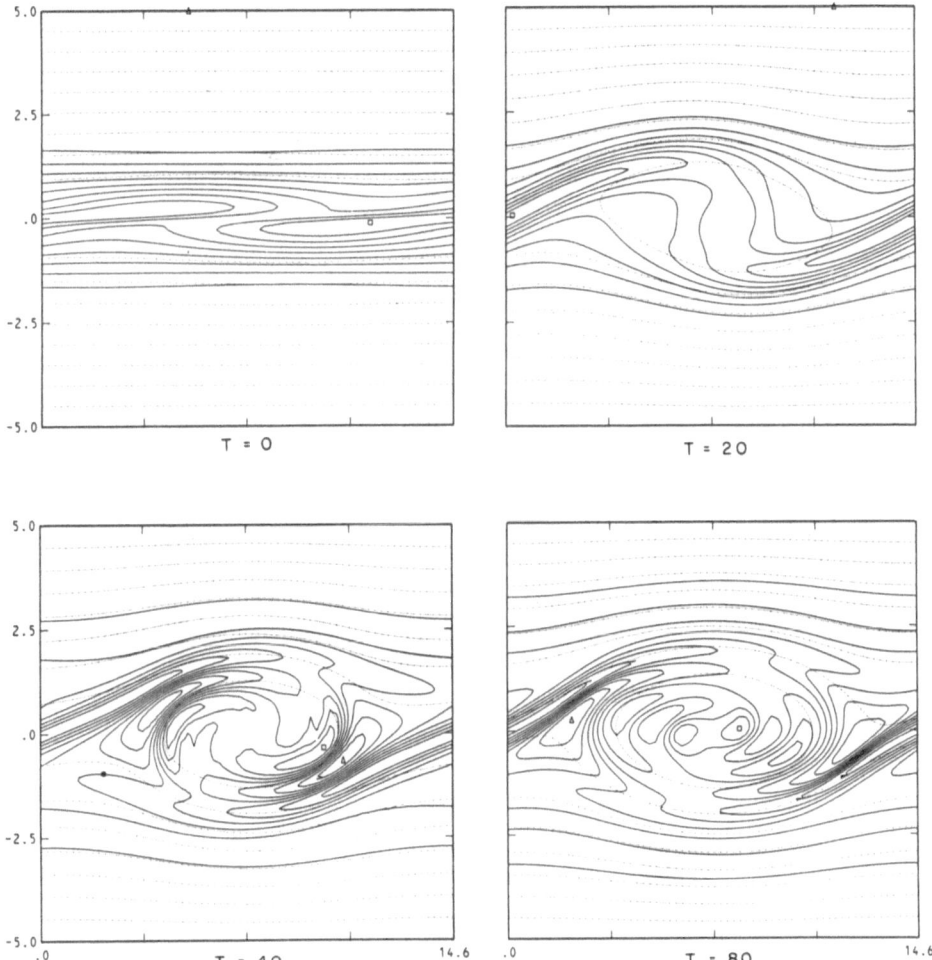

Fig. 14. Constant vorticity contours (solid) and stream-
lines (dotted) for the Boussinesq equations. Parameters
given in Fig. 13.

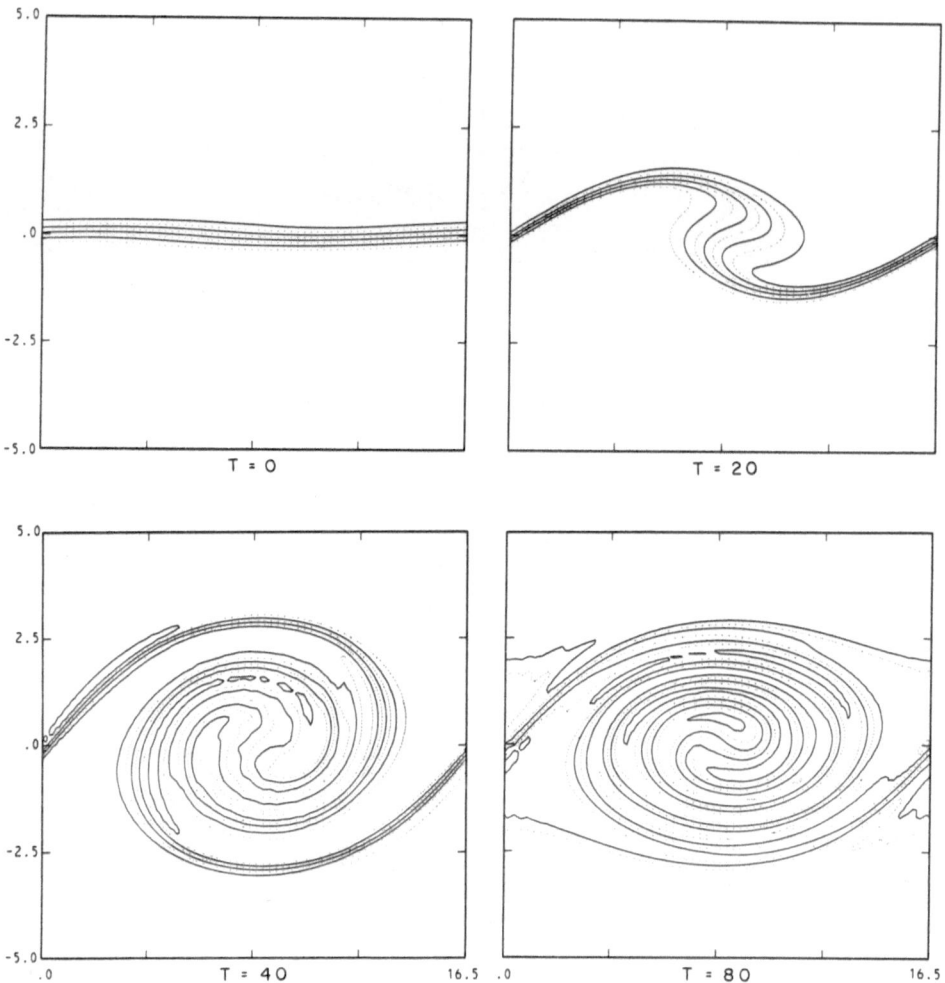

Fig. 15. Constant density contours for the Boussinesq
equations. R = 3.5, J = 0.1, α = 0.38, Re = 100, Pr = 10.
(Deem 1977, Run 3).

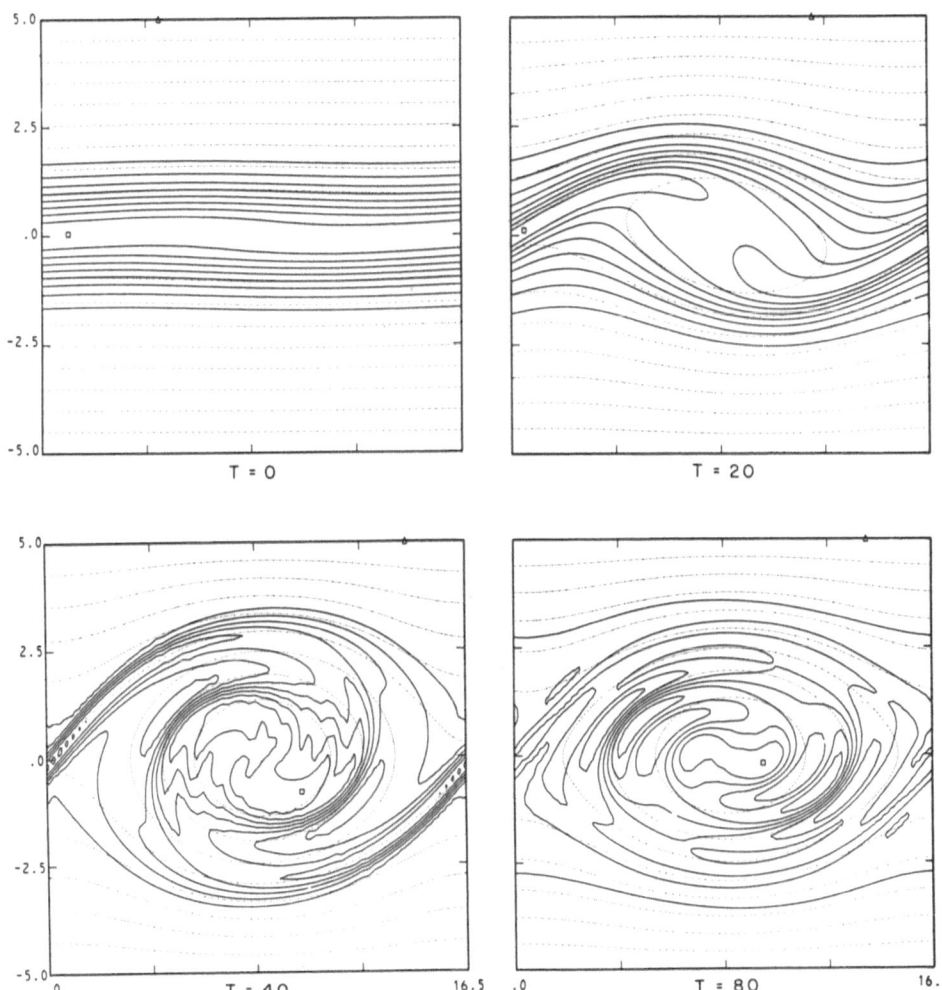

Fig. 16.   Constant vorticity contours (solid) and stream-
lines (dotted) for the Boussinesq equations.   Parameters
given in Fig. 15.

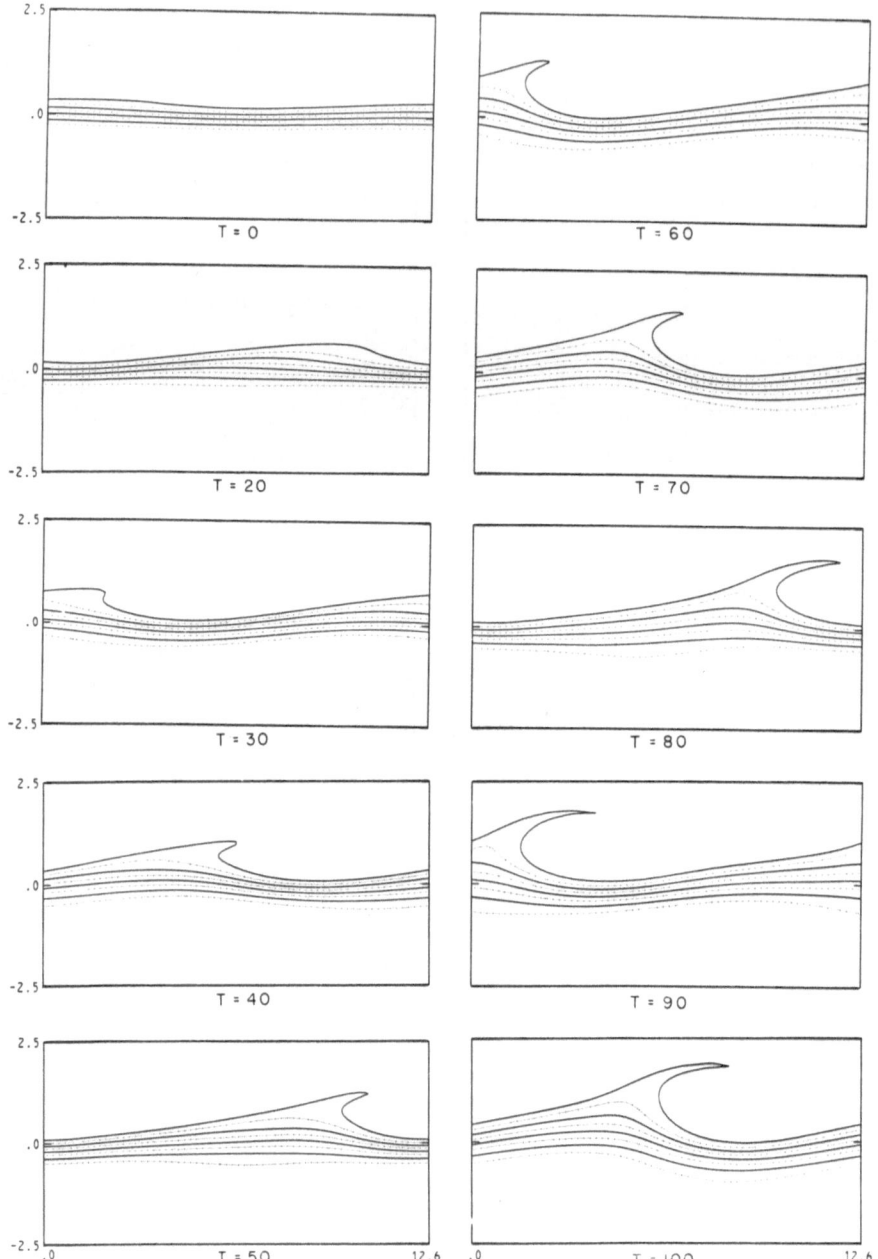

Fig. 17.   Constant density contours for the Boussinesq
equations.   R = 3.5, J = 0.4, α = 0.5, Re = 100, Pr = 10.
(Deem 1977, Run 7).

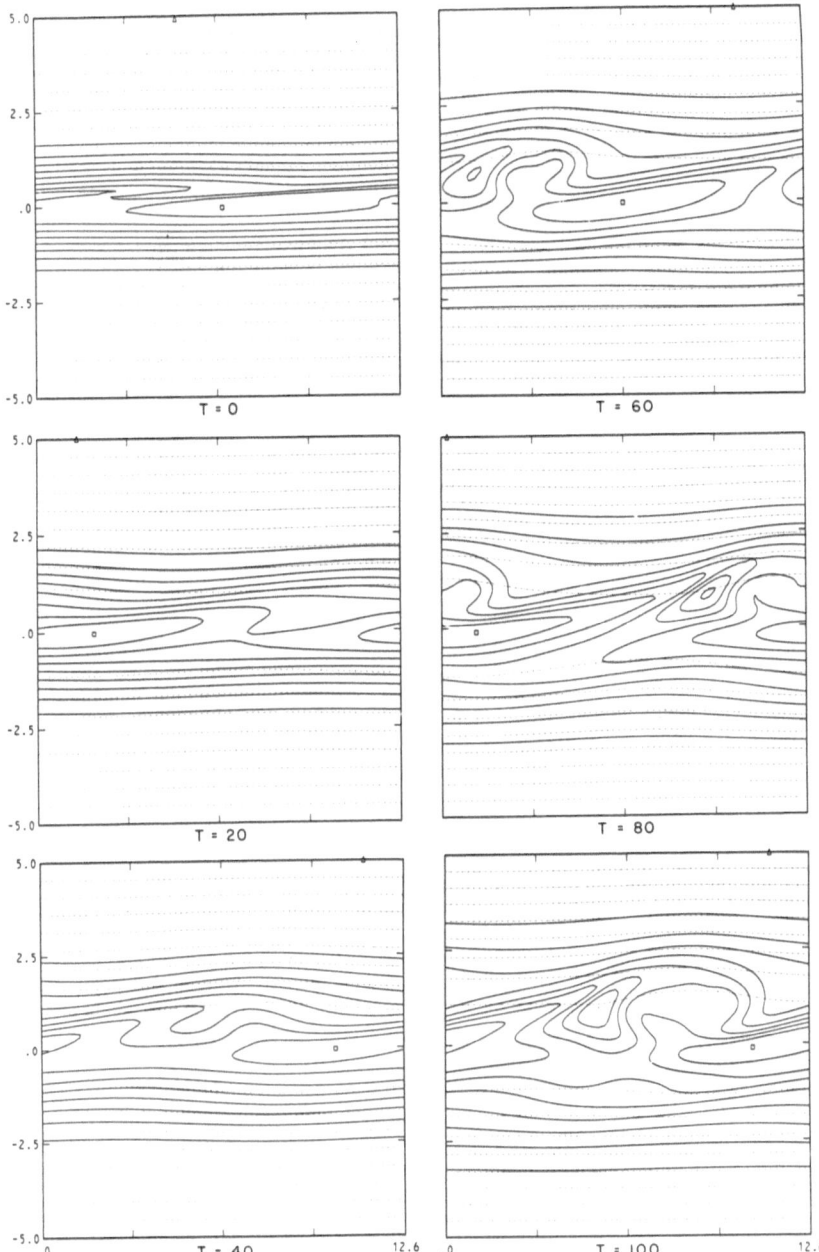

Fig. 18. Constant vorticity contours (solid) and stream-
lines (dotted) for the Boussinesq equations. Parameters
given in Fig. 17.

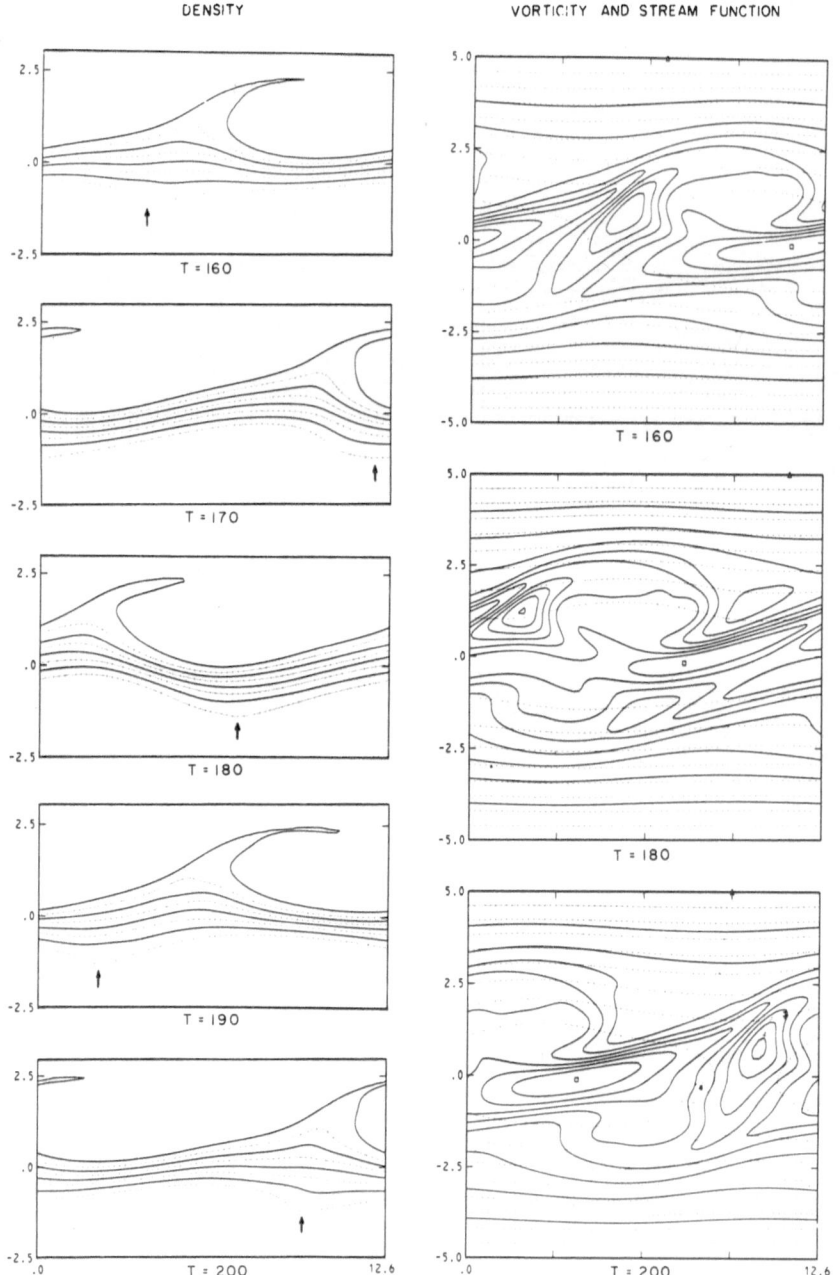

DENSITY   VORTICITY AND STREAM FUNCTION

Fig. 19. Long-time behavior of constant density, vorticity and streamline contours for parameters corresponding
to Figs. 17 and 18. The arrows indicate the position of
a left-moving secondary wave.

in strength (two closed contours at t = 10, three closed
contours at t = 160 and four closed contours at t = 200).
There is a left-moving wave on the lower half of the flow
(see upward pointing arrow, left column of Fig. 19).
The upper vortical region is enhanced as the left-moving
wave approaches or recedes from it. That is, the energy
in fluctuations has <u>twice</u> the period of the propagating
wave. If the Reynolds number is increased to 200, the
cusped will break. In Fig. 20, at t = 60 the density
contour becomes "reentrant" and breaking is evident at
t = 70. This causes an enhanced rate of growth of the
lower left-moving wave at t = 90 and overturning is
evident at t = 100.

Deem also discusses the balance of "overall" energy
sources and sinks

$$\frac{d}{dt} \, \varepsilon_f = S + N - \varepsilon.$$

Here, ε_f is the fluctuation kinetic energy

$$\varepsilon_f = \frac{1}{2} \, <(u'^2 + w'^2)>,$$

S is the Reynolds shear stress

$$S = - \, <u'w' \, \partial_z \bar{u}>,$$

N is the buoyancy flux

$$N = - \, Fr^{-1} \, <\rho'w'>,$$

and ε is the dissipation rate

$$\varepsilon = Re^{-1} \, <(\partial_x u')^2 + (\partial_x w')^2 + (\partial_z u')^2 + (\partial_z w')^2>.$$

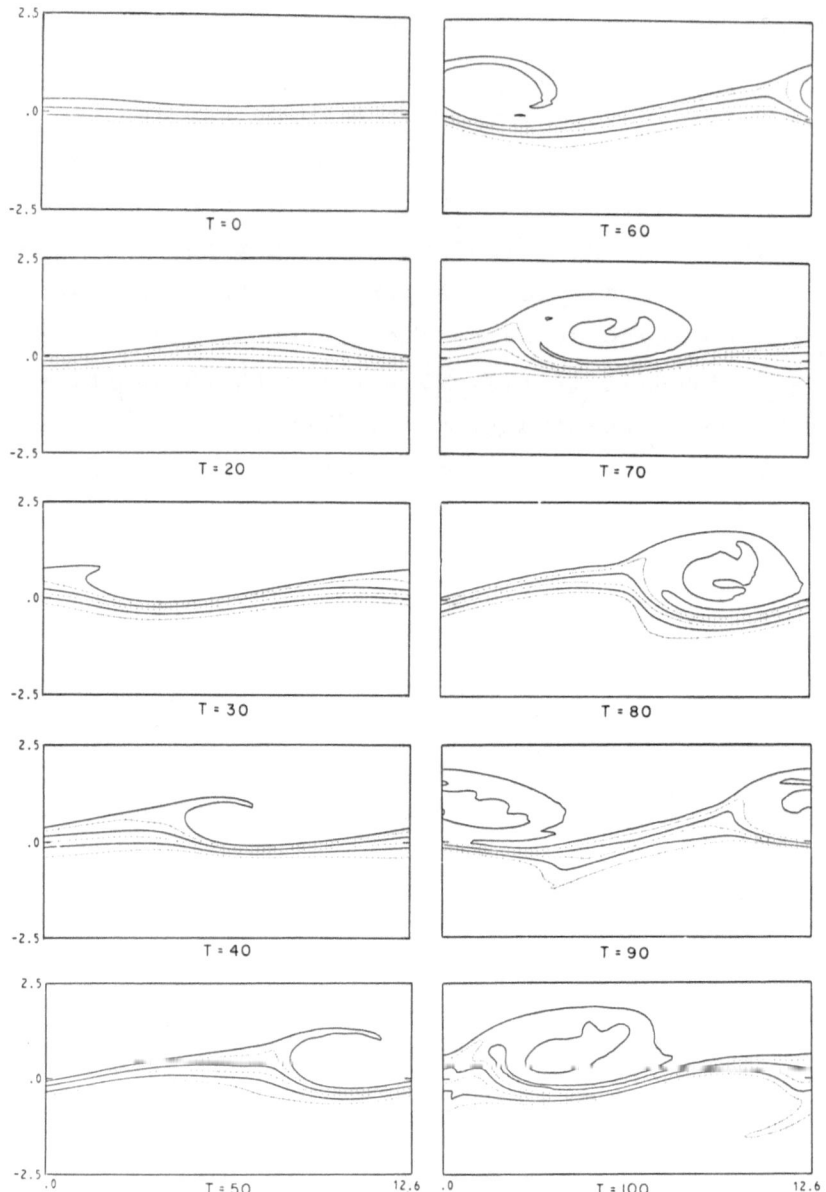

Fig. 20. Constant density contours for the Boussinesq
equations. (Reynolds number increased to allow propagating mode to break). R = 3.5, J = 0.4, α = 0.5, Re =
200, Pr = 10. (Deem 1977, Run 10). Note the enhanced
growth of the left-moving wave after t = 90.

He has divided the flow into its downstream mean and
fluctuating parts (e.g. $u = \bar{u} + u'$), and braces are
used to denote averages over the entire flow (in the
computational box). In Fig. 21 (Deem's Fig. 16) ε_f
grows rapidly at first and then exponentiates at a
slower rate. S grows monotonically. Both curves are
modulated by twice the frequency of the waves travelling
to the left (lower half) and right (upper half).

This nonlinear dispersive system is rich in phenomena
needing mathematical explanation and quantification;
in particular, the breaking of the cusped waves and the
long-time generation of the lower left-propagating wave,
both of which have been ovserved experimentally by
Thorpe (1973). Note that unlike the cat's-eye case,
the growth and breaking of cusped waves are buoyancy-
dominated phenomena strongly dependent on the density
distribution.

NEW DIRECTIONS

Computing weakly dissipative flows with finite-
difference, particle-field, or spectral algorithms is
expensive in two dimensions and prohibitive in three
dimensions. We must look for new methods to investigate
the long-time persistence of two and three-dimensional
coherent states.

The "contour dynamics" approach investigated by
Zaroodny and Greenberg (1973) and Zabusky and Roberts
(1972) provides an example for two-dimensional vorticity
conserving flows. Leonard's (1975, 1977) Lagangian
vortex filament method is useful for vortex rings, wakes
of objects, etc. in three dimensions. We discuss these
numerical methods and conclude with a discussion of recent experiments and analysis of three-dimensional flows.

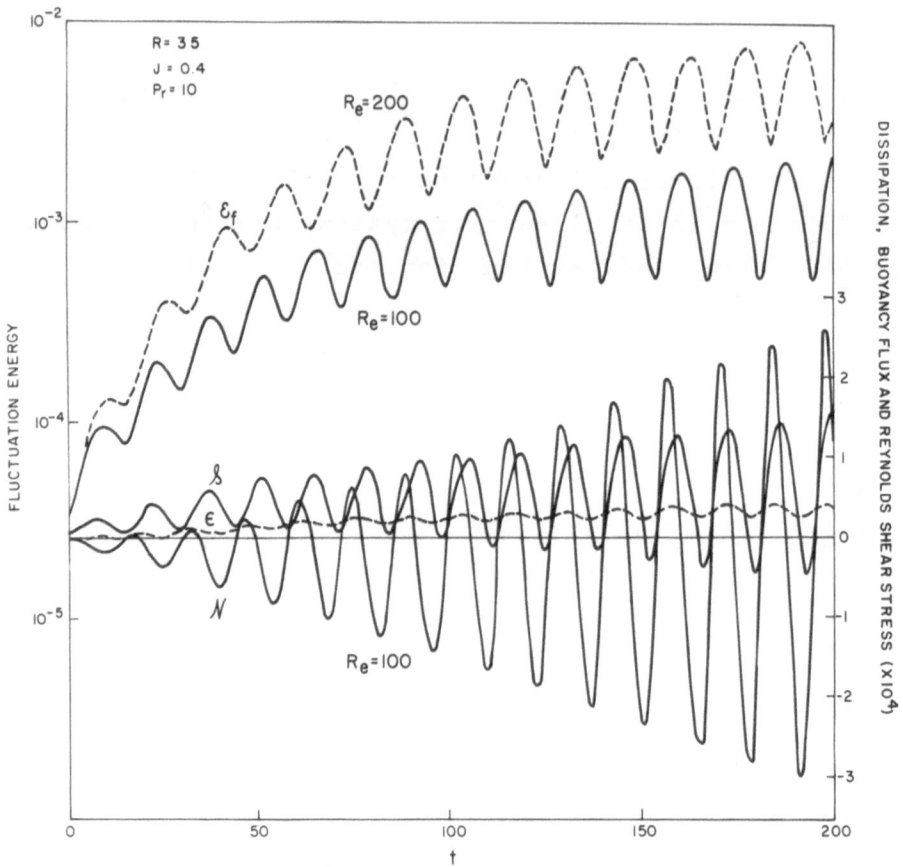

Fig. 21.   Fluctuation energy, Reynolds shear stress,
bouyancy flux and dissipation vs. time.   R = 3.5, J = 0.4,
α = 0.5, Pr = 10, Re = 100 (Run 7).   The upper dashed
curve shows ε_f at Re = 200 (Run 9) for a breaking cusped
wave.

Contour Dynamics

Two facts motivated Zabusky and Roberts to seek an efficient contour dynamics algorithm. First, Lamb (1932, p. 231) and also Batchelor (1970, p. 532) noted that the surface of the Rankine vortex (a circular vortex of radius R_0, with constant vorticity ζ_0) supports infinitesimal azimuthal waves

$$r = R_0 + \varepsilon \cos(m\theta - \omega t),$$

where

$$\omega = (\frac{1}{2} \zeta_0)(m-1).$$

First, these are dispersive waves, because $(\omega/m) \neq d\omega/dm$, (or the larger m, the smaller the rotational period). It seems reasonable that if the amplitude ε were made small but finite, these azimuthal waves would have non-linear-dispersive or perhaps soliton-like properties. Second, the calculations of Christiansen (1973) showed strong surface deformations when two like-signed FAVRs approached, and the calculations of Christiansen and Zabusky (1973) suggested that these azimuthal surface deformations could provide a stabilization mechanism for the von Kármán wake.

The contour-dynamics algorithm follows the motion of closed curves surrounding regions of constant vorticity. Every computational "node" on the surface is advected by the local velocity created by all other surfaces nodes. The algorithm is analogous to the "water-bag" model that Berk and Roberts (1970) used to study the evolution of instabilities of the one-dimensional Vlasov equation of plasma physics.

Fig. 22 shows the geometry associated with one FAVR, where we are determining the velocity at (x,y) from all nodes on the contour, (ξ,η). We write the solution of (5) in terms of the two-dimensional Green's function, or

$$\psi = -(2\pi)^{-1} \int_A d\xi d\eta \; \zeta(\xi,\eta) \log r, \qquad (29)$$

where $\zeta(\xi,\eta)$ is the piecewise constant vorticity distribution and $r^2 = (x-\xi)^2 + (y-\eta)^2$. Then, the velocity is

$$\underset{\sim}{u} = \hat{\underset{\sim}{z}} \times \nabla\psi = \hat{\underset{\sim}{z}} \times (2\pi)^{-1} \int d\xi d\eta \log r(\nabla_{\underset{\sim}{\xi}} \zeta), \qquad (30)$$

where

$$\nabla_{\underset{\sim}{\xi}} = \hat{\underset{\sim}{x}} \, \partial_\xi + \hat{\underset{\sim}{y}} \, \partial_\eta.$$

Since $\nabla_{\underset{\sim}{\xi}}\zeta$ is zero everywhere except on the surface, where it is singular, we can replace the surface integral by the line integral

$$\underset{\sim}{u} = \underset{\sim}{u}(x_n,y_n) = \frac{1}{2\pi} \sum_{A_i} [\zeta_i] \oint \log r[d\xi \; \hat{\underset{\sim}{x}} + d\eta \; \hat{\underset{\sim}{y}}], \qquad (31)$$

where

$$[\zeta_i] = (\zeta_i)_{outside} - (\zeta_i)_{inside} \qquad (32)$$

is the jump in vorticity across the boundary of the ith FAVR. Hence, we have replaced a two-dimensional integral with a one-dimensional line integral. The integral has been discretized and the position of the boundary nodes advanced with a leap-frog algorithm applied to

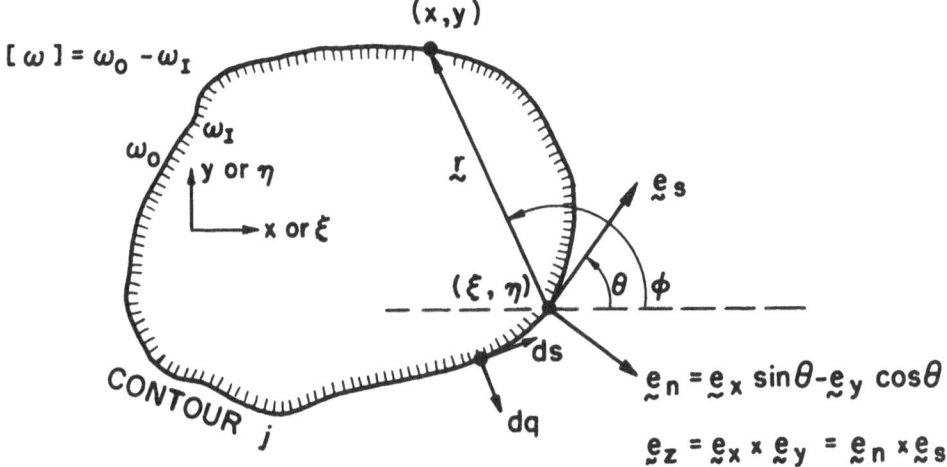

Fig. 22. Schematic illustrating the contour dynamics
algorithm for a two-dimensional incompressible inviscid
fluid. Note, ω is used for vorticity (not ζ) and the
unit normal vectors are e_x, e_y, and e_z.

$$u(x_n,y_n) = \dot{x}_n = \frac{1}{2\pi} \sum_{A_i} [\zeta_i] \oint \log r \, d\xi, \quad (33)$$

$$v(x_n,y_n) = \dot{y}_n = \frac{1}{2\pi} \sum_{A_i} [\zeta_i] \oint \log r \, d\eta. \quad (34)$$

Results for two circular FAVRs with 4 different
initial separations were obtained by Roberts and Hane and
are shown in Fig. 23. The initial separation distance
decreases as one moves from column 1 to column 4 (1.06,
1.022, 1.02, 0.8), and time increases as one moves downward. The first row contains solutions at two times,
0 (two vertically displaced circles) and 4.9. The mutual
interaction induces wave-like azimuthal (m = 2, 3, etc.)
deformations onto the surface, and when close interactions
occur, cusps form.

Case (b) shows the sharing of vortex fluid at t =
19.9 and then a separation at t = 24.9, leaving a single
vortex thread joining the two FAVRs. In case (c) they
seem to have permanently coalesced, and the vortex fluid
of one is flowing around the other in a narrow ridge.
Case (d) shows full coalescence and the ejection of two
arms as the rotation proceeds. Looking across at t =
14.9, one sees the net rotation rate increasing at the
initial separation distance decreases.

This algorithm can be adapted to piecewise-linear
vortex surfaces or finite surface elements having higher-
degree polynomials. In these cases one will follow many
closed curves. In particular, Zabusky has developed an
algorithm (yet untested) for piecewise linear vortex
surfaces. The method could be adapted to other boundary
conditions (e.g. periodic) with an appropriate Green's
function.

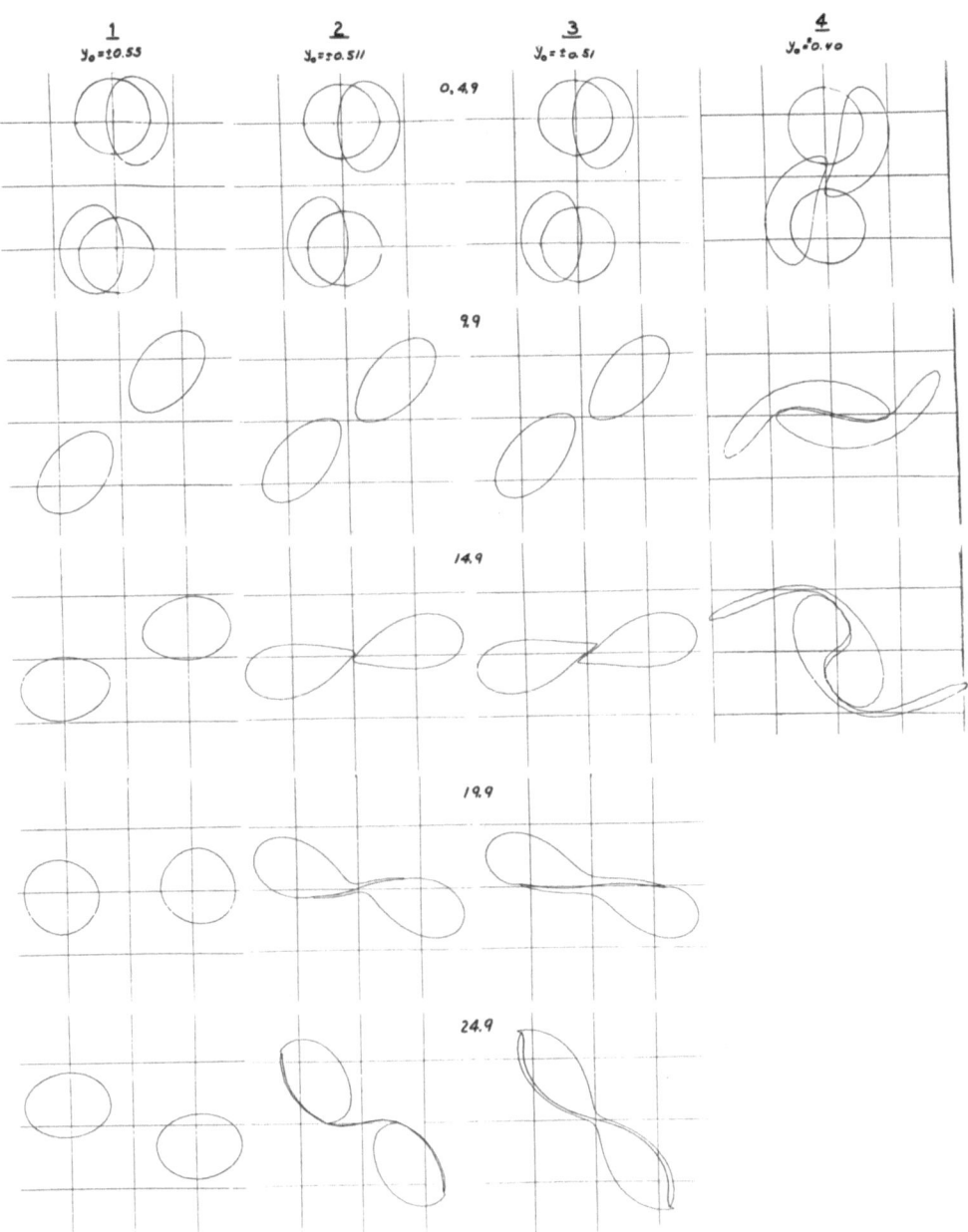

Fig. 23. Contour dynamics solutions for two initially
circular like-signed FAVRs. Initial separation of
centers: (a) 1.06; (b) 1.022; (c) 1.02; (d) 0.08.

If the surface becomes highly structured, one would
have to insert numerous nodes to maintain the level of
initial accuracy. This is evident in case (c) at t =
24.9, where one sees kinks in the surface, and in case
(d), (where we stopped the calculation at t = 14.9).
If two surfaces approach, one should remove or "trim"
segments as discussed by Berk and Roberts (1970). This
disadvantage needs further consideration. Also a
limitation is the algorithm's inability to treat slightly
dissipative or baroclinic (vorticity generating) flows.
Dissipative flows could probably be treated by an adapta-
tion of Chorin's (1973) method. Furthermore, one might
develop rules for the shape change and "life-time" of
very fine arms and very small FAVRs. That is, they would
be removed from the flow after a time, dependent upon
their initial strength.

In another two-dimensional calculation, Morikawa
and Swenson studied the linear stability and nonlinear
evolution of a localized uniformly rotating entity. It
was composed of N "geostrophic" (modified Bessel function)
point vortices of equal strength placed on a circle and
one geostrophic vortex of different strength γ_0 placed
at the center. For all parameters the center vortex
motion was found to be inherently nonlinear. For $\gamma_0 < 0$,
after an oscillatory induction period, some vortices
"move to infinity". It would be interesting to examine
how these "particles" moving in a potential flow field
would interact with each other.

Experimental and Computational Results in Three
Dimensions

In three-dimensional, high Reynolds number wakes
beyond spheres, experiments of Magarvey and MacLatchy
(1965), Bailey (1968), and Achenbach (1972, 1974)

(Re ~ 1000) have shown a regular near-periodic release
of looped and intertwined FAVRs. In Fig. 24 we see
Bailey's (1968) Schlieren photographs of the far wakes
of spheres (Re > 47,000). The remarkable feature is
the frequent appearance of near-periodic looped braids,
sometimes mottled with fine "turbules". In Fig. 25 we
see the near wake of a hypersonic cone and again the
braided loops emerge. The shedding and interaction of
3D FAVRs are being studied by Leonard (1975, 1977) with
his Lagrangian vortex-filament method. Much work remains
to be done validating this technique at long times when
persistent features coalesce, entrain, eject arms and
undergo other fine structuring. Experiments by Kambe
& Takao (1971) have shown the existence of torodial
waves (m = 2 and 3) on ring vortices and the "fission"
and subsequent "fusion" of ring vortices that were
propelled at each other. Oshima (1972) generated non-
toroidal isolated ring vortices and examined oscillations
in their shape and their persistence.

   Analytical Results in Three Dimensions

   Analytically, a large class of 3D hydrodynamic
localized stationary states have been found. Methods
for calculcating them have been developed by Moffatt
(1969), Frankel (1972) Norbury (1972, 1973) and Pekeris
(1972). Such states will prove valuable for validating
and testing the long time behavior of 3D algorithms.
These stationary entities will support wave-like disturbances that may be unstable if their geometry is
simple or if imposed perturbations are large-amplitude
(Widnall, 1975).

   For example, Hasimoto (1972) considered the equations
governing the evolution of curvature and torsion of an
isolated thin vortex filament in an inviscid incompressible

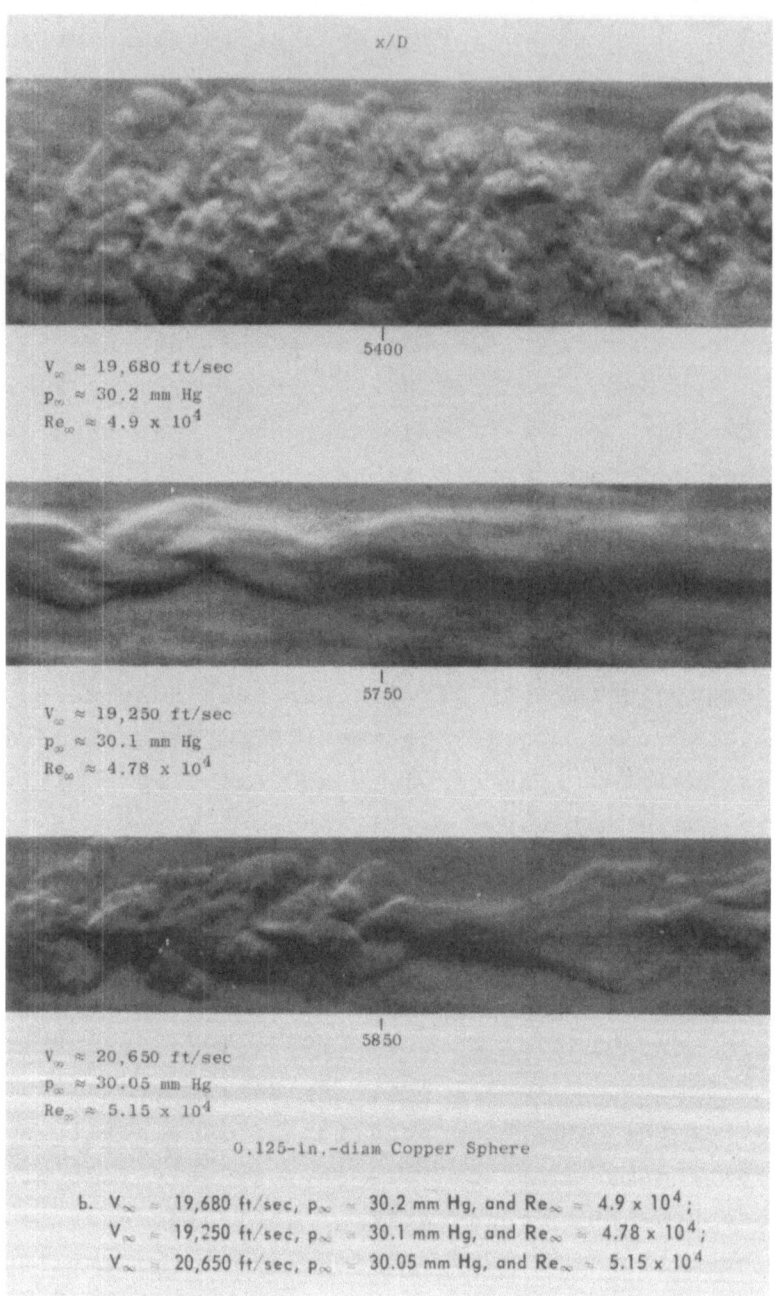

Fig. 24.   Schlieren photographs of wakes of spheres at
high Reynolds number (Bailey, 1968).

fluid assuming no stretching. His equations reduce to
type of nonlinear Schrodinger equation

$$- i \ \psi_t = \psi_{ss} + \frac{1}{2} (A + |\psi|^2) \ \psi,$$

where

$$\psi = \kappa \ \exp[i \int_0^s \tau \ ds']$$

and κ and τ are curvature and torsion and s is arc
length. He finds that a solitary wave describes "the
propagation of a loop or a hump of helical motion along
a line vortex".

I conclude by recalling Basset's pessimistic remarks and von Neumann's prescient optimism. I hope I
have demonstrated my optimism. We are performing more
accurate experiments with a high degree of space-time
localization and multiple correlation; we can process
and correlate great amounts of data; we are developing
new algorithms for two-and three-dimensional motions
that will provide solutions to bridge the theoretical/
experimental gap. All this promises a deeper understanding of the evolution of weakly dissipative or non-
dissipative flows - in neutral fluids and in plasmas.

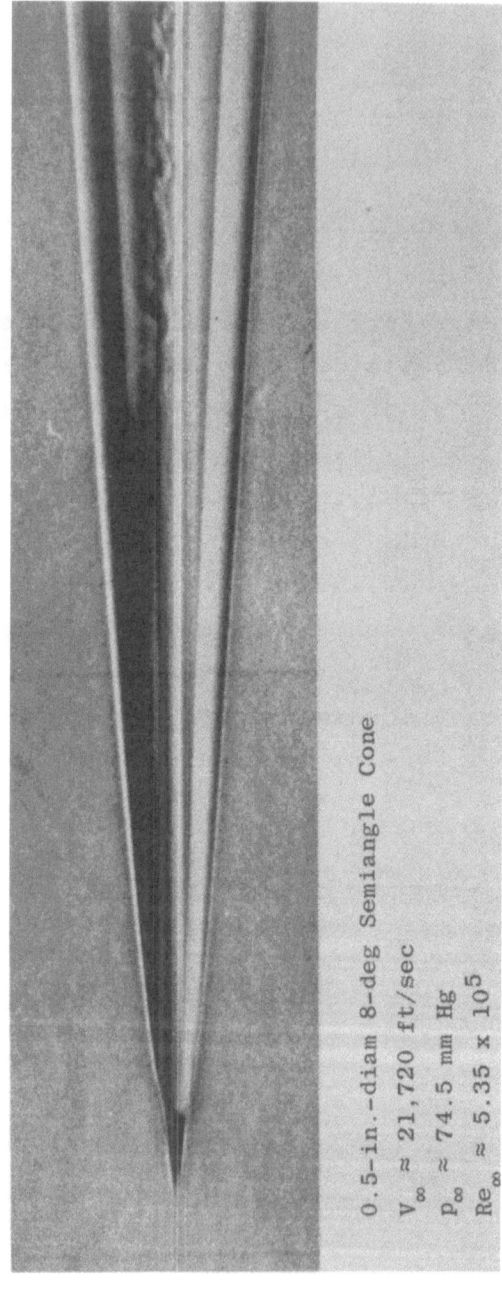

0.5-in.-diam 8-deg Semiangle Cone

$V_\infty \approx 21,720$ ft/sec

$p_\infty \approx 74.5$ mm Hg

$Re_\infty \approx 5.35 \times 10^5$

Fig. 25. Schlieren photographs of the near wake of a hypersonic cone (Bailey, 1968).

REFERENCES

1. Abernathy, F. H. and Kronauer, R. E.   The Formation
    of Vortex Streets, J. Fluid Mech. 13, 1, (1962).

2. Achenbach, E. Experiments on Flow Past Spheres at
    Very High Reynolds Numbers. J. Fluid Mech. 54,
    565-575, (1972).

3. Achenbach, E.   Vortex Shedding from Spheres. J.
    Fluid Mech. 62, 209-221, (1974).

4. Ahlborn, F. Uber den Mechanismus des hydrodynamischen
    Widerstandes, - Hamburg, (1902).

5. Bailey, A.B. Observations of Sphere Wakes Over a
    Wide Range of Velocities and Ambient Pressures.
    Arnold Research Organization, Inc. Report AEDC-TR-
    68-112, (1968).

6. Basset, A.B. 1888. A Treatise or Hydrodynamics.
    Dover, 1961.

7. Batchelor, G.K. Homogeneous Turbulence, Cambridge
    University Press (1953).

8. Bearman, P.W. On Vortex Shedding From A Circular
    Cylinder in the Critical Reynolds Number Regime.
    J. Fluid Mech. 37, 577, (1969).

9. Beavers, G.S. and Wilson, T.A. Vortex Growth in
    Jets. J. Fluid Mech. 44, 97, (1970).

10. Bénard, H. Formation de Centres de Giration a
    l'arriere d'un Obstacle en Mouvement, Comp. Rend
    147, 839, (1908).

11. Berger, E. and Wille, R. Periodic Flow Phenomena,
    Ann. Rev. Fluid Mech. 4, 313, (1972).

12. Berk, H.L. and Roberts, K.V. The Water-bag Model.
    Methods of Computational Physics, Plasma Physics,
    Vol. 9, 88-135, (1970).

13. Betchov, R. and Criminale, W.O. Stability of
    Parallel Flows, Academic Press, New York, (1967).

14. Browand, F.K. and Weidman, P.D. Large Scales in the Developing Mixing Layer. J. Fluid Mech. 76, 127-144, (1976).

15. Brown, G. L. and Roshko, A. On Density Effects and Large Structure in Turbulent Mixing Layers. J. Fluid Mech. 64, 775-816, (1974).

16. Chorin, A.J. Numerical Study of Slightly Viscous Flow. J. Gluid Mech. 57, 785-796, (1973).

17. Christiansen, J.P. Numerical Simulation of Hydro-dynamics by the Method of Point Vortices. J. Comp. Physics. 13, 363-379, (1973).

18. Christiansen, J.P. and Zabusky, N.J. Instability Coalesence and Fission of Finite-Area Vortex Struc-tures. J. Fluid Mech. 61, 219-243, (1973).

19. Davies, P.O.A.L. and Yule, A.J. Coherent Structures in Turbulence. J. Fluid Mech. 69, 513-537, (1975).

20. Deem, G.S. The Origin of Cusped Waves in Layered Fluids. J. Fluid Mech., to be published. (1977).

21. Durgin, W.W. and Karlsson, S.K.F. On the Phenomenon of Vortex Street Breakdown. J. Fluid Mech. 48, 507, (1971).

22. Fetter, A.L. Vortices and Ions in Helium. The Physics of Liquid and Solid Helium, eds. K.H. Bennemann and J.B. Ketterson. J. Wiley and Sons. (1970).

23. Frankel, L.E. Examples of Steady Vortex Rings of Small Cross-Section in an Ideal Fluid. J. Fluid Mech. 51, 119, (1972).

24. Fromm, J.E. and Harlow, F.H. Numerical Solution of the Problem of Vortex Street Development, Phys. Fluids 6, 975, (1963).

25. Goldstine, H.H. and Von Neumann, J. "On the Principles of Large Scale Computing Machines", in Collected

Works *of* John *von* Neumann, ed. A. Taub, Vol. 5, pp. 1-32, Macmillan, New York, 1963. The material in this paper was first given as a talk on May 15, 1946. Also see: "Recent Theories in Turbulence", in *Collected* *Works* *of* John *von* Neumann ed. A. Taub, Vol. 6, pp. 437-472. This paper was issued as a report in 1949.

26. Griffin, O.M. and Votaw, C.W. The Vortex Street in the Wake of a Vibrating Cylinder. J. Fluid Mech. 51, 31, (1972).

27. Griffin, O.M. and Ramberg, S.E. The Vortex-Street Wakes of Vibrating Cylinders. J. Fluid Mech. 66, 553-578, (1974).

28. Hama, F.R. Three-Dimensional Vortex Pattern Behind a Circular Cylinder, J. Aerosp. Sci. 24, 156, (1957).

29. Hama, F.R. Streaklines in a Perturbed Shear Flow. Phys. Fluids 5, 644, (1962).

30. Hama, F.R. Progressive Deformation of a Perturbed Line Vortex Filament. Phys. Fluids 6, 526, (1963).

31. Hardin, R.H., Deem, G.S., Tappert, F.D. and Zabusky, N.J. Visualization of Properties of the Two-Dimensional Navier-Stokes Equation. Unpublished computer generated film produced at Bell Laboratories, Whippany, N.J. 1971.

32. Hasimoto, H. A Soliton on a Vortex Filament. J. Fluid Mech. 51, 477-485, (1972).

33. Kochin, N.E., Compt. Rend. (Doklady) de l'Acadamie des Sciences, l'sssr 24, 18-22 (1939).

34. Kochin, N.E., Kiebel, I.A. and Roze, N.U. *Theoretical* *Hydromechanics*, Interscience, (1964).

35. Lamb, H. *Hydrodynamics*, Cambridge, University Press, 6th ed. (1932).

36. Leonard, A.   Proc. of the IVth Int. Conf. on
    Numerical Methods in Fluid Dynamics, ed. R. D.
    Richtmyer, P. 245. Springer Verlag, (1975).

37. Leonard, A. Simulation of Three-Dimensional Separated
    Flows with Vortex Filaments. Proc. of the Vth Int.
    Conf. on Numerical Fluid Dynamics (1976). Lecture
    Notes in Physics, ed. R.D. Richtmyer, Springer
    Verlag, (1977).

38. Leslie, D.C.   Developments in the Theory of Turbu-
    lence. Clarendon Press, Oxford, (1973).

39. Magarvey, R.H. and MacLatchy, C.S.   Vortices in
    Sphere Wakes. Cand. J. Phys. 43, 1649-1656, (1965).

40. Miura, A. and Sato, T.   Theory of Vortex Nutation
    and Amplitude Oscillation in an Inviscid Shear
    Layer. Submitted to J. Fluid MEch. (1977).

41. Moffatt, H.K.   The Degree of Knottedness of Tangled
    Vortex Lines.   J. Fluid Mech. 35, 117-129, (1969).

42. Montgomery, D.   Individual and collaborative papers
    with G. Joyce, G. Knorr, Y. Salu and C.R. Seyler, Jr.
    in Phys. Rev. Letters, J. Plasma Phys., Phys. Fluids,
    1973-1977. (1977).

43. Morikawa, G.K. and Swenson, E.V.   Interacting Motion
    of Rectilinear Geostrophic Vortices. Phys. Fluids
    14, 1058-1073, (1971).

44. Morkovin, M.V.   Flow Around a Circular Cylinder:
    A Kaleidoscope of Challenging Fluid Phenomena.
    ASME Symposium on Fully Separated Flows, pp. 102-118.
    (A Fluids ENg. Div. Conference). (1964).

45. Norbury, J.   A Steady Vortex Ring Close to Hills
    Spherical Vortex. Proc. Camb. Phil. Soc. 72, 253,
    (1973).

46. Norbury, J.   J. Fluid Mech. 57, 417-431, (1973).

47. Onsager, L. Nuovo Cimento Suppl. $\underline{6}$, 279. (Supplement No. 2, Series 9). (1949).

48. Orszag, S.A. Lecture Notes on the Statistical Theory of Turbulence. Les Houches Lectures, 1973, ed. R. Balian North-Holland, (1976).

49. Papailiou, D.D. and Lykoudis, P.S. Turbulent Vortex Streets and the Entrainment Mechanism of the Turbulent Wake. J. Fluid Mech. $\underline{62}$, 11-31, (1974).

50. Patnaik, P.C., Sherman, F.S. and Corcos, G.M. A Numerical Simulation of Kelvin-Helmholtz Waves of Finite Amplitude. J. Fluid Mech. $\underline{73}$, 215, (1976).

51. Pekeris, C.L. Stationary Spherical Vortices in a Perfect Fluid. Proc. Nat. Acad. Sci. $\underline{69}$, 2460-2462, (1972).

52. Phillips, O.M. The Dynamics of the Upper Ocean. Cambridge University Press, (1969).

53. Prandtl, L. The Generation of Vortices...London. (1927).

54. Roberts, K.V. and Christiansen, J.P. Topics in Computational Fluid Mechanics. Compt. Phys. Comm. $\underline{3}$, (Suppl), 14, (1972).

55. Roberts, P.H. and Donnelly, R.J. Superfluid Mechanics. Ann. Rev. Fluid Mech. $\underline{6}$, 179-225, (1974).

56. Rosenhead, L. Vortex Systems in Wakes. Adv. in Appl. Mech. $\underline{3}$, 185, (1953).

57. Roshko, A. Structure of Turbulent Shear Flows. AIAA J. $\underline{14}$, 1349-1357, (1976).

58. Saffman, P.G. Lectures on Homogeneous Isotropic Turbulence, in Topics in Nonlinear Physics. ed. N.J. Zabusky, pp. 485-614, Springer-Verlag. Saffman advocates a physical-space approach to very high-Re turbulence, pp. 567-581, (1968).

59. Saffman, P.G. Structure of Turbulent Line Vortices, Phys. Fluids 16, 1181-1188, (1973).

60. Saffman, P.G. On the Formation of Vortex Rings. Studies in Appl. Math. 54, 261-268, (1975).

61. Sato, H. and Kuriki, K. The Mechanism of Transition in the Wake of a Thin Flat Plate Placed Parallel to a Uniform Flow. J. Fluid Mech. 11, 321, (1961).

62. Strouhal, V. Uber eine besondere Art der Tonerregung. Ann. Phys. Chem. 5, 216, (1878).

63. Stuart, J. T. Nonlinear Stability Theory. Ann. Rev. Fluid Mech. 3, 347, (1971).

64. Taneda, S. Experimental Investigation of the Wakes Behind Cylinders and Plates at low Reynolds Numbers. J. Phys. Soc. Japan 11, 302, (1956).

65. Taneda, S. Oscillations of the Wake Behind a Flat Plate Parallel to the Flow. J. Phys. Soc. Japan 13, 418, (1958).

66. Taneda, S. Downstream Development of Wakes Behind Cylinders. J. Phys. Soc. Japan, 14, 834, (1959).

67. Taneda, S. Experimental Investigation of Vortex Streets. J. Phys. Soc. Japan, 20, 1714, (1965).

68. Thorpe, S.A. Turbulence in Stably Stratified Fluids. Boundary Layer Meterology 5, 95, (1973).

69. Thorpe, S.A. Experiments on Instability and Turbulence in a Stratified Shear Flow. J. Fluid Mech. 61, 731, (1973).

70. Townsend, A.A. The Structure of Turbulent Shear Flow. Cambridge University Press, (1956).

71. Ulam, S.M. A Collection of Mathematical Problems, Wiley, (Interscience), (1960).

72. von Kármán, Th. "Über der Mechanisms des Widerstands, den ein bewegter Körper in einer Flüssigkeit erfährt, Gottinger Nachrichten, Math. Phys. Kl. 509-519 and 547-556, (1911).

73. von Kármán, Th. and Rubach, H. Über den Mechanisms des Flüssigkeits - und Luftwiderstands Phys. Zeitschrift 13, 49-89, (1912).

74. Widnall, S.E. The Structure and Dynamics of Vortex Filaments. Ann. Rev. Fluid Mech. 7, 141-165,(1975).

75. Wille, R. Karman Vortex Street, Adv. in Appl. Mech. 6, 237, (1960).

76. Winant, C.D. and Browand, F.K. Vortex Pairing: The Mechanism of Turbulent Mixing Layer Growth at Moderate Reynolds Number. J. Fluid Mech. 63, 237, (1974).

77. Woods, J.D. and Wiley, R.L. Billow Turbulence and Ocean Microstructure. Deep-sea Res. 19, 87, (1972).

78. Zabusky, N.J. A Synergetic Approach to Problems of Nonlinear Dispersive Wave Propagation and Interaction Nonlinear Partial Differential Equations, ed. W. Ames, p. 223, Academic Press, Inc. (1967).

79. Zabusky, N.J. and Deem, G.S. Dynamical Evolution of Two-Dimensional Unstable Shear Flows. J. Fluid Mech. 47, 353-379, (1971).

80. Zabusky, N.J. and Roberts, K.V. Contour Dynamics for Incompressible Dissipationless Fluids, I. Algorithms, (preprint), (1972).

81. Zabusky, N.J. Solitons and Energy Transport in Non-linear Lattices. Comp. Phys. Comm. 5, 1-10, (1973).

82. Zaroodny, S.J. and Greenberg, M.D. On a Vortex Street Approach to the Numerical Calculation of Water Waves. J. Comp. Phys. 11, 440-446, (1973).

83. Zdravkovich, M.M. Smoke Observations on the Formation of Karman Vortex Street, J. Fluid Mech. 37, 491 (1969).

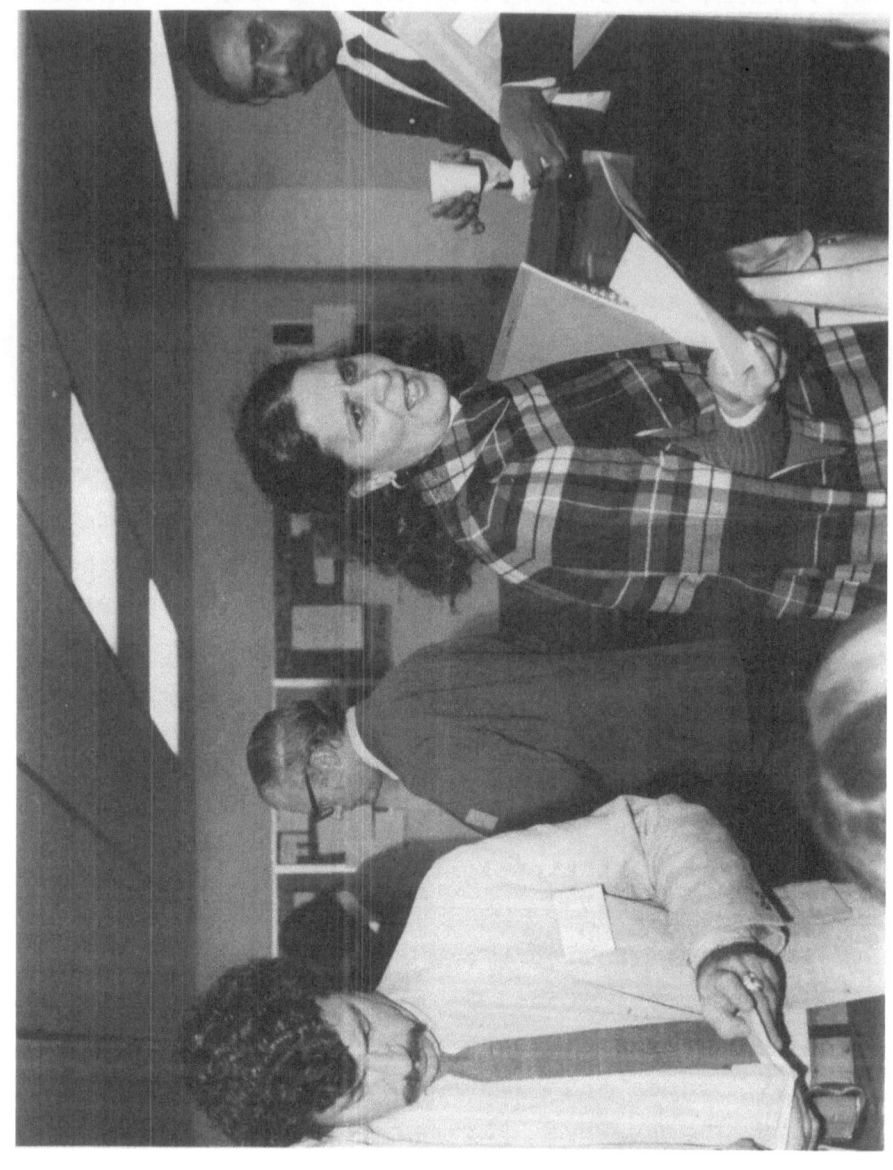

SEVDA KURSUNOGLU, ORBIS SCIENTIAE HONORARY "PARTICIPANT"

COLLISIONLESS MULTIPHOTON DISSOCIATION OF SF_6:
A STATISTICAL THERMODYNAMIC PROCESS

Eli Yablonovitch

Division of Engineering and Applied Physics

Harvard University, Cambridge, Mass. 02138

ABSTRACT

An experiment, which combines picosecond CO_2 laser pulses with opto-acoustic techniques, proves that the collisionless multiphoton dissociation SF_6 is a statistical thermodynamic process. The RRKM theory of unimolecular reactions, familiar to most physical chemists, provides a <u>quantitative</u> explanation of the dissociation effect. The reaction rate is primarily limited by transitions in the quasicontinuum rather than by the anharmonicity of the first few discrete levels.

I. INTRODUCTION

The interest in laser dissociation of molecules[1] became widespread when it was first shown[2] to be isotopically selective. Figure 1, the early work[3,4] of Letokhov and Ambartsumyan, shows that the isotopic selectivity is simply due to the mass shift in the infra-red spectrum of SF_6.

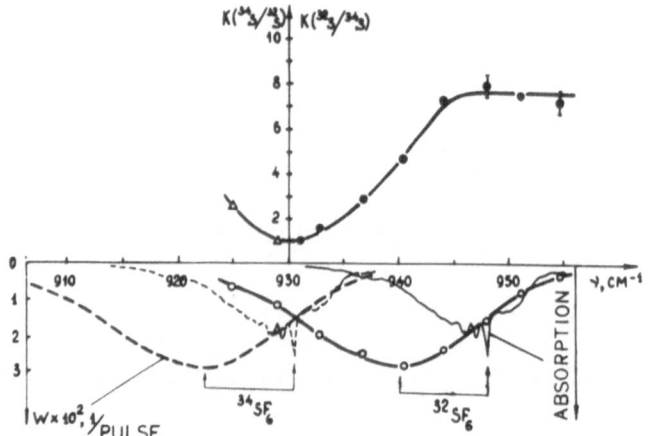

Fig. 1. This is early data by Ambartsumyan et al.,
(ref.4).   The upper part of the graph shows the isotopic
selectivity, while the lower part compares the dissociation yield spectrum with the linear absorption spectrum
for the two isotopes ^{34}S and ^{32}S.   Notice the red shift
in the dissociation spectrum.

   It was quickly established[5] that the process is
essentially collisionless, involving only the interaction
of an isolated molecule with the intense radiation field.
As such, it presents an interesting and novel physical
problem: How can an isolated molecule absorb the 40 or
more infrared quanta needed for dissociation?
   This paper will stress the physics of the multiphoton
dissociation mechanism. We will find that statistical
thermodynamics plays a major role in the effect. The
paper will be divided into four sections:
   The first section will describe the quantum levels
of the SF_6 molecule, from the ground state, through the
quasicontinuum, up to dissociation. The changing spectral
properties of each region will be emphasized.

The second section of the paper will describe measurements of the dissociation yield as a function of laser energy and pulse duration down to 500 picoseconds. We will show why the pulse duration measurements are able to distinguish between the "anharmonicity bottleneck" and other dissociation inhibiting mechanisms.

In the third section of the paper we will show that a statistical thermodynamic approach is appropriate for the quasicontinuum. Therefore we will introduce the RRKM theory of unimolecular reactions.

The fourth section of the paper will describe direct opto-acoustic measurements of the vibrational temperature induced by ultrashort CO_2 laser pulses. These measurements are in good agreement with the statistical thermodynamic theory of dissociation yield. They permit a direct estimate of the fraction of molecules stuck in the "anharmonicity bottleneck." The measurements also show a rapidly falling absorption cross-section as the molecule ascends the quasicontinuum.

II. QUANTUM LEVEL STRUCTURE OF SF₆

A physical understanding has to be based on a knowledge of the energy spectrum of the SF_6 molecule. As a result of the upsurge of interest in the past year[6] we now have rough idea of the SF_6 molecular energy levels, as shown in Figure 2. The detailed structure[7] is omitted both for simplicity and also because much of it is not yet known. Nevertheless, the important overall features are present in Fig. 2. In particular the energy levels are divided into three regions,[8] as indicated by the large Roman numerals.

Region I, the lowest excited states, consists of

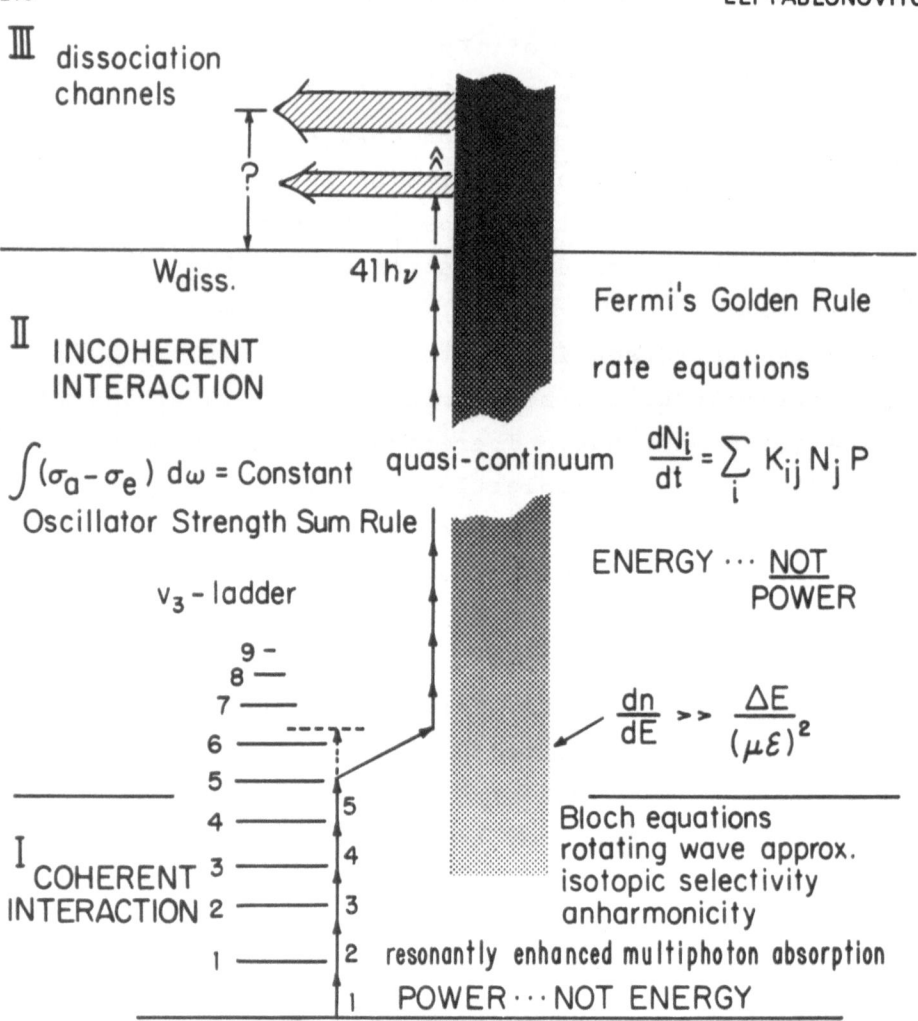

Fig. 2. The quantum level structure of the SF_6 molecule
(simplified!). The energy levels are divided into three
regions; Region I, the discrete levels in which the
interaction is coherent, Region II the quasicontinuum
of levels where incoherent interaction predominates
and Region III, the true continuum where dissociation
actually occurs. The quasicontinuum sets in when Eq. (1)
becomes valid. Transitions in region I are primarily
peak power dependent, while those in region II are
energy fluence dependent.

discrete energy levels. It is symbolized by the ladder of anharmonic oscillator levels of the ν_3 vibrational mode. Of course, the other vibrational modes are not shown. With increasing energy, the phase space volume of a polyatomic molecule rises very rapidly, resulting in a high density of quantum levels. In SF_6 this quasicontinuum already sets in at a level of excitation corresponding to only 5 or 6 CO_2 laser photons (10.6µ wavelength). This we will call region II. Finally there is the true continuum of energy levels lying beyond the dissociation energy, W_{diss}, about 40 laser photons. This is region III.

There is much to be learned by following a laser excited molecule up through Fig. 2, from the ground state to dissociation. In region I, containing discrete levels, the molecular evolution is best described in terms of Bloch equations.[9] This is the region of <u>coherent interaction</u>. The rotating-wave approximation is appropriate. Anharmonicity of the vibrational ladder tends to limit the sequential absorption of photons. This is called the "anharmonicity bottleneck". Resonantly enhanced multiphoton absorption[6] tends to overcome the "anharmonicity bottleneck." But the peak absorption is then red-shifted a few wave numbers from the linear absorption peak, as shown in Fig. 1. Due to the multiphoton character[6] of the absorption it is the peak power of the laser pulse rather than its energy which is important in driving the molecule through region I. Finally, the most important property of region I is the isotopic shift of the discrete levels, which permits the selectivity. The other isotopic species is not excited and remains near the bottom of region I.

At a level of excitation corresponding to about 5 laser photons, the density of states rises and the oscillator strength of the ν_3 mode becomes spread over a number of vibrational levels. When the density of states is large enough and the matrix elements are small enough, Fermi's Golden Rule takes over as the correct description of the molecular evolution. From the conditions of validity[9] of Fermi's Golden Rule, region II begins when the density of states is:

$$\frac{dn}{dE} \gg \frac{\Delta E}{(\mu E)^2} \quad , \qquad (1)$$

where ΔE is the energy width of the band into which the oscillator strength is smeared and μE is the dipole matrix element from the ground state. Equation (1) becomes valid at a level of excitation of 5 or 6 photons, marking the onset of region II.

Fermi's Golden Rule implies that Rate equations rather than Bloch equations are appropriate for the temporal evolution of the molecule in the quasicontinuum. Therefore <u>incoherent interaction</u> describes region II. A typical rate equation is

$$\frac{dN_i}{dt} = \sum_j K_{ij} N_j P - K_{ji} N_i P \quad , \qquad (2)$$

where N_i is the probability of occupation of the i-th stationary state and $K_{ij} P$ is the transition rate between i and j which is explicitly proportional to the laser power P. The important property of Eq. (2) is that the right hand side is explicitly proportional to the laser power P which can therefore be divided out. Equation (2) becomes:

$$\frac{dN_i}{d(Pt)} = \sum_j K_{ij}N_j - K_{ji}N_i \quad . \qquad (3)$$

Equation (3) depends only on (Power) × (time) or in other words laser energy. Only the laser energy, rather than the peak power, is important for transitions in the quasicontinuum.

Another important property, the oscillator strength sum rule,[10] is valid for any state in region II:

$$\int (\sigma_a - \sigma_e) \, d\omega = \text{Constant} \quad . \qquad (4)$$

The net absorption cross section $(\sigma_a - \sigma_e)$ is the difference between the absorption and stimulated emission cross-sections. We may anticipate that further smearing of the oscillator strength in the quasicontinuum will reduce the net absorption cross-section, hindering dissociation.

Finally we come to region III of figure 2. Here the important question is how much excess energy, above W_{diss}, would be required to produce dissociation in a finite time.

III. PICOSECOND MEASUREMENTS OF DISSOCIATION YIELD

A typical measurement[11] of dissociation yield is shown in Fig. 3. The absolute dissociation probability rises rapidly from a threshold at 1.4Joules/cm^2. At a laser fluence of 10J/cm^2 the probability is already near unity. There are two possible "bottlenecks" which would tend to limit the dissociation:

(i) Anharmonicity of the first few discrete levels.[12] This may prevent excitation of the molecule while it is in region I.

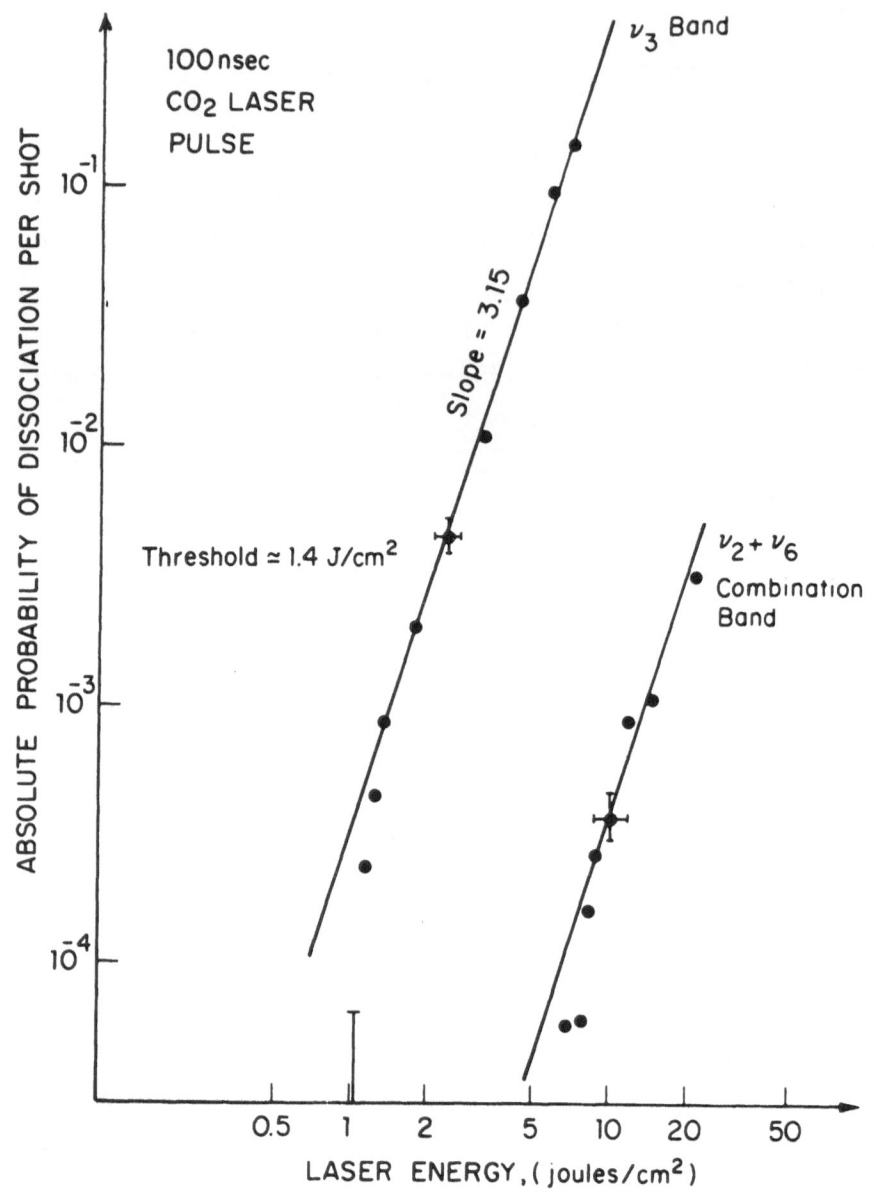

Fig. 3. The absolute dissociation yield vs. energy fluence for irradiation in the ν_3 band. There is a sharp threshold at $J = 1.4$ J/cm^2 followed by a rapid rise to near unity probability. Also shown is the yield for irradiation in the $\nu_2 + \nu_6$ combination band.

(ii) Smearing of the oscillator strength in the
quasicontinuum. This would lead to a reduced absorption
cross-section and less efficient heating of the molecule
in region II.

Pulse duration dependence of the dissociation yield
is able to discriminate between these two possible
bottlenecks. Overcoming the "anharmonicity barrier"
is quite sensitive to the peak power of the laser
beam. Multiphoton absorption in region I is rather
dependent on the magnitude of the Rabi-precession
frequency $\mu E/\hbar$. On the other hand, only the laser
energy, not the peak power, is important for transitions
in the quasicontinuum, region II.

Harvard's short pulse CO_2 laser facility[13] was
available for these measurements. It generates 500
psec pulses by the optical free induction decay
technique[14] and amplifies them up to the 0.15 J level.
For a given energy, these ultrashort pulses have 200
times higher peak power than the typical 100 nsec
TEA CO_2 laser pulses. Therefore, if mechanism (i) is
at all important, the reduction of pulse duration, for
fixed energy, should have a powerful effect on the
dissociation yield. Figure 4 shows the experimental
result. While the peak power is changing 2 orders
of magnitude, the dissociation increases only 30 per
cent!

Therefore mechanism (i) plays only a minor role
and it is mechanism (ii), transition in the quasicontinuum,
which acts to limit the dissociation yield. Region II,
of the quantum level structure (Fig. 2), is where the
important rate-limiting physics is taking place.

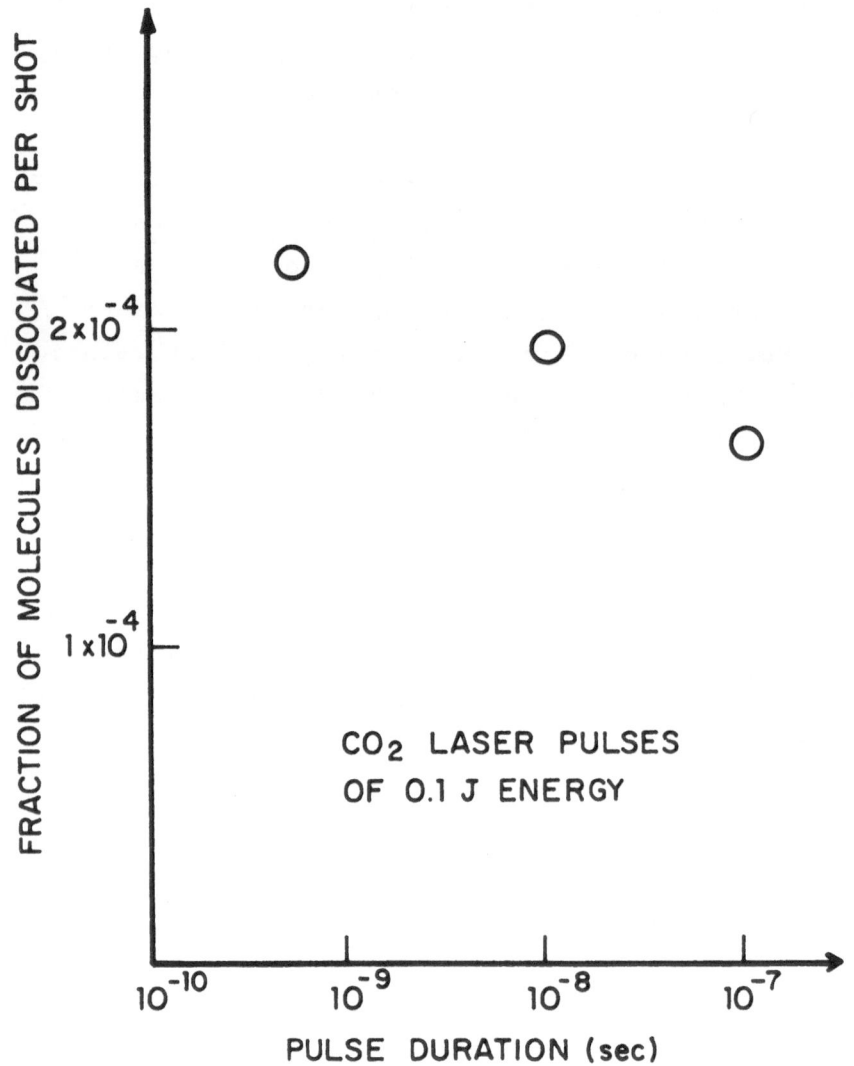

Fig. 4.   A comparison of the dissociation yield for
pulses of widely varying duration, but fixed energy.
For a 200-fold increase in peak power the fraction of
molecules dissociated increased only 30%. This proves
that interactions in the quasicontinuum are more
important than the "anharmonicity bottleneck" in the
discrete levels.

IV. STATISTICAL DESCRIPTION OF THE QUASI-CONTINUUM

We have already mentioned that transitions in the quasicontinuum are of an <u>incoherent</u> nature. That is, the molecular evolution is best described in terms of kinetic rate equations (2), rather than the coherent Bloch-type[9] equations. Two conclusions follow directly from this:

(a) In spectral region II only the energy fluence (Joules/cm^2) is important. This was proved earlier in this paper.

(b) It is known in quantum electronics that an <u>incoherently</u> driven oscillator assumes[15] a "Planck" distribution at a well-defined temperature. This is shown in Reference 15 or may be proven directly from the rate equations for an oscillator.

Conclusions (a) and (b) follow from the kinetic rate equations (2). They rely on the assumption of ergodicity, which is justified if the rate of intra-molecular VV relaxation is faster than the rate of absorption of photons. A large body of experimental evidence[16] shows that intramolecular VV times are in the one picosecond range for polyatomics. Absorption cross-section measurements, to be described later in this paper, indicate that the rate of photon absorption is indeed slower than this.

Conclusion (b) is very important. Together with ergodicity it implies that all the oscillators in the SF$_6$ molecule have a "Planck" distribution at a well defined temperature. Therefore statistical thermo-dynamics may be employed in the description of the molecular evolution. A statistical approach is well known to physical chemists and is described, for example, in the text-book[17] by Robinson and Holbrook, entitled "Uni-Molecular Reactions."

In the RRKM theory, the accepted statistical model[17] for molecular dissociation, the molecular energy must exceed the dissociation energy by a finite amount in order to get reasonable dissociation rates. Physically, this is because the phase space of a highly excited polyatomic is so huge. The fraction of phase space which leads directly to dissociation is relatively small unless the dissociation energy is significantly exceeded. A typical formula[17] for the rate of dissociation is:

$$\text{Rate} = \omega \, \frac{n! \, (n-m+s-1)!}{(n-m)! \, (n+s-1)!} \qquad . \qquad (5)$$

Here ω is an average vibrational frequency, s is the no. of vibrational degrees of freedom (15 for SF_6), m is the minimum no. of absorbed photons needed for dissociation and n is the number actually absorbed. Formula (5) predicts that n must exceed m by 10 to 15 photons in order to produce dissocation during a typical laser pulse. This is in rough agreement with the molecular beam measurements of Lee[18] et al.

When formula (5) is averaged over a statistical distribution of SF_6 states, the result for the average rate of dissociation is:

$$\text{Average Rate} = \omega \exp \left\{ -\frac{ms}{\langle n \rangle} \right\} \equiv \omega \exp \left\{ - \frac{W_d}{kT} \right\}, \qquad (6)$$

where $\langle n \rangle$ is the average number of photons absorbed per molecule. The temperature of the molecule is:

$$kT \equiv \frac{\langle n \rangle \, \hbar \omega}{s} \qquad . \qquad (7)$$

Formula (6) is the familiar Arrhenius equation, which is valid for many kinetic processes.

The theoretical arguments given in this section, imply that a statistical thermodynamic approach is appropriate to describe the multiphoton dissociation of SF$_6$. An experimental confirmation of this approach demands a direct measurement of the molecular temperature, as well as good agreement with the Arrhenius equation (6). There measurements are described in the next section.

V.   PICOSECOND, OPTO-ACOUSTIC, TEMPERATURE MEASUREMENT

As discussed in Section III of this paper, ultrashort CO$_2$ laser pulses are particularly suited for measuring the quasicontinuum properties of SF$_6$. The high peak power of an ultrashort pulse, implies a high Rabi precession frequency, which implies that all molecules in the rotational manifold overcome the anharmonicity barrier and interact in the quasicontinuum. Figure 5 shows a simple opto-acoustic[19] apparatus used for measuring the quasicontinuum temperature of SF$_6$.

The microphone responds to the pressure rise induced in the gas. The system must be carefully calibrated since a major portion of the energy is conducted to the walls without contributing to the acoustic impulse. A laser power meter determined the energy deposited per molecule at 5 torr pressure. Leaving the illumination conditions unchanged, the pressure was reduced to 0.2 torr and the acoustic signal was determined for the given energy deposition. (This technique is similar to that employed by Letokhov[19] except that the ultrashort pulses obviated the need for a long optically thick cell).

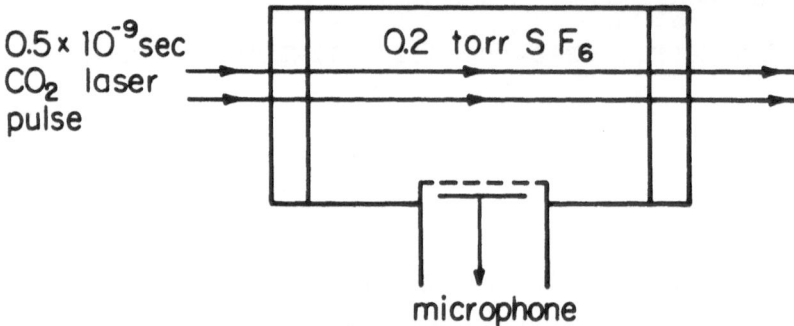

Fig. 5.   The opto-acoustic experiment for the measurement of the energy deposited per molecule.   This directly
determines the vibrational temperature to be used in
the Arrhenius Eq. (6).   The ultra-short CO_2 laser pulse
ensures that there is no bottlenecking in the discrete
levels.

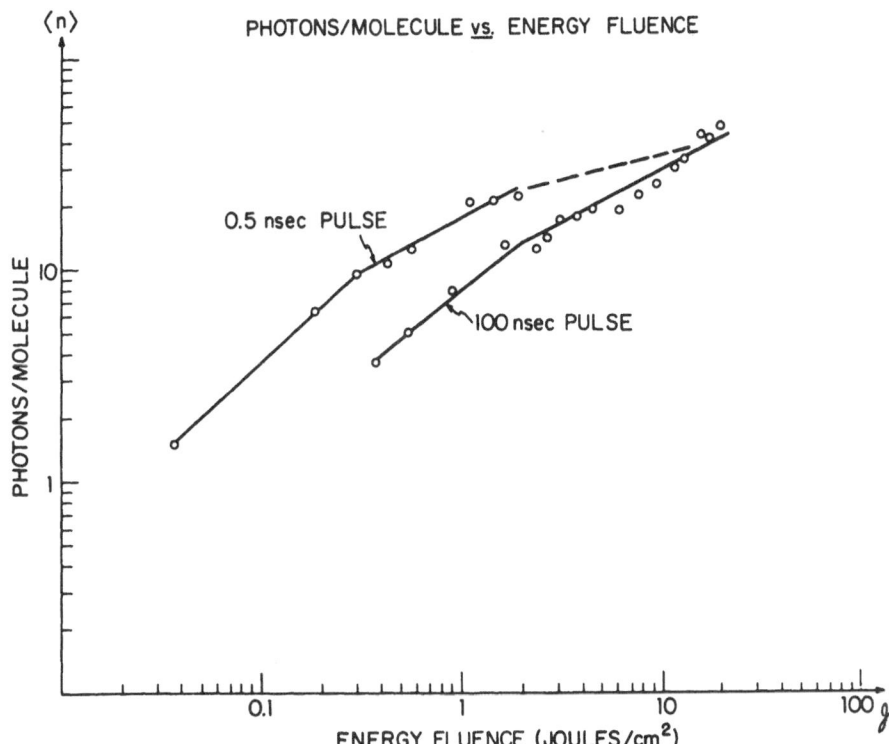

Fig. 6. The number of photons per molecule vs. the energy fluence (J/cm^2). The difference between the two curves permits am estimate of the fraction, f, of molecules stuck in the discrete energy levels. The vibrational temperature as measured by the 0.5 nsec pulses, is in good quantitative agreement with the Arrhenius eq. (6), confirming the statistical thermodynamic approach.

A comparison of the number of photons/molecule deposited by an ultrashort pulse and a conventional TEA laser pulse is shown in Figure 6. At dissociation threshold, 1.4 J/cm^2, the 100 nsec pulse deposits $\langle n' \rangle$ = 10 photons/molecule while the ultrashort pulse deposits $\langle n \rangle$ = 20 photons/molecule. The difference[20] between $\langle n \rangle$ and $\langle n' \rangle$ must be attributed to anharmonicity bottlenecking in the long TEA laser pulse. Therefore the fraction, f, of bottleneck molecules may be readily estimated;

$$ f \sim \frac{\langle n \rangle - \langle n' \rangle}{\langle n \rangle - 3} \, , \qquad (8) $$

assuming 3 is the average excitation of the molecules which remain bottlenecked in the discrete levels. The fraction, f, which is about 2/3 at the dissociation threshold, 1.4 J/cm^2, eventually goes to zero as the energy fluence rises to 10 J/cm^2.

The number $\langle n \rangle$ = 20 photons/molecule implies a vibrational temperature of the isolated molecule, T = 1800°K at threshold. This absolute measurement is in good quantitative agreement with the Arrhenius equation (6) estimate, $\approx$ 2000°K, of the threshold temperature required for dissociation under the experimental conditions. The dissociation yield rises rapidly above threshold (see Fig. 3) even for the small observed vibrational temperature increase (see Fig. 6), again consistent with the Arrhenius equation. Therefore the statistical theory (RRKM) provides good quantitative agreement between measurements of dissociation yield and vibrational temperature.

The measurement of $\langle n \rangle$, the number of photons absorbed per molecule, may also be interpreted in terms

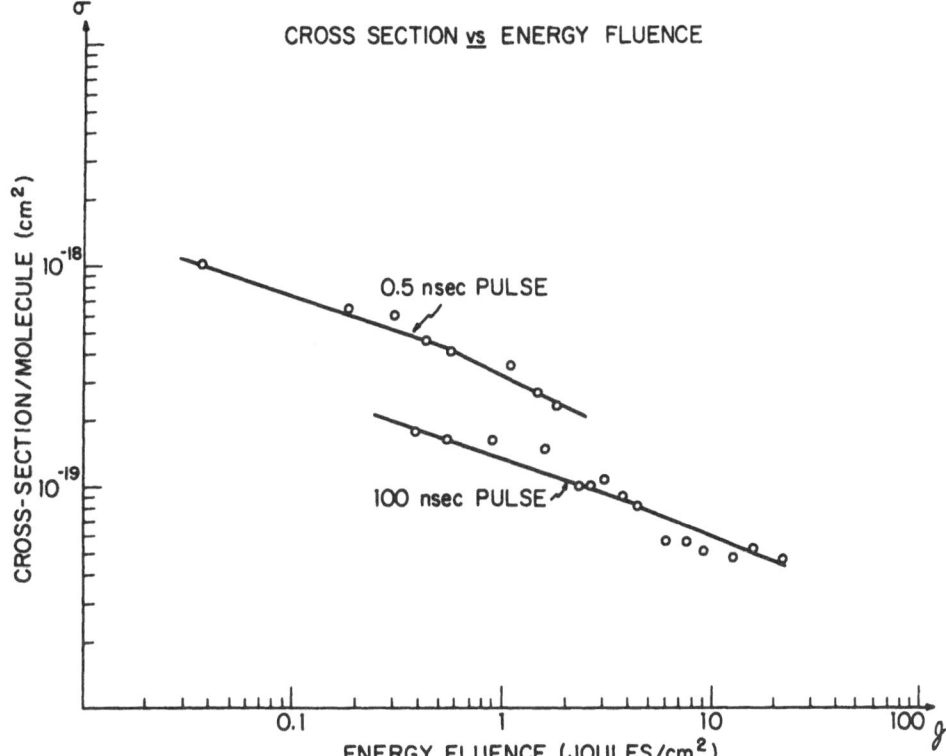

Fig. 7.   The cross section falls rapidly as the molecule
is heated.   This is the major dissociation limiting mechanism.

of the absorption cross section, σ, of the SF_6 molecule.

$$\sigma = \frac{\langle n \rangle \hbar \omega}{J} \quad , \tag{9}$$

where $\hbar\omega$ is the photon energy and J is the energy fluence in Joules/cm^2. This data is plotted in Figure 7. Of perhaps greater interest is the absorption cross section σ' of the excited molecule in the quasicontinuum, which is given by the differential version of formula (9):

$$\sigma' = \frac{d}{dJ} \left(\langle n \rangle \hbar\omega \right) \quad . \tag{10}$$

Obviously, σ and σ' differ only by a factor α, which is the logarithmic slope of the curve in Figure 6.

Figure 7 shows that the absorption cross-section falls rapidly to less than 10^{-19} cm^2, as the molecule is heated in the quasicontinuum. This is consistnet with the oscillator-strength sum rule[10], Eq. (4), provided that the oscillator strength is smeared over 1000 cm^{-1}. Alternatively, the drop in cross-section may be due to a shift in the absorption spectrum as the molecule is heated. Clearly, more work is needed on the spectral properties[21] in the quasicontinuum of polyatomic molecules.

REFERENCES

1. N. R. Isenor and M. C. Richardson, Appl. Phys. Lett. 18, 225 (1971).

2. R. V. Ambartsumyan, V. S. Letokhov, E. A. Ryabov and N. V. Chekalin, JETP Lett. 20, 273 (1974). J. L. Lyman, R. J. Jensen, J. Rink, C. P. Robinson and S. D. Rockwood, Appl. Phys. Lett. 27, 87 (1975).

3. Proceedings of the Nordfjord Conference on Tunable Lasers and Applications, ed. by A. M. Mooradian (Springer-Verlag, 1976).

4. See the article by R. V. Ambartsumyan in Ref. 3.

5. See the article by Kompa in Ref. 3.

6. See the article by N. Bloembergen, C. D. Contrell and D. M. Larsen in Ref. 3.

7. C. D. Cantrell and H. W. Galbraith, Opt. Comm. 18, 513 (1976).

8. S. Mukamel and J. Jortner, Chem. Phys. Lett. 40, 150 (1976).

9. R. H. Pantell and H. E. Puthoff, Fundamentals of Quantum Electronics, (Wiley, New York, 1969).

10. E. Merzbacher, Quantum Mechanics, (Wiley, New York, 1961).

11. P. Kolodner, C. Winterfeld and E. Yablonovitch, Opt. Comm. to be published.

12. N. Bloembergen, Opt. Comm. 15, 416 (1975).

13. H. S. Kwok and E. Yablonovitch, Rev. Sci. Instrum. 46, 814(1975).

14. E. Yablonovitch and J. Goldhar, Appl. Phys. Lett. 25, 580 (1974).

15. W. H. Louisell, Quantum Statistical Properties of Radiation, (Wiley, New York, 1973).

16. J. D. Rynbrandt and B. S. Rabinovitch, J. Phys. Chem. 75, 2164 (1971).

17.   P. J. Robinson and K. A. Holbrook, Uni-Molecular
      Reactions (Wiley, New York, 1972).

18.   M. J. Coggiola, P. A. Schulz, Y. T. Lee and Y. R.
      Shen, Phys. Rev. Lett. 38, 17 (1977).

19.   V. N. Bagratashvili, I. N. Knyazev, V. S. Letokhov
      and V. V. Lobko, Opt. Comm. 18, 525 (1976).

20.   The departure from RRKM theory observed by D. F.
      Dever and E. Grumwald, J. Am. Chem. Soc. 98,
      5055 (1976) was probably due to the fraction, f,
      of molecules which were unable to overcome the
      "anharmonicity barrier". A similar departure
      would be observed here if ⟨n'⟩ rather than ⟨n⟩
      were used to determine the vibrational temperature.

21.   R. V. Ambartsumyan, N. P. Fuzikov, Yu. A. Gorokhov,
      V. S. Letokhov, G. N. Marakhov and A. A. Puretzky,
      Optics Comm. 18, 517 (1976).

STRUCTURE OF THE VIBRATIONAL STATES IN THE ν_3-FUNDAMENTAL AND ITS OVERTONES IN SF_6 AND MULTIPHOTON ABSORPTION EFFECTS*

H. W. Galbraith and C. D. Cantrell

(Presented by H. W. Galbraith)

Los Alamos Scientific Laboratory

Los Alamos, New Mexico 87544

I. INTRODUCTION

In a previous work[1], we have put forth the thesis that the experimentally observed[2,3,4] rapid absorption of several CO_2 photons of like frequency, is due to the vibrational octahedral splitting of the vibrational states of overtones of the triply degenerate ν_3 mode. Up to this point, all other treatments of collisionless multiple-photon dissociation of SF_6 by CO_2 laser radiation[5,6,7,8,9,10,11,12] have utilized a simple one dimensional anharmonic oscillator (or equivalent) model for the vibrational states of SF_6. In such a model, a fundamental difficulty is immediate: how is the rapid absorption of 25-30 laser photons of approximately like frequency possible, when the (strong) anharmonicity of the molecular potential shrinks the spacing of successive

* Work performed under the auspices of the United States Energy Research and Development Administration

vibrational states too rapidly and by too large an
amount? Several explanations have appeared on the
scene. The power broadening and stark shifting of the
molecular states due to the high fields involved <u>is</u>
an effect which surely plays a role in the absorption
process, but SF_6 has only an induced dipole moment
and this effect is therefore too small quantitatively;
another popular explanation is compensation of the
vibrational anharmonicities through rotational
broadening. Once again this effect <u>is</u> present and
<u>does</u> aid in overcoming anharmonicities, but such
compensation can only explain resonant absorption of
approximately three photons.[13] It is our viewpoint that
this anharmonicity bottleneck is a consequence of an
insistence on modeling octahedral SF_6 by a simple
oscillator. In our recent analysis of high resolution
laser diode spectra of SF_6,[14,15,16,17] we have found
that the interpretation of such spectra <u>required</u>
(even for low J-values) the introduction of nonspherically
symmetric, fourth-rank, vibration-rotation tensor
operators. The existence and effects of these operators
in spectra of spherical top molecules has been the
subject of much investigation by the spectroscopy
community, dating back to the early work on CH_4 by
Jahn.[18,19,20,21,22,23] The molecular vibration-rotation
Hamiltonian for spherical top molecules has a long and
interesting history. The Darling-Dennison Hamiltonian[24]
was first written out to second order (including up to
fourth degree polynomials in the normal coordinates)
by Shaffer, Nielsen and Thomas. In this work, a Van Vleck
contact transformation[26] was applied to the Hamiltonian,
the effect of which is to <u>remove all terms which are</u>
<u>nondiagonal in the total vibrational quantum number</u>.

The Shaffer et al Hamiltonian is extremely lengthy and, consequently, it has never been used for calculations. Later Hecht[20] and (independently) Louck[27] expressed the Shaffer Hamiltonian in terms of irreducible spherical tensor operators and, utilizing the Racah-Wigner angular momentum theory, Hecht was able to explain very complex high resolution grating spectra of the ν_3 fundamental of CH_4. Moret-Bailly[21,22,23] extended the Hamiltonian to fourth order (which involved a second contact transformation) and recognized that the resulting Hamiltonian could be used for <u>any</u> <u>spherical</u> <u>top</u> <u>molecule</u>.[28,29] Throughout these works it has proven essential, for the interpretation of the experimental spectra of the fundamentals, that tensor splittings of vib-rot levels be taken into account. It is our opinion that meaningful multiphoton calculations should be consistent with at least the dominant features of the current spherical top theory as it applies to SF_6.

In this work, we first introduce a model Hamiltonian for the vibrating-rotating SF_6 molecule which incorporates all of the known qualitative, as well as quantitative, features which we have found from our spectroscopy work on the ν_3-fundamental. We then discuss (and contrast) the two most important approaches for constructing energy levels from this model. Two kinds of splittings are apparent:vib-rot and pure vibrational. From our empirically known (and calculated) parameter values, we see that the pure vibrational splittings are dominant, and hence we are led to consider the nonrotating molecule first. We find that the second order pure vibrational Hecht-Hamiltonian can be considerably simplified and a comprehensible model of vibrational splittings is thereby introduced. This model is solved

completely and explicitly in the Cartesian basis set
for the triply degenerate oscillator, and the energies
and relative intensities are readily computed. The
treatment of the full second order vib-Hamiltonian can
be simplified with the introduction of a set of basis
functions which diagonalize our simple model and also
carry octahedral representation labels. Such functions
are constructed by standard group theoretic techniques
and are tabulated in the Appendix. Finally the dynamical
response of our model to a strong laser pulse of very
long duration is calculated. Our calculations include
39 vibrational states (up to six photon absorption) and
indicate a maximum energy deposition in the ν_3 overtone
states, provided that the driving laser is tuned to
944 cm^{-1}. Comparison of our model with one in which
the tensor parameter is set equal to zero indicates the
fundamental qualitative, as well as quantitative,
difference between such models and ours. Finally, the
change in absorption by one state of $6\nu_3$ with changing
field intensity (power broadening) is calculated. The
results of this computation are in agreement with our
speculations on the magnitude of such effects. The
striking agreement between our results and the recent
two frequency experiments of Ambartzumian et al.[13] is
noted.

II. MODEL OF VIBRATING ROTATING SF$_6$

In order to obtain a good mathematical model of
the vibrating rotating SF$_6$ molecule in its ν_3 overtone
states, one need only consider those pieces of the
second order Hecht-Hamiltonian which have nonzero matrix
elements in the ν_3 ladder. Our experience in analyzing
the recent laser diode spectra of the ν_3 fundamental
indicates that the following model should suffice:

$$H = H_o + x_{33}H_o^2/\omega^2 + T_{33}T_{404} + G_{33}L_{\sim}^2$$

$$+ BJ_{\sim}^2 - 2B\zeta_3(J_{\sim}\cdot L_{\sim}) + t_{224}T_{224} \; .$$

(2.1)

In Eq. (2.1):

(1) H_o is the triply degenerate oscillator Hamiltonian, $\frac{\omega}{2}(p_{\sim}^2 + q_{\sim}^2)$, for the ν_3 normal coordinate $q_{\sim}$, with $\omega \sim 951.6$ cm^{-1};

(2) $x_{33}H_o^2/\omega^2$ is the quartic scalar anharmonicity and carries a calculated[29] coefficient of -2.8cm^{-1};

(3) T_{404} is a quartic tensor anharmonicity operator; $t_{404} = -0.44$cm^{-1}. This operator is a combination of tensors of fourth rank with respect to the vibrational angular momentum such that the composite operator is an octahedral invariant.

(4) $L_{\sim}^2$ is the square of the vibrational angular momentum operator of the ν_3 mode and carries a calculated[29] coefficient of 1.34cm^{-1}.

(5) $BJ_{\sim}^2$ is the familiar rigid rotator Hamiltonian, and the $0 \to 1\nu_3$ analysis[15] gives $B \cong 0.0929$cm^{-1}.

(6) $J_{\sim}\cdot L_{\sim}$ is the first order Coriolis vibration-rotation interaction term and has a parameter of value -0.13cm^{-1} as determined from the $0 \to 1$ spectra.

(7) Finally T_{224} is an octahedral vibration-rotation fourth-rank tensor operator carrying a coefficient value of $\sim 4 \times 10^{-5}$cm^{-1}. This operator is responsible for the splitting of various R-states into their octahedral components and leads to the rich rotational structure observed in the fundamental.

Based upon this model, one calculates a theoretical spectrum from the time independent Schrödinger equation

$$H\Psi_n = E_n\Psi_n \quad . \tag{2.2}$$

By now the standard approach[30] derives from the
"Coriolis stabilized basis," of Hecht. This basis set
is the vector coupling of spherical oscillator functions
with rigid rotor wavefunctions

$$\Psi^{nJ}_{RK_R} = \sum_{K+m=K_R} \langle J\ell\, K\, m \mid RK_R\rangle\, \phi^*_{n\ell m}\, \Psi^J_{KM} \quad . \tag{2.3}$$

In Eq. (2.3): $\phi^*_{n\ell m}$ is the complex conjugate of the
spherical oscillator wavefunction[20]; n is the number
of ν_3 quanta; ℓ, the vibrational angular momentum
quantum number ($\ell=n,n-2,\ldots,0/1$); m, its projection
onto the body-fixed z-axis; J is the total orbital
angular momentum of the molecule and K and M are its
z-projections onto body and lab-fixed frames, respectively;
$\langle J\ell Km \mid RK_R\rangle$ is a standard Clebsch-Gordan coefficient
for coupling J with ℓ to obtain R, which has molecule-
fixed z-component K_R. This basis set could equivalently
have been defined as that set of functions which
diagonalize the operator set $\{H_o,\, J^2,\, L^2,\, J{\cdot}L\}$. The
irreducible tensor operators T_{404} and T_{224} must be
dealt with separately. These operators have, however,
been written as irreducible tensors with respect to
ℓ and J, as well as R, so that the matrix elements take
on the standard form:[31]

$$\langle n\ell'JR'K_R' \mid T_{k_1 k_2}^{\ 4} \mid n\ell JRK_R \rangle = D$$

$$\times \ \langle n\ell' \| t_{k_1} \| n\ell \rangle \langle J \| t_{k_2} \| J \rangle \begin{Bmatrix} \ell' & \ell & k_1 \\ J & J & k_2 \\ R' & R & 4 \end{Bmatrix}$$

$$\times \left[\sqrt{70} \begin{pmatrix} R & 4 & R' \\ K_R & 0 & -K_R' \end{pmatrix} + 5 \begin{pmatrix} R & 4 & R' \\ K_R & 4 & -K_R' \end{pmatrix} + 5 \begin{pmatrix} R & 4 & R' \\ K_R & -4 & -K_R' \end{pmatrix} \right], \qquad (2.4)$$

where D is a dimension factor, $\langle \| \ \| \rangle$ are vibrational and rotational reduced matrix elements,[20,21,22,23] $\begin{Bmatrix} \vdots & \vdots & \vdots \\ \vdots & \vdots & \vdots \end{Bmatrix}$ is a 9-J symbol and $(\ldots)$ are standard 3-J symbols. Notice that the tensors are generally __nondiagonal in__ $\underline{\ell}$, $\underline{R}$, and $\underline{K_R}$.

The Coriolis basis set lends itself to various perturbation schemes. For example, in the $\nu_3=1$ manifold there is only one ℓ-value available, namely $\ell=1$. It happens there, also, that the separation in energy between R-states (belonging to the same J) is large compared to the fourth rank tensor coupling of the states. Hence one may diagonalize $\langle R \mid T_{k_1 k_2}^{\ 4} \mid R \rangle$, initially and deal with the off-diagonal matrix elements in perturbation theory. This approximation may also prove valid for the rotational levels in excited states, but such knowledge awaits further empirical evidence.

For the sake of this discussion, let us suppose that interactions, off-diagonal in ℓ, are also small, and focus upon $2\nu_3$.[30] The Hamiltonian matrix then may be approximated by

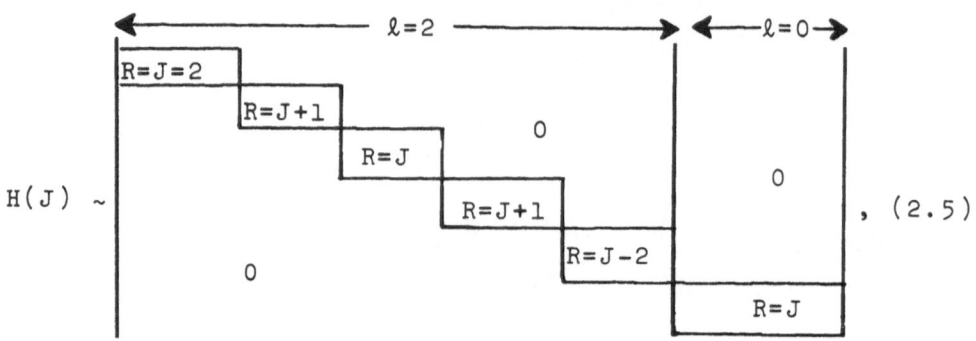

$$H(J) \sim \qquad , \quad (2.5)$$

for each J-value. The eigenvalues of the separate
R-blocks are known, and are proportional to the $F^{(4RR)}_{A_1 \alpha\alpha}$-
coefficients of Moret-Bailly.[32] This scheme corresponds
to an energy level diagram as shown in Fig. 1,

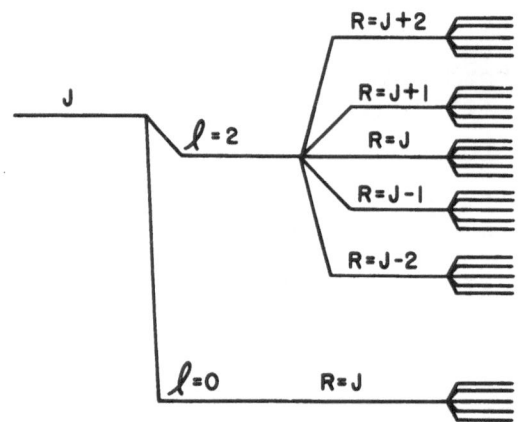

Fig. 1 Energy level diagram corresponding to perturbations
in the Coriolis basis set for $2\nu_3$.

which is patently <u>not</u> the case for the parameter values
presented earlier. Since, at present, we are not capable
of dealing with the full matrix of H for up to ~ $6\nu_3$,
we abandon the Coriolis scheme and present a new one which

appears more appropriate for our approximations.

Such a model is based upon large vibrational splittings and correspondingly smaller rotational structure. For example, in $2\nu_3$ a given J-state would be split qualitatively as shown in Fig. 2,

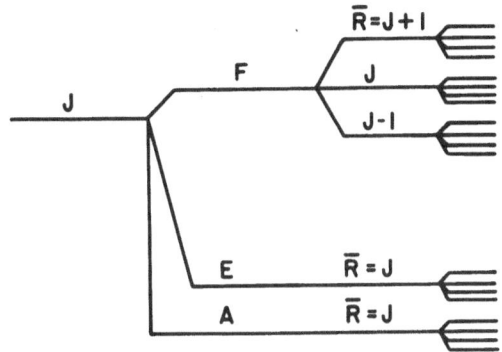

Fig. 2 Energy level diagram corresponding to
        perturbations on the vibrational basis set for
        $2\nu_3$.

This scheme is based upon the inequalities

$$T_{33}T_{404} + G_{33} \underset{\sim}{L}^2 >> -2B\zeta \underset{\sim}{J} \cdot \underset{\sim}{L} >> t_{224},$$

and appears suitable for our SF_6 work. In the next section, we will study the nonrotating molecule and obtain a model for the vibrational states of SF_6.

III. SIMPLIFIED MODEL OF OCTAHEDRAL SPLITTING

Let us then study the nonrotating model and set J=0 in Eq. (2.1). (We will in a future work relax this condition). One has, after some algebra, the following form:

$$H = 951.6 \, N - 4.4M + 0.46\underset{\sim}{L}^2 - 0.16N^2 \quad , \quad (3.1)$$

with

$$N = \sum_{i=1}^{3} a_i^+ a_i \qquad (3.2)$$

and

$$M = \sum_{i=1}^{3} (a_i^+ a_i)^2 \; . \qquad (3.3)$$

We have defined standard boson creation, annihilation operators in arriving at Eqs. (3.1)-(3.3). We get the commutation relations

$$[N,M] = 0 \; , \qquad (3.4)$$

$$[N,L^2] = 0 \; , \qquad (3.5)$$

and

$$[M,L^2] \neq 0 \; . \qquad (3.6)$$

It is easy to see that N and M are diagonal in the standard cartesian direct product space of functions

$$|n_1 n_2 n_3 > = \frac{(a_1^+)^{n_1}(a_2^+)^{n_2}(a_3^+)^{n_3}}{\sqrt{n_1! n_2! n_3!}} \; | \, 0 > \; , \qquad (3.7)$$

and carry eigenvalues as follows:

$$N \, |n_1 n_2 n_3 > = (n_1 + n_2 + n_3) \, |n_1 n_2 n_3 > \; , \qquad (3.8)$$

and

$$M \, |n_1 n_2 n_3 > = (n_1^2 + n_2^2 + n_3^2) \, |n_1 n_2 n_3 > \; . \qquad (3.9)$$

Since the eigenvalues of M and L^2 are of comparable magnitude (for $N \leq 6$), and the coefficient of L^2 is down by a factor of ten, we may as a first approximation study the simplified Hamiltonian

$$H^{(0)} = 952.4N - 4.4M .\qquad (3.10)$$

From Eqs. (3.8) and (3.9), the eigenvalues of $H^{(0)}$ may be written out directly,

$$E_{n,m} = 948 \, (n_1 + n_2 + n_3) + 4.4 \, (n_1 + n_2 + n_3 - n_1^2 - n_2^2 - n_3^2). \quad (3.11)$$

The Hamiltonian $H^{(0)}$ and corresponding eigenvalues $E_{n,m}$ give a simple and yet not unrealistic model of octahedral splitting.[1]

From the form of $E_{n,m}$, it is evident that if $| \, n_1 n_2 n_3 >$ carries energy $E_{n,m}$, then $|n_1 n_3 n_2 >$ etc., do as well, i.e. $E_{n,m}$ is invariant to permutations of the n_i. Furthermore different m-values will correspond to different choices of numerical values for the n_k. Therefore the degeneracy spaces of $H^{(0)}$ may be put into correspondence with the ordered partitions of n into three, $\{[n_1 n_2 n_3] : n_1 + n_2 + n_3 = n \; ; \; n_1 \geq n_2 \geq n_3\}$. This correspondence is not strict, but has direct utility in Section IV.

Transitions between energy states $n \rightarrow n+1$ occur via an induced dipole moment. The dipole is expanded in the molecular normal coordinates and since SF_6 has no permanent dipole moment, the leading term for ν_3-ladder transitions is proportional to $\underset{\sim}{q}$. But since

$$q = \frac{1}{\sqrt{2}} \, (\underset{\sim}{a} + \underset{\sim}{a}^+), \qquad (3.12)$$

we see that (to this order of approximation) the field can transfer only one quantum to any of $n_1 n_2$ or n_3. Hence we obtain the selection rules

$$\Delta n = \pm 1 \; , \tag{3.13}$$

and

$$\Delta m = 2n_s + 1, \tag{3.14}$$

where n_s is the shifted n-value in the transition. If we define by $S^n_{[\lambda]}$ the degeneracy subspace of $H^{(0)}$ having energy $E_{n,m}$ and corresponding to partition $[\lambda]$,

$$[\lambda] = [\lambda_1 \lambda_2 \lambda_3] = \lambda_k + \lambda_2 + \lambda_3 = n, \; \lambda_1 \geq \lambda_2 \geq \lambda_3 \; , \tag{3.15}$$

then the intensity of a transition is proportional to the reduced B-value

$$B\left(S^n_{[\lambda]} \to S^{n+1}_{[\lambda]}\right) = \sum_{\substack{A, \{n_i\} \epsilon S^n_{[\lambda]} \\ \{n_i'\} \epsilon S^{n+1}_{[\lambda']}}} |\langle n_1' n_2' n_3'| \; q_A \; |n_1 n_2 n_3\rangle|^2 \; , \tag{3.16}$$

which may be evaluated directly in our case to give

$$B\left(S^n_{[\lambda]} \to S^{n+1}_{[\lambda']}\right) = d\begin{pmatrix}[\lambda]\\ [\lambda]\end{pmatrix}(n_s+1)/2 \; , \tag{3.17}$$

where d is the larger of the dimensions of $S^n_{[\lambda]}$ or $S^{n+1}_{[\lambda']}$. Finally, from Eq. (3.11) the frequency of allowed transitions is found to be

$$\nu_{m \to m'} = 948 - 8.8 n_s \tag{3.18}$$

in cm^{-1}. Equations (3.10)-(3.18) give a complete and explicit model for octahedral splitting in SF_6. This model yields a picture of the vibrational states and allowed transitions up to $4\nu_3$ as shown in Fig. 3.

Plotted along the vertical axis is the number of
quanta in the ν_3 mode, and along the horizontal, we
have the energy of the vibrational levels minus 948n.
Dipole allowed transitions are indicated by $\}$, and the
relative intensities are given by Eq. (3.17). Transitions
occurring vertically in the figure correspond to $n_s=0$,
and have a frequency of $948cm^{-1}$. Those involving a
change in $n_s=1$ involve a change in off-resonance energy
by $8.8cm^{-1}$, etc. From the figure, we may also trace
out various <u>routes</u> in arriving at $4\nu_3$, for example, there
is only one route to the triply degenerate manifold

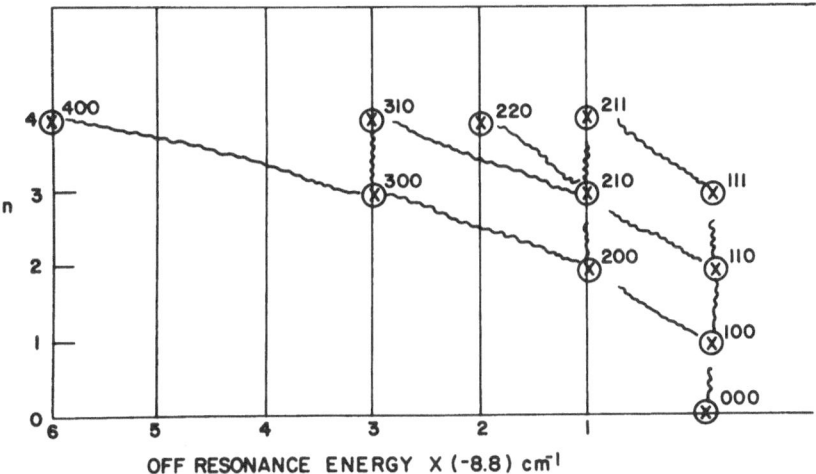

OFF RESONANCE ENERGY X (-8.8) cm⁻¹

Fig. 3 Off-resonance energy versus n for the simplified
vibrational Hamiltonian of Eq. (3.10).

$S^4_{[400]}$, and that is via $S^2_{[200]}$. There are two routes to
the states $S^4_{[200]}$ and three to $S^4_{[211]}$, and $S^4_{[310]}$.
three to $S^4_{[211]}$, and $S^4_{[310]}$.

In order to improve this model of the vibrational
states of SF_6, we return to the full second order
vibrational Hamiltonian of Eq. (3.1) and observe that

it is only $\underset{\sim}{L}^2$ with which we need be concerned. Since

$$L = -i \, \underset{\sim}{a}^+ \times \underset{\sim}{a} \quad , \qquad\qquad (3.19)$$

we could construct the matrices of $\underset{\sim}{L}^2$ directly and diagonalize them. However, a simplification occurs if we note that $\underset{\sim}{L}^2$ is also an octahedral invariant operator, so that, if $|n,\Gamma,n_\Gamma>$ is an n-state which also carries symmetry species (irrep) label Γ under transformations of the octahedral group O_h, having components labelled by n_Γ, then the Wigner Eckart theorem implies that

$$< n,\Gamma,n_\Gamma \, |\underset{\sim}{L}^2| \, n,\Gamma',n_\Gamma' > \, \propto \, \delta_{\Gamma\Gamma'} \, \delta_{n_\Gamma n_{\Gamma'}} \quad . \qquad (3.20)$$

By Eq. (3.20), the matrix of $\underset{\sim}{L}^2$ is block diagonalized in the <u>symmetry</u>-<u>adapted</u> <u>basis</u> set, the dimensions of the blocks being the multiplicity of each Γ in the set of all functions of the same n-value. (Of course this applies to all further perturbations which may be added to H and which also carry octahedral symmetry).

The problem which we solve in the next section then is: Reduce the space of n-quanta in a three-dimensional oscillator into irreducibles with respect to O_h, the oscillator being such that its displacement vector carries the F_{1u} representation of O_h.

IV. O_h-SYMMETRY ADAPTED FUNCTIONS

The space of n-quanta in a three-dimensional oscillator is denoted

$$V_n = \{|n_1 n_2 n_3 > \, : \, \Sigma n_i = n\} \, , \qquad (4.1)$$

of dimension

$$(n)_D = (n+1)(n+2)/2; \qquad (4.2)$$

and our problem is to decompose this space into
irreducibles with respect to the octahedral group, O_h.
We first outline the steps in the solution of this
problem:

(1) Introduce transformations of O_h as linear
operators mapping V_n onto itself; $V_n \rightarrow V_n$.

(2) Show that M is O_h invariant, which implies that
operators of O_h map <u>each</u> degeneracy subspace onto itself;
$O_h: S^n_{[\lambda]} \rightarrow S^n_{[\lambda]}$.

(3) Reduce the one, three or six dimensional
representations of O_h carried by each degeneracy space
$S^n_{[\lambda]}$.

While the octahedral group O_h is usually defined as
that set of forty-eight rotations/inversions of real
Euclidean three-space which carry a regular octahedron
into itself, an alternative definition given by
Biedenharm, et al.[33] shows that each operation of O_h
may be written as a product of only three, called
generators of the group. The generators satisfy

$$R^4 = S^2 = (RS)^3 = E, \qquad (4.3)$$

$$[*,R] = [*,S] = 0 , \qquad (4.4)$$

and

$$** = E . \qquad (4.5)$$

Equations (4.3)-(4.5) are a definition of abstract O_h.
Our three dimensional oscillator wavefunctions in
the cartesian basis are of the form

$$\Psi_{n_1 n_2 n_3} (\underset{\sim}{q}) = \psi_{n_1} (q_1) \psi_{n_2} (q_2) \psi_{n_3} (q_3) \ , \quad (4.6)$$

where each $\psi_{n_i}(q_i)$ is a one dimensional oscillator state vector satisfying

$$\psi_{n_i} (-q_i) = (-1)^{n_i} \psi_{n_i} (q_i) \qquad (4.7)$$

and each n_i may be any positive integer (including zero).

It is assumed furthermore, that $\underset{\sim}{q}$ carries the F_{1u} irreducible representation of O_h. With our geomentrical octahedron as shown in Fig. 4,

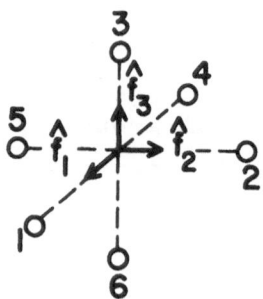

Fig. 4   The equilibrium model of SF_6.

the operators R and S are chosen to be

$$R = R \left(\frac{\pi}{2}, \ \hat{f}_1 \right) \ , \qquad (4.8)$$

and

$$S = R \left[\pi, \ (\hat{f}_1 + \hat{f}_2 / \sqrt{2}) \right] \ , \qquad (4.9)$$

with * the operator taking each vector into its negative. In (4.8) and (4.9), $R(\theta, \hat{n})$ is a rotation by θ about axis

$\hat{n}$ in the right-handed sense.[34] With such definitions, $\underset{\sim}{q}$ transforms as a standard vector in the space, i.e.

$$R \begin{pmatrix} q_1 \\ q_2 \\ q_3 \end{pmatrix} = \begin{pmatrix} 1 & 0 & 0 \\ 0 & 0 & -1 \\ 0 & 1 & 0 \end{pmatrix} \begin{pmatrix} q_1 \\ q_2 \\ q_3 \end{pmatrix} = \begin{pmatrix} q_1 \\ -q_3 \\ q_2 \end{pmatrix} , \qquad (4.10)$$

$$S \begin{pmatrix} q_1 \\ q_2 \\ q_3 \end{pmatrix} = \begin{pmatrix} 0 & 0 & 1 \\ 0 & -1 & 0 \\ 1 & 0 & 0 \end{pmatrix} \begin{pmatrix} q_1 \\ q_2 \\ q_3 \end{pmatrix} \begin{pmatrix} q_3 \\ -q_2 \\ q_1 \end{pmatrix} , \qquad (4.11)$$

and

$$* \begin{pmatrix} q_1 \\ q_2 \\ q_3 \end{pmatrix} = \begin{pmatrix} -1 & 0 & 0 \\ 0 & -1 & 0 \\ 0 & 0 & -1 \end{pmatrix} \begin{pmatrix} q_1 \\ q_2 \\ q_3 \end{pmatrix} = \begin{pmatrix} -q_1 \\ -q_2 \\ -q_3 \end{pmatrix} . \qquad (4.12)$$

In order to obtain a linear representation of O_h on V_n, we define operators T_R, T_S, T_* by

$$[T_R \, \Psi_{n_1 n_2 n_3}] \, (\underset{\sim}{q}) \equiv \Psi_{n_1 n_2 n_3} \, (R^{-1} \underset{\sim}{q})$$

$$= \psi_{n_1} \, (q_1) \, \psi_{n_2} \, (q_3) \, \psi_{n_3} \, (-q_2) = (-1)^{n_3} \, \Psi_{n_1 n_3 n_2} (\underset{\sim}{q}),$$
$$(4.13)$$

where use has been made of Eq.'s (4.10), (4.6) and (4.7). Similarly we have

$$[T_S \, \Psi_{n_1 n_2 n_3}] \, (\underset{\sim}{q}) = (-1)^{n_2} \Psi_{n_3 n_2 n_1} (\underset{\sim}{q}) , \qquad (4.14)$$

and

$$[T_* \ \Psi_{n_1 n_2 n_3}] \ (\underset{\sim}{q}) = (-1)^{n_1 + n_2 + n_3} \Psi_{n_1 n_2 n_3} (\underset{\sim}{q}). \quad (4.15)$$

These results suggest a definition of permutation/inversion operators as follows:

(1) (ik) interchanges the ith and k^{th} n-values counting in $\Psi_{\alpha\beta\gamma}$, $n_1 = \alpha$, $n_2 = \beta$, $n_3 = \gamma$, from the left.

(2) i_p has eigenvalue $(-1)^{n_p}$ on $\psi_{\alpha\beta\gamma}$.

We then have

$$T_R \ \Psi_{\alpha\beta\gamma} = (-1)^\gamma \ \Psi_{\alpha\gamma\beta} , \quad (4.16)$$

or

$$T_R = i_2 (23) = (23) i_3 ; \quad (4.17)$$

$$T_S \ \Psi_{\alpha\beta\gamma} = (-1)^\beta \Psi_{\gamma\beta\alpha} , \quad (4.18)$$

or

$$T_S \ i_2 (13) = (13) i_2 , \quad (4.19)$$

and

$$T_* \ \Psi_{\alpha\beta\gamma} = (-1)^{\alpha+\beta+\gamma} \Psi_{\alpha\beta\gamma}, \quad (4.20)$$

or

$$T_* = i_1 i_2 i_3 . \quad (4.21)$$

The definitions (4.17), (4.19) and (4.21) are not enough since, as Eq. (4.17) shows, the permutation operators in general do not commute with inversions. The way out is given in terms of the standard theory of semidirect products.[35,36,37,38]

Firstly, we consider the two groups D_{2h}, of order eight, and S_3 of order six, out of which we will construct

O_h, of order $8 \times 6 = 48$.

The set of operators $\{E, i_2 i_3, i_3 i_1, i_1 i_2, i_1 i_2 i_3, i_1, i_2, i_3\}$ is an operator realization of D_{2h}, an Abelian group of order eight. For this discussion, we label operators as follows:

$$A_i = i_j \, i_k \qquad , \qquad ijk \text{ cyclic in } 1,2,3 \quad (4.22)$$

and

$$B_\ell = i_\ell \qquad , \qquad \ell = 1,2,3 \quad . \qquad (4.23)$$

An automorphism of D_{2h} is a mapping, α, of the group onto itself. Consider the set of correspondences

$$A_1' = A_2 \qquad , \qquad B_1' = B_2 \quad ,$$

$$A_2' = A_3 \qquad , \qquad B_2' = B_3 \quad ,$$

$$A_3' = A_1 \qquad , \qquad B_3' = B_1 \quad ,$$

$$E' = E \qquad , (i_1 i_2 i_3)' = (i_1 i_2 i_3) \; ; \quad (4.24)$$

i.e.

$$A_k' = (123) A_k \quad ,$$

$$B_k' = (123) B_k \quad . \qquad (4.25)$$

One easily shows that A' and B' satisfy the D_{2h} multiplication table and it is clear that there is such an automorphism for each operator of S_3.

$$S_3 = \{E, (12), (13), (23), (123), (321)\} \quad . \quad (4.26)$$

Hence S_3 is an automorphism group of D_{2h}, and operators in the semidirect product $D_{2h} \wedge S_3$ satisfy the product rule

$$(\alpha, A) \wedge (\beta, B) = (\alpha\beta, \beta(A)B) , \qquad (4.27)$$

for $\alpha, \beta \in S_3$ and A and B $\in D_{2h}$. The operator which multiplies B in the product is the image of A under automorphism β.

Having obtained the operators of O_h in permutation/ inversion form, it is now evident that M is an invariant operator, so that we have arrived at the important result (2). Each degeneracy subspace $S^n_{[\lambda]}$ is mapped onto itself by the operators of O_h.

$$O_h: S^n_{[\lambda]} \to S^n_{[\lambda]} . \qquad (4.28)$$

Hence we have decomposed V_n into a direct sum of O_h invariant subspaces

$$V_n = \bigoplus_{[\lambda]} S^n_{[\lambda]} , \qquad (4.29)$$

and we must only deal with one ($[\lambda]=[ppp]$, three ($[\lambda]=[p,q,q]$ or $[p,q,q]$) or six ($[\lambda]=[p,q,r]$) dimensional representations of O_h.

The first use of this machinery is the construction of the decomposition table (Table I) for the $S^n_{[\lambda]}$ spaces. For example, $S^2[200]$ falls into category II, since two entries are equal and it has p=2 (+) and q=o (+) so that

$$S^2[200] = A_1 + E. \qquad (4.30)$$

In order to construct the explicit SAF basis set, the following (unnormalized) projection operators are used:

$$P^{A_2^1} = \{e+(132)+(123)\pm[(12)+(13)+(23)]I\}\ I_1\ ,$$

$$P_{aa}^{E} = \{e-1/2(132)-1/2(123)+[1/2(12)+1/2(13)-(23)]I\}\ I_1\ ,$$

$$P_{ba}^{E} = \{(132)-(123)+[(13)-(12)]I\}\ I_1\ ,$$

$$P_{xx}^{F_2^1} = \{e\mp(23)I\}\ I_3\ ,$$

$$P_{yx}^{F_2^1} = \{(132)\mp(12)I\}\ I_3\ ,$$

$$P_{zx}^{F_2^1} = \{(123)\mp(13)I\}\ I_3\ ;$$

$$I_1 = e+i_1i_2+i_1i_3+i_2i_3\ ,$$

$$I_3 = e+i_2i_3-i_1i_3-i_1i_2\ ,\ \text{and}\ I = i_1i_2i_3\ . \qquad (4.31)$$

Application of these operators gives the SAF basis states listed in the Appendix.

The final states may be labeled by $|\ n,m,\Gamma,n_\Gamma >$ and in the case of p,q,r, all even or odd, the two E states are designated by ($\pm$) labels and are constructed to be mutually perpendicular. The transformation matrices from the Cartesian basis set to the SAF set are orthogonal:

$$|\ n,m,\Gamma,n_\Gamma > = \sum_{\{n_i\}\epsilon S_{[\lambda]}^n} G_{n_\Gamma;\{n_i\}}^{\Gamma;n,m}\ |\{n_i\}>\ , \qquad (4.32)$$

$$\sum_{\Gamma,n_\Gamma\epsilon S} G_{n_\Gamma;\{n_i\}}^{\Gamma;n,m}\ G_{n_\Gamma;\{n_i'\}}^{\Gamma;n,m} = \delta_{\{n_i\}\ \{n_i'\}}\ , \qquad (4.33)$$

$$\sum_{\{n_i\} \epsilon S} G_{n_\Gamma;\{n_i\}}^{\Gamma;n,m} \quad G_{n_{\Gamma'}}^{\Gamma';n,m} = \delta_{\Gamma\Gamma'} \, \delta_{n_\Gamma n_{\Gamma'}} \quad . \qquad (4.34)$$

Equations (4.32)-(4.34) are useful for finding dipole sum rules for transitions[39,40] between SAF states (recall that we have completely solved this problem for the simple model).

Using the SAF basis set and including the full second order vibrational Hamiltonian, one has the energy level structure as shown in Table II; with energies relative to 948N.

These calculations illustrate:

(1) Very considerable spread in the energies of eigenvectors of V_n due to octahedral splitting (over $100cm^{-1}$ in $6\nu_3$).

(2) The effects of $\underset{\sim}{L}^2$ become important as N increases, particularly when adjacent $S_{[\lambda]}^n$ have states of like symmetry.

(3) There is a net anharmonic shift to the low frequency side.

For a positive coefficient of $\underset{\sim}{L}^2$ in H, we derive the splitting rules from a first order perturbation treatment as shown in Fig. 5.
The [420] states of $6\nu_3$ are split according to +++ above (-52.6 is A_2).

In this section, we have obtained a full diagonalization of the second order vibrational Hamiltonian and constructed eigenstates and eigenvalues for the most current model of SF_6. In the next section, we will consider the application of a powerful laser field, and solve the time dependent Schrödinger equation for the dynamical response of this model.

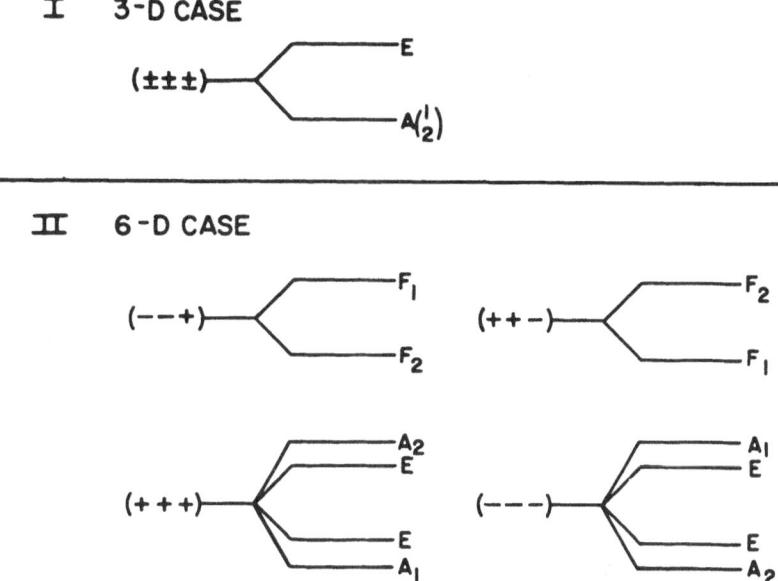

Fig. 5   Splittings due to $\underset{\sim}{L}^2$ obtained to first order. Notation is as in Table I.

V.   MULTIPLE PHOTON ABSORPTION EFFECTS

In this section, we calculate the response of our full second order vibrational states model of SF_6 to a powerful laser pulse of long duration. The time dependent probability amplitudes for finding the model in several excited vibrational states is calculated in the effective Hamiltonian formalism.[1,7,8,10,11] Thirty-nine vibrational states were included in the calculations; all states of (0→4) quanta and the states nearest to resonance of the n=5 and 6 manifolds. Our results indicate that a maximum energy deposition in the ν_3-ladder is obtained for laser frequencies around $944 cm^{-1}$ in excellent agreement with recent two-frequency experiments.[13]

The time dependent transition amplitude for
populating the α-th excited vibrational state is
obtained from the following integral[11]

$$C_\alpha(t) = \frac{-1}{2\pi i} \int dE e^{-iEt} \langle \alpha | (E-H_{eff})^{-1} | 0 \rangle \quad , \quad (5.1)$$

where $|\alpha\rangle$ is one of the molecular states (say of $n-\nu_3$)
multiplied by a state of (P-n) photons for P, a large
integer). H_{eff} is the effective Hamiltonian
corresponding to our model:

$$H_{eff} = H_{Mol} - N\omega + \underset{\sim}{\mu} \cdot \underset{\sim}{E} \quad . \quad (5.2)$$

In Eq. (5.2) H_{Mol} is the second order vibrational
Hamiltonian, N is the number operator, giving the number
of laser quanta absorbed at laser frequency ω; $\underset{\sim}{\mu}=\mu_o \underset{\sim}{q}$ is
the first order dipole moment operator and $\underset{\sim}{E}$ the electric
field vector. Since we are not interested in computing
a dissociation probability but are calculating energy
deposition into various molecular levels, we have not
introduced width or decay parameters into the theory.
Coupling between the various n-manifolds occurs via the
$\underset{\sim}{q}$ operator and we have chosen a value of $\mu_o E=3cm^{-1}$
corresponding to $\sim 1GW/cm^2$ laser power in a 100 ns pulse.
It is hoped that this large value for the laser electric
field will broaden the vibrational states sufficiently
to compensate the effects of rotation. We also assume
a pulse length of ~100 ns which corresponds to ~10^6
oscillations in ν_3 (of an equivalent classical oscillator)
and integrate $|C_\alpha(t)|^2$, over the pulse to obtain an
average excitation of the model. If we wish to interpret
the two-frequency experiments with our model, it is
indeed this averaged excitation which the second

dissociative pulse "sees."

Figure 6 shows the results of such a calculation. We find a definite tendency for stabilization of the probability functions in the range of 943-947 cm^{-1}, with a peak absorption occurring at 944cm^{-1}. The various resonaces seen are a consequence of the different routes available in getting to a given manifold of states. Figure 7 represents a similar calculation, but now with the tensor splitting operator, T_{33}, set equal to zero. Comparison of these results shows the enormous qualitative difference between a three-dimensional anharmonic oscillator and a three dimensional anharmonic oscillator with splitting. Without splitting ($\underset{\sim}{L}^2$ splitting is still included), we see that more quanta will be absorbed if the laser is tuned farther and farther below 948 cm^{-1} in agreement with the results of other models;[12] there appears to be no stabilization in this model.

Finally, Fig. 8 shows the effect of power broadening the [222] absorption of $6\nu_3$. These results appear in rough agreement with the $I^{1/2}$ dependence[13] of the response with laser intensity. It is also very significant that a reduction in the laser field amplitude gave no reduction in the absorption probability but only narrowed (and shifted) the peak. The vertical axis represents the absolute probability for finding our model in the indicated states in each of our figures; indicating a very strong response to the applied field; e.g., in Fig. 8, even at $\mu E=1.5cm^{-1}$, we find 14% of molecules in $6\nu_3$ with a laser frequency of 944.5cm^{-1}. In our future work, we hope to be able to make more direct comparison with experiments on multiphoton absorption in SF_6.

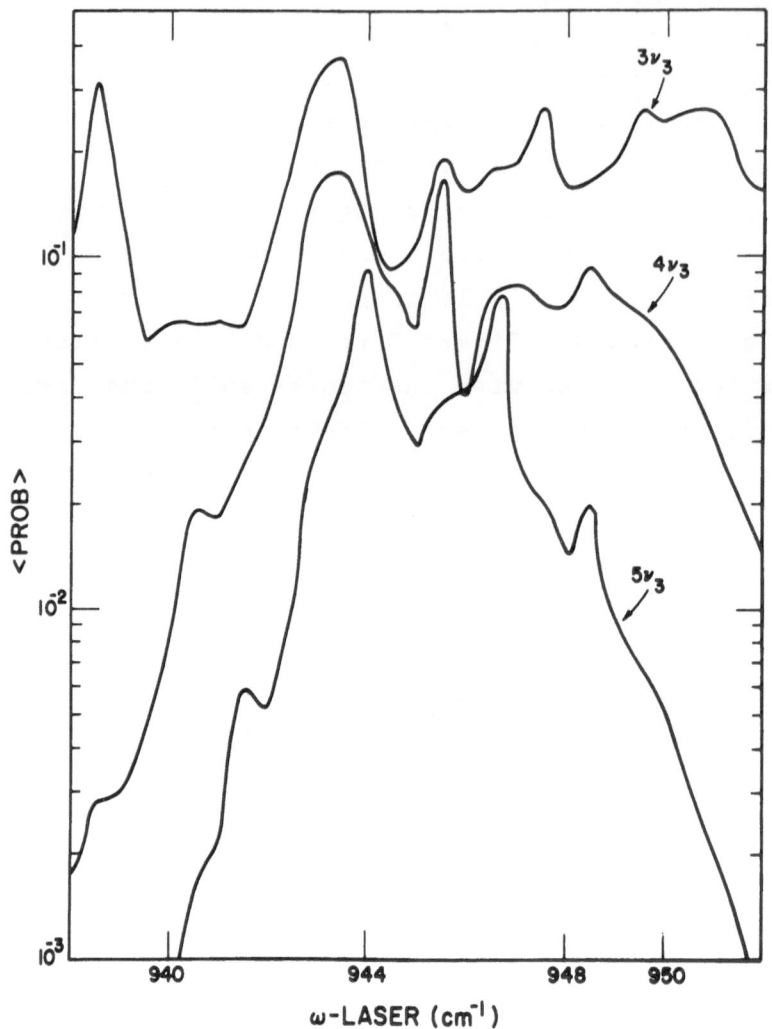

Fig. 6 Averaged probability for absorbing three, four, or five laser photons (only the manifold of states nearest to resonance were included in five and six[*]ν_3) versus laser frequency. Parameters for the calculation are $\mu E = 3 \text{cm}^{-1}$; $x_{33} = -2.8 \text{cm}^{-1}$; $T_{33} = -.44 \text{cm}^{-1}$ $G_{33} = 1.34 \text{cm}^{-1}$.

[*]Results for six ν_3 are not included in Figures 6 and 7, but for clarity, are given in Fig. 8 for two laser energies.

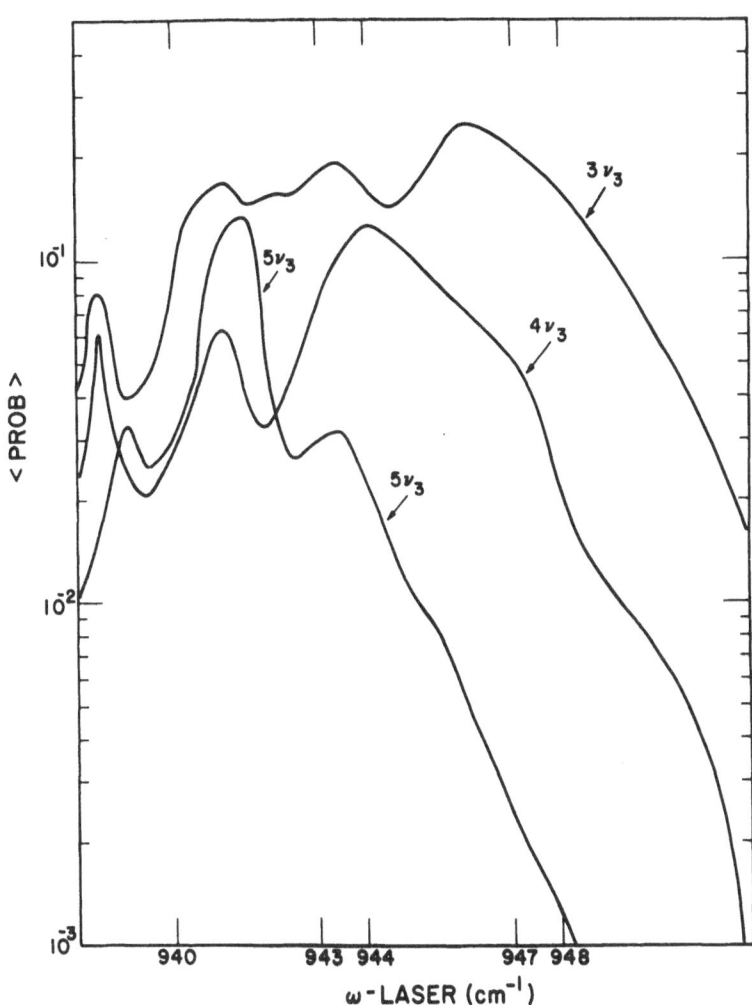

Fig. 7 Averaged probability for absorbing three, four or five photons with no tensor splittings. Other parameters are the same as in Fig. 6.

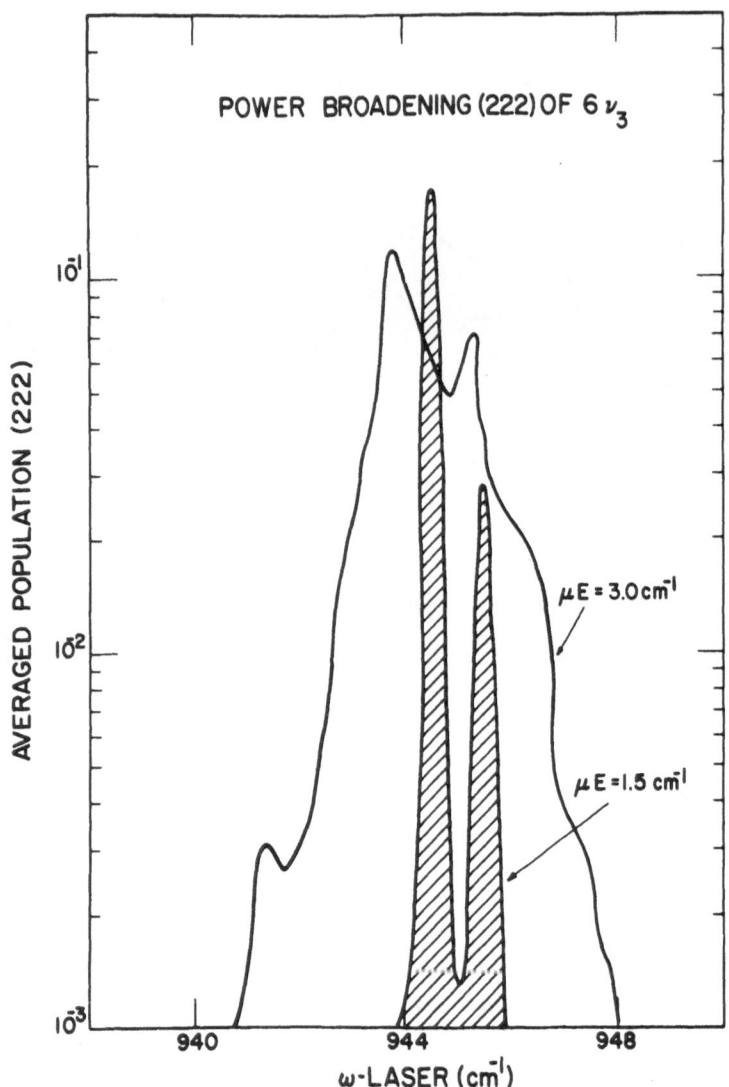

Fig. 8 Power broadening of the absorption into [222] of
$6\nu_3$. The narrower (hatched) curve corresponds to
$\mu E=1.5 cm^{-1}$ while the broader feature is for $\mu E=3.0 cm^{-1}$.
Molecular parameters of the calculations are as in Fig. 6.

TABLE I

DECOMPOSITION TABLE FOR V_n

I. 6-D , $[\lambda] = [p,q,r]$*

+ + +	$A_1 + A_2 + 2E$**		
+ + -	$F_1 + F_2$		
+ - -	$F_1 + F_2$		
- - -	$A_1 + A_2 + 2E$		

II. 3-D , $[\lambda] = [p,q,q]$*

p q	
+ +	$A_1 + E$
+ -	F_2
- +	F_1
- -	$A_2 + E$

III. 1-D $[\lambda] = [ppp]$

+	A_1
-	A_2

---

*The order of the entries is immaterial here.

**($\pm$) refers to the parity of the integer.

TABLE II

VIBRATIONAL ENERGIES OF SF_6 TO SECOND ORDER*

.600	[110]	F_2	
$\left.\begin{array}{c} -\ 8.200 \\ -10.960 \end{array}\right\}$	[200]	$E + A_1$	$2\nu_3$
1.800	[111]	A_2	
$\left.\begin{array}{c} -\ 7.000 \\ -\ 8.570 \end{array}\right\}$	[210]	$F_1 + F_2$	$3\nu_3$
-27.630	[300]	F_1	
-\ 5.894	[211]	F_2	
$\left.\begin{array}{c} -14.793 \\ -17.139 \end{array}\right\}$	[220]	$E + A_1$	
$\left.\begin{array}{c} -22.800 \\ -28.546 \end{array}\right\}$	[310]	$F_1 + F_2$	$4\nu_3$
$\left.\begin{array}{c} -54.847 \\ -55.261 \end{array}\right\}$	[400]	$E + A_1$	
-12.996	[221]	F_1	
$\left.\begin{array}{c} -20.400 \\ -28.680 \end{array}\right\}$	[311]	$A_2 + E$	
$\left.\begin{array}{c} -31.237 \\ -32.858 \end{array}\right\}$	[320]	$F_1 + F_2$	$5\nu_3$
$\left.\begin{array}{c} -53.043 \\ -53.410 \end{array}\right\}$	[410]	$F_1 + F_2$	
-91.496	[500]	F_1	

Table II con.
next page

TABLE II (continued)

-19.494	[222]	A_1	
$\left.\begin{array}{c}-26.982\\-30.659\end{array}\right\}$	[321]	$F_1 + F_2$	
-48.126	[330]	F_2	
-51.042	[411]	F_2	
$\left.\begin{array}{c}-52.600\\-53.708\\-64.053\end{array}\right\}$	[420]	$A_1 + A_2 + 2E$	$6\nu_3$
$\left.\begin{array}{c}-88.338\\-89.654\end{array}\right\}$	[510]	$F_1 + F_2$	
$\left.\begin{array}{c}-137.119\\-137.156\end{array}\right\}$	[600]	$A_1 + E$	

*The energies above are measured relative to n
times the 0→1 energy.

REFERENCES

1. C. D. Cantrell and H. W. Galbraith, Optics Commun.
    18, (1976) 513.

2. N. R. Isenor and M. C. Richardson, Appl. Phys.
    Lett. 18, (1971) 224.

3. R. V. Ambartzumian, V. S. Letokov, E. A. Ryabov and
    V. Chekalin, ZhETF Pis. Red. 20 (1974) 597 [JETP
    Lett. 20 (1974) 273].

4. J. L. Lyman, R. J. Jensen, J. Rink, C. P. Robinson
    and S. D. Rockwood, Appl. Phys. Lett. 27 (1975) 87.

5. V. S. Letokov and A. A. Makarov, Optics Commun. 17
    (1976) 250; Coherent Excitation of Multilevel
    Molecular Systems in Intensen Quasiresonant Laser
    IR Field (Moscow, 1976).

6. N. Bloembergen, Optics Commun. 15 (1975) 416.

7. D. M. Larsen and N. Bloembergen, Optics Commun. 17
    (1976) 254.

8. D. M. Larsen, Optics Commun. 19 (1976) 404.

9. D. P. Hodgkinson and J. S. Briggs, Chem. Phys. Lett.
    43 (1976) 451.

10. C. J. Elliott and B. J. Feldman, paper OA-2, 28th
    Annual Gaseous Electronics Conference, Rolla,
    Missouri (1975).

11. S. Mukamel and J. Jortner, Chem. Phys. Lett. 40
    (1976) 150.

12. T. P. Cotter, W. Fuss, K. L. Kompa and H. Stafast,
    Optics Commun. 18 (1976) 220.

13. R. V. Ambartzumian, N. P. Furzikov, Yu. A. Gorokhov,
    V. S. Letokhov, G. N. Makarov and A. A. Puretzky,
    Optics Commun. 18 (1976) 517.

14. R. S. McDowell, H. W. Galbraith, B. J. Krohn,
    C. D. Cantrell and E. D. Hinkley, Optics Commun.
    17 (1976) 178.

15. "Methods of Rotational Assignment of the P and R Branches of the ν_3 Fundamental Bank of SF_6", H. W. Galbraith, R. S. McDowell and C. D. Cantrell, submitted to J. Mol. Spectrosc (1977).

16. "The ν_3 Q-Branch of SF_6 at High Resolution: Assignment of the Levels Pumped by P(16) of the CO_2 Laser," R. S. McDowell, H. W. Galbraith, C. D. Cantrell, N. G. Nereson, and E. D. Hinkley submitted to J. Mol. Spectrosc. (1977).

17. "Coriolis Splitting Constant for UF_6 from Laser Diode Spectroscopy," J. P. Aldridge, H. Filip, K. Fox, H. W. Galbraith, R. S. McDowell and D.F. Smith, presented at "Lasers for Isotope Separation" conference (Albuquerque, N.M., 1976).

18. The literature on this subject is rather extensive and we cite only the classic works. H. A. Jahn, Proc. Roy. Soc. A168, 469 and 495 (1938).

19. W. H. J. Childs and H. A. Jahn, Proc. Roy. Soc. A169, 451 (1939).

20. K. T. Hecht, J. Mol. Spectrosc. 5, 335 and 390 (1960).

21. J. Moret-Bailly, Cah. Phys. 73, 476 (1959).

22. J. Moret-Bailly, Cah. Phys. 15, 237 (1961).

23. J. Moret-Bailly, J. Mol. Spectrosc. 16, 344 (1965).

24. B. T. Darling and D. M. Dennison, Phys. Rev. 57, 128 (1940).

25. W. H. Shaffer, H. H. Nielsen, and L. H. Thomas, Phys. Rev. 56, 895 and 1051 (1939).

26. J. H. van Vleck, Rev. Mod. Phys. 23, 213 (1951).

27. J. D. Louck [Dissertation Abst. 19, 840 (1958)].

28. See also, F. Michelot, J. Moret-Bailly, and K. Fox, J. Chem. Phys. 60, 2606 and 2610 (1974).

29. The parameters G_{33}, T_{33} and X_{33} have been calculated by C. C. Jensen, W. B. Person, B. J. Krohn, and

J. Overend, "Anharmonic Splittings of Vibrational Energy Levels of Octahedral Molecules: Application to the $n\nu_3$ manifold of $^{32}SF_6$", using explicit expressions for the vibration energy levels found in Hecht. For the purposes of our calculations, the eigenvectors are also needed as well as higher states; 5, and 6 ν_3, so that the results of Hecht are not sufficient. The remaining parameters were determined in ref. 15.

30. See K. Fox, J. Mol Spectrosc. $\underline{9}$, 381(1962), and the following paper of this volume, for an excellant and comprehensive discussion of 2 ν_3.

31. D. M. Brink and G. R. Satchler, "Angular Momentum" Oxford University Press, London, 1962.

32. J. Moret-Bailly, L. Gautier, J. Montagutelli, J. Mol. Spectrosc. $\underline{15}$, 355 (1965).

33. L. C. Biedenharn, W. Brouwer, and W. T. Sharp, 1968, "The Algebra of Representations of Some Finite Groups" (Rice University Studies, Houston, Texas).

34. J. D. Louck and H. W. Galbraith, Rev. Mod. Phys. $\underline{48}$, 69 (1976).

35. J. S. Lomont, 1959, "Applications of Finite Groups" (Academic Press, New York).

36. S. L. Altman, Rev. Mod. Phys. $\underline{35}$, 641 (1963).

37. H. V. McIntosh, J. Mol. Spectrosc. $\underline{5}$, 269 (1960).

38. H. V. McIntosh, J. Mol. Spectrosc. $\underline{10}$, 51 (1963).

39. H. W. Galbraith and C. D. Cantrell, "Vibrational States, Dipole Moments, and Dipole Sum Rules for F_1 and F_2 Ladders in Octahedral Spherical Top Molecules," to be submitted to J. Mol. Spectrosc..

40. J. Overend, H. W. Galbraith, W. B. Person and C. D. Cantrell, "The Collisionless Multi-Photon Laser Dissociation of SF_6: Relative Dipole

Strengths of One-Quantum Transitions in the ν_3 Vibrational Manifold," to be submitted for publication in J. Mol. Spectrosc..

41. L. Mower, Phys. Rev. 142, 799 (1966).

APPENDIX

CASE I $n_1 \neq n_2 \neq n_3$

(i) $(n_1, n_2, n_3) = (\text{odd, even, even})$.

$$|F_{(\frac{1}{2})}x> = \frac{1}{\sqrt{2}} \, [|n_1 \, n_2 \, n_3> \pm |n_1 \, n_3 \, n_2 >] \, ,$$

$$|F_{(\frac{1}{2})}y> = \frac{1}{\sqrt{2}} \, [|n_3 \, n_1 \, n_2> \pm |n_2 \, n_1 \, n_3 >] \, ,$$

$$|F_{(\frac{1}{2})}z> = \frac{1}{\sqrt{2}} \, [|n_2 \, n_3 \, n_1> \pm |n_3 \, n_2 \, n_1 >] \, ,$$

(ii) $(n_1, n_2, n_3) = (\text{even, odd, odd})$

$$|F_{(\frac{1}{2})}x> = \frac{1}{\sqrt{2}} \, [|n_1 \, n_2 \, n_3> \mp |n_1 \, n_3 \, n_2 >] \, ,$$

$$|F_{(\frac{1}{2})}y> = \frac{1}{\sqrt{2}} \, [|n_3 \, n_1 \, n_2> \mp |n_2 \, n_1 \, n_3 >] \, ,$$

$$|F_{(\frac{1}{2})}z> \quad \frac{1}{\sqrt{2}} \, [|n_2 \, n_3 \, n_1> \mp |n_3 \, n_2 \, n_1 >] \, ,$$

(iii) $(n_1, n_2, n_3) = (\text{even, even, even})$

$$|A_{(\frac{1}{2})}> = \frac{1}{\sqrt{6}} \, [|n_1 n_2 n_3> + |n_3 n_1 n_2> + |n_2 n_3 n_1> \pm$$
$$\{|n_2 n_1 n_3> + |n_3 n_2 n_1> + |n_1 n_3 n_2>\}]$$

$$|E_a^+> = \frac{1}{2} \, [|n_1 n_2 n_3> + |n_3 n_2 n_1> - |n_3 n_1 n_2> - |n_1 n_3 n_2>]$$

$$|E_b^+> = \frac{1}{\sqrt{12}} [-2|n_2 n_1 n_3> -2|n_2 n_3 n_1> + |n_3 n_2 n_1> +$$
$$|n_3 n_1 n_2> + |n_1 n_2 n_3> + |n_1 n_3 n_2>]$$

$$|E_a^-> = \frac{1}{\sqrt{12}} [2|n_2 n_1 n_3> -2|n_2 n_3 n_1> + |n_3 n_1 n_2 -$$
$$|n_3 n_1 n_2> - |n_1 n_3 n_2>]$$

$$|E_b^-> = \frac{1}{2} \, [|n_3 n_1 n_2> + |n_3 n_2 n_1> - |n_1 n_2 n_3> - |n_1 n_3 n_2>]$$

(iv) $(n_1, n_2, n_3) = (\text{odd, odd, odd})$

$$|A_{\binom{1}{2}}> = \frac{1}{\sqrt{6}} [|n_1 n_2 n_3> + |n_3 n_1 n_2> + |n_2 n_3 n_1> \mp$$
$$\{|n_2 n_1 n_3> + |n_3 n_2 n_1> + |n_1 n_3 n_2>\}]$$

$$|E_a^+> = \frac{1}{\sqrt{12}} [-2|n_2 n_1 n_3> -2|n_2 n_3 n_1> + |n_3 n_2 n_1> +$$
$$|n_3 n_1 n_2> + |n_1 n_2 n_3> + |n_1 n_3 n_2>]$$

$$|E_b^+> = \frac{1}{2} [|n_1 n_2 n_3> + |n_3 n_2 n_1> - |n_3 n_1 n_2> - |n_1 n_3 n_2>]$$

$$|E_a^-> = -\frac{1}{2} [|n_3 n_1 n_2> + |n_3 n_2 n_1> - |n_1 n_2 n_3> - |n_1 n_3 n_2>]$$

$$|E_b^-> = \frac{1}{\sqrt{12}} [2|n_2 n_1 n_3> -2|n_2 n_3 n_1> + |n_1 n_2 n_3> +$$
$$|n_3 n_1 n_2> - |n_3 n_2 n_1> - |n_1 n_3 n_2>]$$

CASE II

(i) $(m, n, n) = (\text{even, odd, odd})$

$$|F_{2x}> = |m\,n\,n>,$$
$$|F_{2y}> = |n\,m\,n>,$$
$$|F_{2z}> = |n\,n\,m>;$$

(ii) $(m, n, n) = (\text{odd, even, even})$

$$|F_{1x}> = |m\,n\,n>,$$
$$|F_{1y}> = |n\,m\,n>,$$
$$|F_{1z}> = |n\,n\,m>,$$

(iii) $(m, n, n) = (\text{even, even, even})$

$$|A_1> = \frac{1}{\sqrt{3}} [|m\,n\,n> + |n\,m\,n> + |n\,n\,m>],$$

$$|E_a> = \frac{1}{\sqrt{2}} [|n\,n\,m> - |n\,m\,n>];$$

$$|E_b> = \frac{1}{\sqrt{6}} [2|m\,n\,n> - |n\,m\,n> - |n\,n\,m>];$$

(iv) $(m, n, n) = (\text{odd}, \text{odd}, \text{odd})$

$$|A_2\rangle = \frac{1}{\sqrt{3}} \left[|m \, n \, n\rangle + |n \, m \, n\rangle + |n \, n \, m\rangle \right],$$

$$|E_a\rangle = \frac{1}{\sqrt{6}} \left[2|m \, n \, n\rangle - |n \, m \, n\rangle - |n \, n \, m\rangle \right]$$

$$|E_b\rangle = \frac{1}{\sqrt{2}} \left[|n \, n \, m\rangle - |n \, m \, n\rangle \right];$$

CASE III

(i) $n = \text{even}$

$$|A_1\rangle = |n \, n \, n\rangle \; ;$$

(ii) $n = \text{odd}$

$$|A_2\rangle = |n \, n \, n\rangle \; .$$

REVIEW OF ROTATIONAL STRUCTURE IN EXCITED VIBRATIONAL STATES OF SPHERICAL-TOP MOLECULES*

KENNETH FOX

DEPARTMENT OF PHYSICS AND ASTRONOMY

UNIVERSITY OF TENNESSEE, KNOXVILLE, TENN. 37916

I. INTRODUCTION

The subject of this review has been a recurring theme through much of the history of molecular spectroscopy, especially in the infrared spectral range. Methane has been the classic prototype molecule in this area of research. Advances in our theoretical understanding of rotational-vibrational spectra of CH_4 have followed, to some extent, developments in infrared technology.

In several instances, theoretical hypotheses and speculations concerning methane have prompted the extension or development of experimental techniques. Two recent examples will illustrate this point dramatically.

In 1971 a theoretical suggestion was made by Fox[1,2] concerning conventionally nonpolar molecules like methane. He predicted that pure rotational transitions

---

*Research supported by ERDA, NASA, and NSF.

in the vibrational-electronic ground state of
molecules with tetrahedral symmetry would be activated
by rotational-vibrational interactions. The predictions
for both the far-infrared[2] and microwave[3] spectral
regions were corroborated quantitatively in experiments
on CH_4 and other tetrahedral molecules. The subject
of this kind of forbidden transition has been reviewed
elsewhere.[4,5]

Standard available microwave techniques were not
sufficient to obtain that type of pure rotational
spectrum in CH_4. It was necessary to go beyond state-
of-the-art to measure transitions weaker than in any
previous microwave lines.[6] Such spectroscopic
measurements yield accurate ground-state rotational
constants for tetrahedral molecules. Comparable
transitions do not appear to exist[7] for octahedral
molecules like SF_6.

A spectrum of Uranus at 6800 Å was observed by
Owen[8] in 1966, and was tentatively attributed to $5\nu_3$ of
CH_4. This work suggested the need for high-resolution
laboratory spectra of that rotational-vibrational band.
However, its very weakness required (at low pressure)
ultra-long path lengths which were beyond state-of-the-
art White-cell methods.[9] Consequently, a time-of-flight
technique was developed[10,11] utilizing an evacuated
pipe of 1 kilometer base length. A total path of 20 km
was achieved. Spectral resolution was not sufficient
to obtain definitive results, however.

Fortson and Fox[12] are currently developing an
alternative approach utilizing moderate path lengths
and ultrasenstivity for absorption. This high-resolution
technique involving a tunable laser is expected to yield
well-resolved rotational-vibrational features in the

band of CH_4 near 6800 Å. The importance of this
and related spectra to our understanding of the
atmospheres of the outer planets Jupiter, Saturn,
Uranus, and Neptune has been reviewed elsewhere.[13]

Such spectra teach us much about the structure
of energy levels in excited vibrational states of
molecules like CH_4. The experimental techniques like
those mentioned above[10-12] may be utilized to obtain
spectra of $3\nu_3$ and $5\nu_3$ of SF_6 which have been modeled
recently in studies of collisionless multiple-photon
dissociation.[14]

The present review will discuss the rotational
structure of several excited vibrational states of
spherical-top molecules. It is not intended to be
exhaustive. Rather, several important problems will
be highlighted. A principal theme will be the
following one:

The impact of rotational structure on the
elucidation of vibrational structure. Patterns
discernable in high-resolution infrared spectra of
spherical-top molecules are intriguing in their own
way. They have been analyzed to yield precise
spectroscopic constants which in turn provide information
on molecular geometry, etc. However, these patterns
vary, depending on the vibrational energy levels on
which they are built. We shall discuss some of these
relationships and their implications in what follows.

II. ROTATIONAL STRUCTURE

Hecht[15] developed an important systematic approach
in the theory of rotational-vibrational perturbations
in tetrahedral XY_4 molecules. He applied the techniques
of group theory and angular-momentum coupling to

greatly simplify the calculations. Fox and coworkers[16-26]
later extended the applicability of this approach.

The simplest starting point for rotational energy
levels is, of course, the term

$$B \vec{J}^2, \tag{1}$$

where[27] $\vec{J}$ is the total angular momentum of the molecule
and B is essentially the reciprocal of the principal
moment of inertia, all three of which are equal for a
spherical top. The quantum-mechanical expectation
value of $\vec{J}^2$ is $J(J+1)$ with $J = 0,1,2,\ldots$ for a rigid
rotor. In the usual situation where electric-dipole
transitions prevail in infrared spectra of molecular
gases, the selection rules for absorption in a rotational-
vibrational band are $\Delta J = 0, \pm 1$. Then the approximate
displacement of a line from the center of a band would
be

$$B'J'(J'+1) - BJ(J+1). \tag{2}$$

Here the primes refer to the final state of the transition.
The rotational constant B' in the excited vibrational
state is usually different from B because of centrifugal
distortion effects, etc. If we specialize to the case
$J'=J+1$, and assume for simplicity that $B'=B$, then Eq. (2)
becomes

$$2B(J+1). \tag{3}$$

As a consequence of this result, it appears that a
sequence of such features in a rotational-vibrational
band should be separated by an amount 2B. However, this
quantity is not what is observed in actual spectra.

For a triply-degenerate normal vibration in a
spherical-top molecule, like ν_3 of CH_4 or SF_6, there is
an essential Coriolis interaction term

$$-2B'\zeta_3 \ (\vec{J}\cdot\vec{\ell}_3) \ . \tag{4}$$

The analogue of this term in classical mechanics resembles a degenerate harmonic oscillator moving on a rotating frame, such as a turntable. For convenience, the subscript 3 in Eq. (4) will be dropped for a while. A useful way to evaluate the contribution of Eq. (4) to the energy is to introduce the rotational angular momentum $\vec{R}$ through the relation

$$\vec{J} = \vec{R} + \vec{\ell} \tag{5}$$

or

$$\vec{R} = \vec{J} - \vec{\ell} \ . \tag{6}$$

Then

$$\vec{R}^2 = \vec{J}^2 + \vec{\ell}^2 - 2\vec{J}\cdot\vec{\ell} \tag{7}$$

and

$$-2B'\zeta(\vec{J}\cdot\vec{\ell}) = B'\zeta(\vec{R}^2 - \vec{J}^2 - \vec{\ell}^2) \tag{8a}$$

$$= B'\zeta[R(R+1) - J(J+1) - \ell(\ell+1)]. \tag{8b}$$

Now, according to Eq. (6),

$$R = J+\ell, \ J+\ell-1, \ldots, \ |J-\ell| \ . \tag{9}$$

For a triply-degenerate normal vibration with principal quantum number v, the possible values of ℓ are

$$\ell = v, \ v-2, \ v-4, \ldots, \ 1 \ \text{or} \ 0. \tag{10}$$

In the ground state, $v = \ell = 0$ so that the Coriolis term Eq. (4) is zero. In the first excited vibrational state, $v = \ell = 1$ so that the Coriolis term takes the values

$$-2B'\zeta(\vec{J}\cdot\vec{\ell}) = B'\zeta[R(R+1) - J(J+1)-2] \qquad (11a)$$

$$= 2B'\zeta J \qquad \text{for } R = J+1 \qquad (11b)$$

$$= -2B'\zeta \qquad \text{for } R = J \qquad (11c)$$

$$= -2B'\zeta(J+1) \quad \text{for } R = J-1. \qquad (11d)$$

As v increases, the Coriolis splittings in excited vibrational states become more complicated.

However, let us briefly examine the effect of the simplest Coriolis term on the infrared-active fundamental transition $v = 0 \to 1$. To do this, we have to anticipate a selection rule $\Delta R = 0$ for the fundamental. It is convenient to label the states involved in a transition by J_R. Clearly for the ground state $R = J$ so that the possible transitions are

$$J_J \to J+1_J, \; J_J, \text{ or } J-1_J. \qquad (12)$$

The corresponding energy differences for the terms

$$B\vec{J}^2 - 2B\zeta(\vec{J}\cdot\vec{\ell}) \qquad (13)$$

are

$$B'(J+1)(J+2) - 2B'\zeta(J+2) - BJ(J+1) \approx 2B(1-\zeta)(J+1) - 2B\zeta, \qquad (14a)$$

$$B'J(J+1) - 2B'\zeta - BJ(J+1) \qquad \approx -2B\zeta \;, \qquad (14b)$$

$$B'(J-1)J + 2B'\zeta(J-1) - BJ(J+1) \qquad \approx -2B(1-\zeta)J-2B\zeta. \qquad (14c)$$

The important result in Eqs. (14) is that the spacing between rotational "multiplets" in the $\Delta J = \pm 1$ sub-bands is not $2B$ as in Eq. (3), but $2B(1-\zeta)$. This effect has been observed repeatedly in infrared spectra of spherical tops, most recently[28-30] in SF_6.

The Coriolis term in Eq. (4) has a significant
effect on the rotational structure of higher excited
vibrational states. However, before considering
those effects in detail, let us turn briefly to the
general aspect of cubic, that is tetrahedral and
octahedral, fine-structure splittings. Hecht[15] has
shown that, even to third order in perturbation theory,
the splittings of a rotational-vibrational level into
its tetrahedral sublevels is governed only by perturbation terms of one basic symmetry in all states in
which vibrational quanta of ν_1, ν_3, and ν_4 are excited,
and to a certain approximation in many other infrared-
active states. The effects carry over to octahedral
sublevels as well because of the cubic symmetry shared
by these two classes of molecules. The perturbation
term is identified as the tetrahedrally (or, equivalently,
octahedrally) invariant linear combination of fourth-rank
spherical tensor operators.[31]

In a recent theoretical study of level splitting,
Fox et al.[25] have made an extensive analysis of the
spectrum of the octahedrally invariant fourth-rank
tensor operator. One physical realization of this
operator in pure angular-momentum space is

$$T \equiv -(24/5)(\vec{J}^2)^2 + (8/5)\,\vec{J}^2 + 8(J_x{}^4 + J_y{}^4 + J_z{}^4)\;. \tag{15}$$

The eigenvalue spectrum of this operator has been
examined in detail.[25] Calculations were carried out
with double precision[26] on a CDC 7600 computer through
values of $J = 100$. Unexpected periodic symmetries and
asymptotic degeneracy features were found in the
eigenvalue spectrum. Although these results are
significant from the mathematical-physics viewpoint,

their further discussion must be deferred.

　　　We now turn to a brief consideration of selection rules prevailing in the rotational-vibrational spectra of molecules of cubic symmetry. For electric-dipole transitions, $\Delta J = 0, \pm 1$ has already been mentioned. This selection rule can be understood in the conventional way, as follows. Let the space-fixed Z component of the electric-dipole-moment operator be expressed as

$$\mu_Z = \lambda_{Zx} \, \mu_x + \lambda_{Zy} \, \mu_y + \lambda_{Zz} \, \mu_z \quad , \qquad (16)$$

where $\lambda_{Z\alpha}$ is the direction cosine between the space-fixed Z axis and the molecule-fixed α axis,[32] and μ_α is the molecule-fixed α component of the dipole-moment operator. The quantum-mechanical matrix elements involving the $\lambda_{Z\alpha}$ and the total angular momentum eigenfuctions ψ_{JK} lead to $\Delta J = 0, \pm 1$.

　　　There are two other selection rules of significance to be considered. The first of these has to do with cubic symmetry, and can be described in terms of the irreducible representations ("irreps") of the cubic point groups. For the present, we will use only the irrep labels A, E, and F without any subscripts 1 or 2, and g or u. In the space of normal coordinates for molecular vibrations, the operator in Eq. (16) is of type A. Consequently, the group multiplication rules for irreps require that the initial and final rotational-vibrational states be of the same symmetry species. This selection rule may be characterized as Δ symmetry = 0. More specifically it is A ⟷ A, E ⟷ E, and F ⟷ F.

　　　The second significant selection rule has to do with the rotational angular-momentum quantum number defined in Eqs. (5) and (6). The coupling scheme introduced by

Hecht[15] is

$$\Psi_{RK_R} = \sum_m (\ell J m K | \ell J R K_R) \phi_{\ell m}^* \psi_{JK} , \qquad (17)$$

where ψ_{JK} and $\phi_{\ell m}$ are the eigenfunctions for total and vibrational angular momentum, respectively. These functions are coupled by standard Clebsch-Gordan coefficients.[31]

For the infrared-active fundamentals ν_3 and ν_4, the transition matrix elements are dominated by the terms

$$\mu_z = A_3 z_3 + A_4 z_4 , \qquad (18)$$

where $A_3 = (\partial \mu_z / \partial z_3)_0$ and $A_4 = (\partial \mu_z / \partial z_4)_0$. In this situation, the overall electric-dipole-moment operator in R space is a tensor of rank zero. Consequently, the selection rule on R is $\Delta R = 0$. This leads to three strong "branches" in the spectrum: $J_J \rightarrow J+1_J$ ("R-branch"), $J_J \rightarrow J_J$ ("Q-branch"), and $J_J \rightarrow J-1_J$ ("P-branch"). Their basic structures, without cubic fine-structure splittings, were indicated in Eq. (14). In addition there are relatively forbidden $\Delta R \neq 0$ transitions which have been described in ν_3 of CH_4, for example.[15]

Energy level diagrams for the $v_3 = 0$ and $v_3 = 1$ states are depicted in Fig. 1. Representative transitions are shown for the case $J_J \rightarrow J_{R'}' = J+1_J$ which correspond to $\Delta R = 0$. The splittings shown are not necessarily to scale. Note that the ordering of fine-structure components need not be the same in the ground and first-excited vibrational states.

The cubic splitting patterns comprise an intriguing area of study by themselves. Their physical origin is from a wide variety of effects. Some examples for a

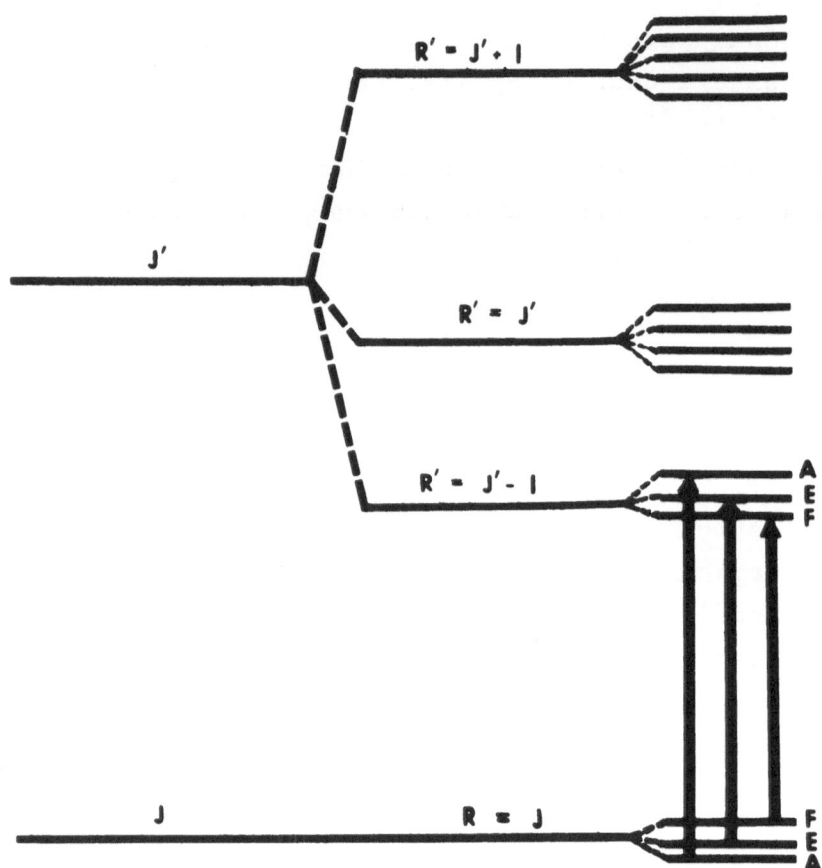

Fig. 1. Schematic representation of Coriolis and fine-
structure splittings in $v_3 = 0$ and $v_3 = 1$ states. The
transitions shown here follow the selection rules
$\Delta J = +1$, $\Delta R = 0$, and Δ symmetry $= 0$.

tetrahedral XY_4 molecule like CH_4 are the following:[15]

$$T_{33}O_{33}(\text{tensor})=T_{33}[(p_{3x}^2+x_3^2)^2+(p_{3y}^2+y_3^2)^2+(p_{3z}^2+z_3^2)^2$$

$$-3(p_{3x}^2+x_3^2)(p_{3y}^2+y_3^2)$$

$$-3(p_{3y}^2+y_3^2)(p_{3z}^2+z_3^2)-3(p_{3z}^2+z_3^2)(p_{3x}^2+x_3^2)$$

$$+2\vec{\ell}_3^2+6] \quad , \tag{19}$$

and

$$F_{3t}O_{PPP3}(\text{tensor}) = F_{3t}[10(J_x^3\ell_{3x}+J_y^3\ell_{3y}+J_z^3\ell_{3z})$$

$$-6\vec{J}^2(\vec{J}\cdot\vec{\ell}_3)+2(\vec{J}\cdot\vec{\ell}_3)] \quad . \tag{20}$$

The coefficients T_{33} and F_{3t} are complicated functions of molecular parameters such as cubic and quartic anharmonic constants. The functional forms in Eqs. (15), (19), and (20) have the same spherical tensor character:[15,25]

$$(14/5)^{1/2}T_{4,0} + T_{4,4} + T_{4,-4} \quad , \tag{21}$$

where the factor $(14/5)^{1/2}$ is required for the cubic symmetry. The power of this insight is that the theory of level splitting of many rotational-vibrational states (including the ground state) in spherical-top molecules may be developed within the framework of the standard Racah-Wigner tensor algebra.

This formalism has recently been extended in significant ways.[23] The rotational energy for spherical-top molecules in the vibrational-electronic ground state has been given in a general form to **all** orders in perturbation theory. The sixth-order result was developed explicitly. The Hamiltonian for a triply-degenerate

fundamental (like ν_3) of spherical-top molecules
was extended to sixth order of approximation. Formulas
for the energy levels were also given.

A striking example of octahedral fine-structure
splitting patterns occurs in ν_3 of SF_6. This stretching
fundamental band contains rotational-vibrational
transitions that are nearly coincident with carbon
dioxide laser lines near 10.6 µm. For this reason there
has been considerable interest in SF_6 as a nonlinear
absorber. Recently Aldridge et al.[28] reported a high-
resolution spectrum in which extensive portions of the
rotational fine structure in the P and R branches were
resolved for the first time. A preliminary analysis
of the spectrum was made, and rotational and
octahedral quantum numbers were assigned to the spectrum.

These spectra were recorded using a PbSnSe
tunable semiconductor diode laser operating
at a resolution of better than $10^{-4} cm^{-1}$. The SF_6
was cooled to about 135°K to reduce interference from
transitions arising from excited vibrational states.
Fig. 2 gives an overall view of the ν_3 band. The
splitting patterns of the J multiplets P(17), P(18),
and P(19) are shown in more detail in Fig. 3, where
the fine-structure components are identified according
to the notation of Moret-Bailly.[33]

The number of cubic fine-structure components
of each type A_1, A_2, E, F_1, and F_2 (with additional
subscripts g and u for O_h symmetry) for a given value
of J is uniquely specified by a standard group-
theoretical algorithm.[21,34] To first approximation,
the _relative_ splittings in each J-multiplet are
determined by the eigenvalue spectrum of operators like
those in Eqs. (15) and (19)-(21). The _absolute_

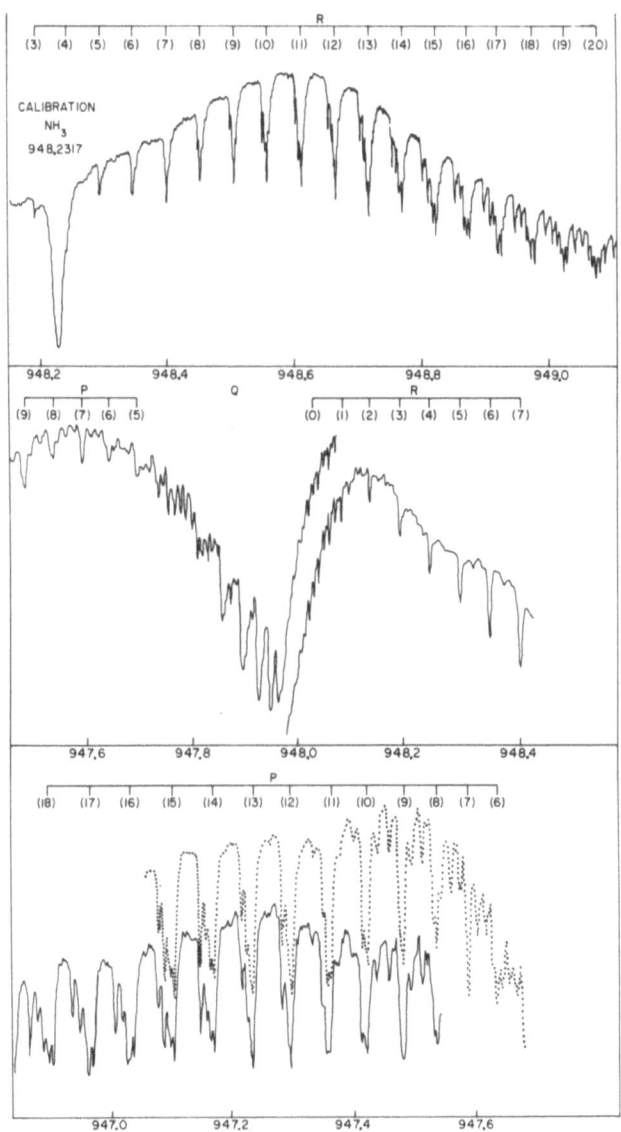

Fig. 2.   Rotational-vibrational fundamental ν_3 of SF_6 from P(18) to R(20), recorded using a tunable diode laser at resolution better than $10^{-4} cm^{-1}$. The SF_6 was cooled to about 135°K at which temperature 91% of the molecules are in the ground vibrational state (after Aldridge et al.[28]).

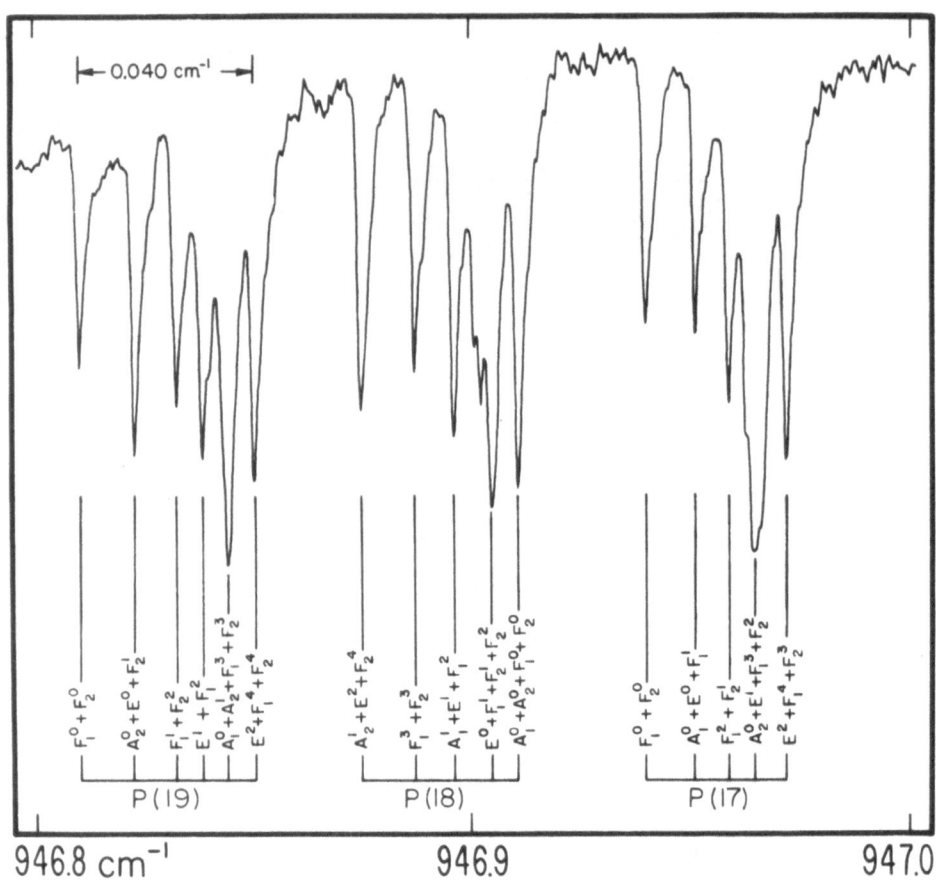

Fig. 3. Detail of P(17), P(18) and P(19) of ν_3 of SF$_6$ with the octahedral fine-structure component lines identified. Gas temperature 135°K, pressure ≈ 0.4 torr, path length 10 cm (after Aldridge et al.[28]).

splittings depend on physical parameters like T_{33} and F_{3t} in Eqs. (19) and (20), respectively. These parameters vary, of course, from molecule to molecule. The relative line <u>intensities</u> within a particular J-multiplet are determined by nuclear-spin statistical weight factors. For example,[35] for CH_4 we have $A_1:A_2:E:F_1:F_2 = 5:5:2:3:3$; and for SF_6 the corresponding values[36] are $2:10:8:6:6$. The relative splitting and intensity patterns are often enough to make it possible to identify one or more J-multiplets in a rotational-vibrational band. This in turn may facilitate a complete analysis of such a spectrum.

Thus far we have concentrated, explicitly or implicitly, on rotational structure in the ground and first-excited vibrational states of triply-degenerate fundamentals of spherical-top molecules. Next we move on to the more difficult problem of more highly excited vibrational states. We shall see how the complexity mounts. In particular, we shall want to note what the observed rotational structure implies about the vibrational states on which it is built. We shall examine several classic cases in high-resolution infrared spectroscopy: $2\nu_3$, $2\nu_4$, and $3\nu_3$ of methane.

For a v = 2 state, the vibrational angular momentum ℓ = 0 or 2. Consequently in the conventional representation[16,17] the rotational angular momentum can take the values R=J±2, J±1, and J. The R = J case can arise from coupling with either ℓ = 0 or 2.   A characteristic splitting scheme is shown in Fig. 4. The selection rules $\Delta J = 0, \pm 1$ and Δ symmetry = 0 hold as before. However, there is a significant change in the selection rule for R.

The transition matrix elements are dominated by

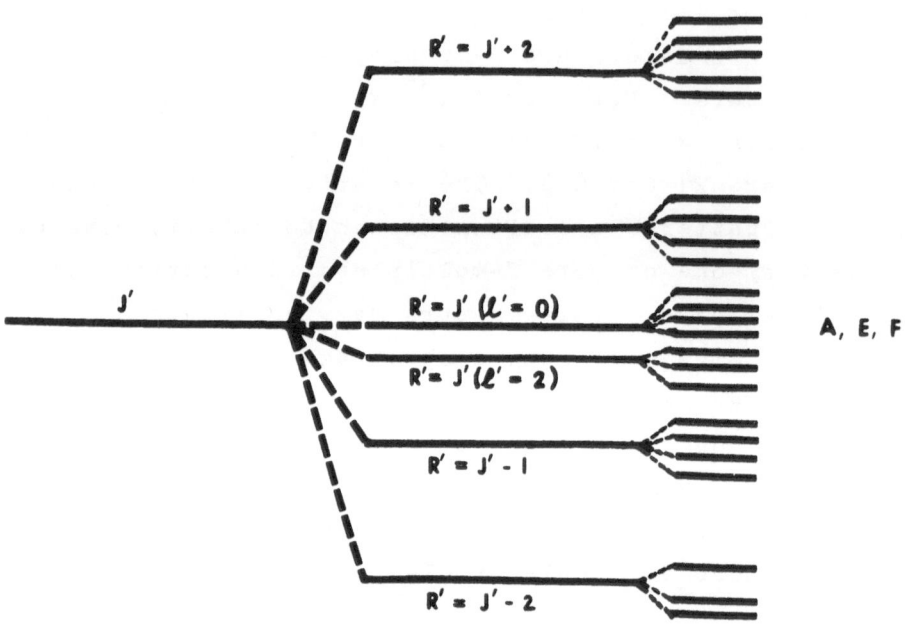

Fig. 4. Schematic representation of Coriolis and fine-structure splittings in $v_4 = 0$ and $v_4 = 2$ states (after Fox[16,17]).

the terms

$$\mu_z = A_3' \, x_3 y_3 + A_4' \, x_4 y_4 \quad , \tag{22}$$

where $A_3' = (\partial^2 \mu_z / \partial x_3 \partial y_3)_0$ and $A_4' = (\partial^2 \mu_z / \partial x_4 \partial y_4)_0$. In this situation, the overall electric-dipole-moment operator in R space is a tensor of rank three. Consequently, the selection rule on R is $\Delta R = 0, \pm 1, \pm 2,$ or ± 3. Referring to Fig.4, let us consider the case $\Delta J = +1$. Then $J' = J+1$, and $R' = J+3$, $J+2$, $J+1(\ell'=0)$, $J+1(\ell'=2)$, J, and $J-1$ all of which are consistent with $|\Delta R| \leq 3$. Thus, a much more complicated spectrum may be expected than for a fundamental as depicted in Figs. 1-3. The $2\nu_4$ band of CH_4 apparently exhibits[37,24] all the complexity implied by Fig.4. Much of this spectrum is depicted in Fig.5. The fine-structure splittings in this band have not yet been completely analyzed, although recently progress has been made along these lines with the help of a simpler form of the spectrum obtained at low temperature.[24]

In striking constrast to $2\nu_4$ of CH_4 is its counterpart $2\nu_3$ which a priori would be expected to appear as complicated. That this was not the case was dramatically demonstrated in several experiments,[38,39] especially that of Rank et al.[38] The $2\nu_3$ spectra appeared to have the aspect of ν_3 with $\ell_3 = 1$. The observed fine-structure splittings were small, and so another measurement was undertaken[40] on $2\nu_3$ of CD_4. The expectation was two-fold: that a fundamental-like band would appear, thus corroborating the puzzling appearance of $2\nu_3$ of CH_4; and that the fine-structure splittings would be larger than in CH_4, thus giving further credence to the explanation to be described below. A portion of

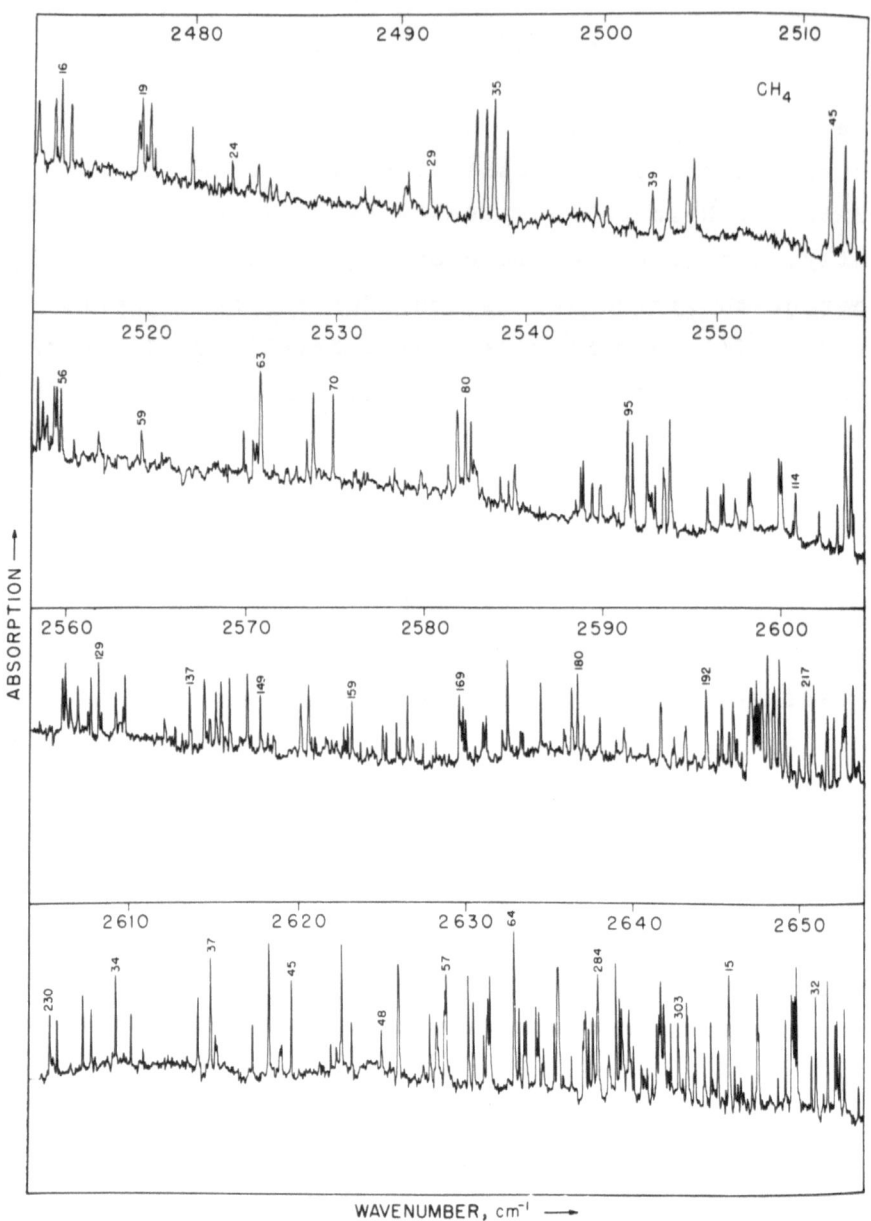

Fig. 5. A portion of the $2\nu_4$ spectrum of CH_4 (after Plyler et al.[37]).

the spectrum of $2\nu_3$ of CD_4 is shown in Fig.6. The
characteristic fine-structure splitting for $\ell_3 = 1$
of a fundamental band is clearly evident.

A new and unconventional representation had to
be introduced[16,17] to rationalize the appearance of
$2\nu_3$ of CH_4 and CD_4. The $\ell_3 = 2$ portion of $2\nu_3$ consists
of two- and three-dimensional substates of symmetry
E and F_2, respectively. These states are split by the
purely vibrational term $T_{33}O_{33}$(tensor) in Eq. (19)
by[15] an amount $20\ T_{33}$. It was estimated[16,17] that this
term value was sufficiently larger than the contribution
of the Coriolis term $-2B'\zeta_3(\vec{J}\cdot\vec{\ell}_3)$ in Eq. (4), that a
new coupling scheme could be invoked. It is shown
schematically in Fig.7.

In this alternative coupling scheme, the F_2
substate of $\ell_3 = 2$ is treated as if it were actually
an $\ell_3 = 1$ state as in the fundamental $v_3 = 1$. Then $\vec{R}$
is replaced by a new angular momentum $\vec{L}$ defined by

$$\vec{L} = \vec{J} - \vec{\ell}\ ,\qquad\qquad (23)$$

so that $L = J \pm 1, J$. Energies and transition moments
are calculated in the new representation using the
following angular-momentum coupled wave functions

$$\Psi_{LK_L} = \sum_m (1JmK \mid 1JLK_L)\phi_m^* \ \psi_{JK}\ ,\qquad\qquad (24)$$

where the ϕ_m are actually $\ell_3 = 2$ type vibrational
functions. Complete computations are carried out in
Refs. 16 and 17. Suffice it to say that the scheme
involving Eqs. (23) and (24) gives a plausible account
of $2\nu_3$ of CH_4 and CD_4. In particular the selection rule
$\Delta L = 0$ which is a natural outcome of the new representation predicts a single P-, Q-, and R-branch with the
usual tetrahedral fine-structure components.

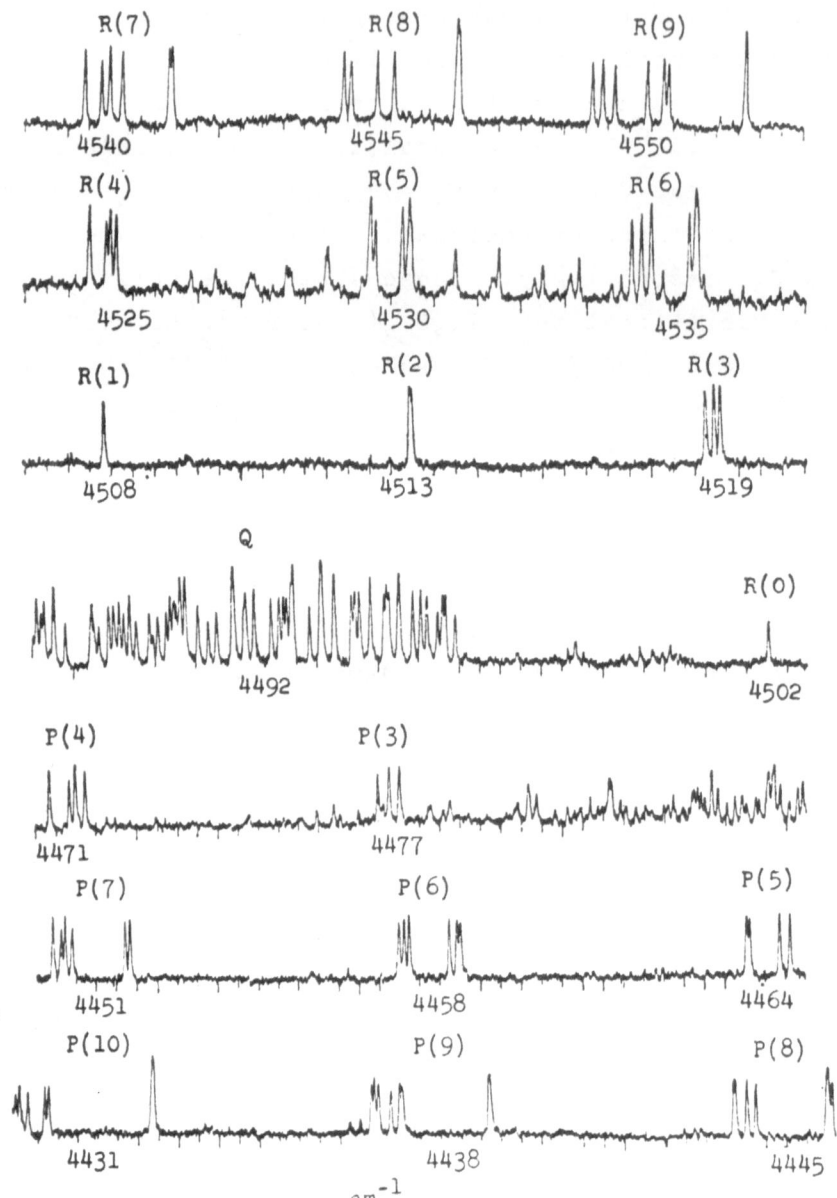

Fig. 6. A portion of the $2\nu_3$ spectrum of CD_4 (after Fox et al.[40]).

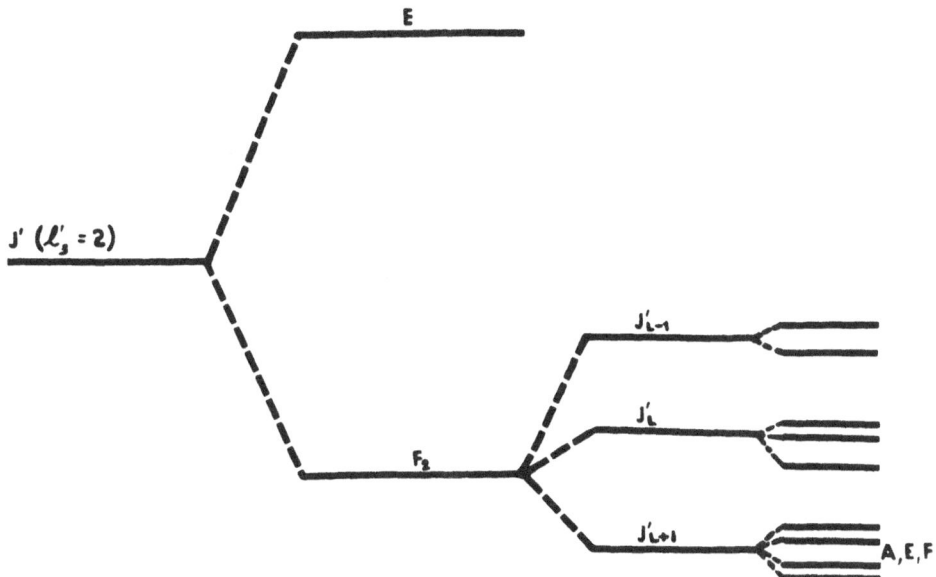

Fig. 7. Schematic representation of successive splittings in the new representation of $2\nu_3$ of methane: pure vibrational tensor, followed by Coriolis term, and tetrahedral fine-structure splitting terms (after Fox[16,17]).

Finally, the $3\nu_3$ band of CH_4 deserves some consideration as part of the pattern of rotational structure observed in excited vibrational states. The R-branch consisting of $\Delta J = +1$ transitions is shown[41] in Fig. 8. One is immediately struck by the fact that this portion of the spectrum appears very much like the ν_3 fundamental, especially with regard to tetrahedral fine-structure splittings. No complete explanation for this rotational-vibrational band has yet been given. However, there are some preliminary qualitative indication of how the simple observed structure arises.[42]

For $v_3 = 3$, the vibrational angular momentum takes the values $\ell_3 = 1$ and 3. The electric-dipole-moment matrix element for a $v_3 = 0 \rightarrow 3$ transition is[42]

$$\langle v_3' = 3,\ \ell_3';\ J'R'K_R' | \mu_Z | v_3 = 0,\ \ell_3 = 0;\ JRK_R \rangle$$

$$= (-1)^{J+R+M'+K_R'+1}\ [(2J'+1)(2R'+1)]^{1/2} \begin{pmatrix} J & 1 & J' \\ M & 0 & -M' \end{pmatrix} \cdot$$

$$\cdot \left\{ -\tfrac{1}{3} C' \delta_{R'R} \delta_{K_R' K_R} \delta_{\ell_3' 1} + \tfrac{3}{20} C (-1)^{J'+J} [3(2J+1)]^{1/2} \cdot \right.$$

$$\cdot \begin{Bmatrix} R' & J & 4 \\ 1 & 3 & J' \end{Bmatrix} \left[(70)^{1/2} \begin{pmatrix} R & 4 & R' \\ K_R & 0 & -K_R' \end{pmatrix} + 5 \begin{pmatrix} R & 4 & R' \\ K_R & 4 & -K_R' \end{pmatrix} \right.$$

$$\left. + 5 \begin{pmatrix} R & 4 & R' \\ K_R-4 & -K_R' \end{pmatrix} \right] \delta_{\ell_3' 3} \right\} \quad . \tag{25}$$

This expression is indeed complicated, but it contains considerable information worth looking at carefully in detail. From the purely technical viewpoint it should be noted that there are four 3-j symbols [instead of Clebsch-Gordan coefficients as in Eqs. (17) and (24)] and one 6-j symbol. The leading 3-j symbol contains the usual selection rule $\Delta J = 0, \pm 1$.

The first term inside the curly braces implies the

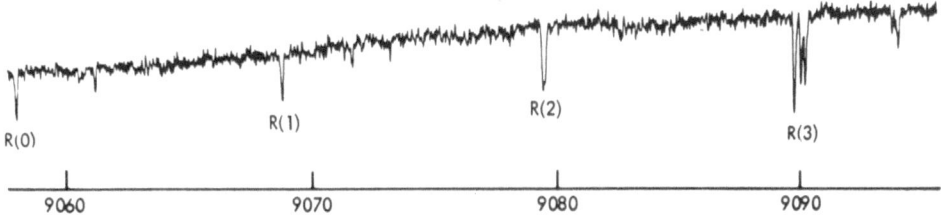

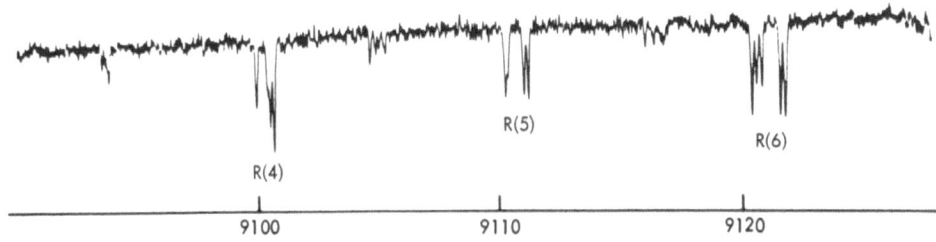

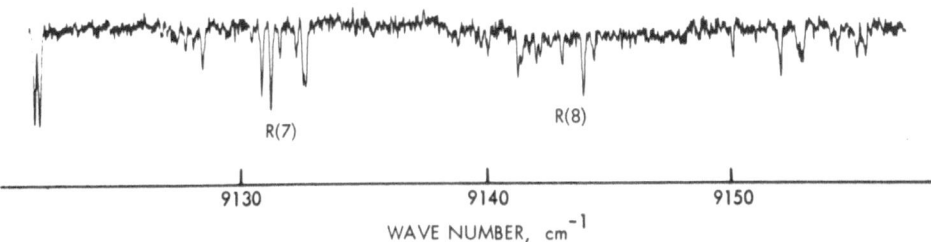

Fig. 8. A portion (R-branch, $\Delta J = +1$) of the $3\nu_3$ spectrum of CH_4 (after Margolis and Fox[41]).

selection rules $\Delta R = 0$, $\Delta K_R = 0$, and $\Delta \ell_3 = 1$. These
are strikingly similar to the systematics for ν_3 and
$2\nu_3$ of CH_4. The coefficient C' is a linear combination
of third derivatives of the electric-dipole-moment
operator with respect to normal coordinates. (Also,
the coefficient C is a similar quantity.) The relatively
simple selection rules may eventually comprise part of
the explanation for the fundamental-like structure of the
R-branch of $3\nu_3$ of CH_4.

The second term inside the curly braces implies
the selection rules $|\Delta R| \leq 4$, $\Delta K_R = 0$, ± 4, and $\Delta \ell_3 = 3$.
These are reminiscent, but not identical to, the systematics
for $2\nu_4$ of CH_4. It may be that $C'/C \gg 1$ so that the term
with the simpler selection rules dominates at least a
portion of the spectrum. It should also be noted that
the $\Delta R = 0$ term has the characteristic of a spherical
tensor of rank zero, that is, a scalar. The $\Delta R \neq 0$ term,
however, has the structure of a fourth-rank tensor
previously discussed in Eqs. (15) and (21). How this
latter term reflects itself in the spectrum remains to be
analyzed.[42]

A relevant point concerning our understanding of
$2\nu_3$ and $3\nu_3$ of CH_4 is whether these quantum identifications
are indeed correct. In fact they are, as has recently
been demonstrated by measurements of isotope shifts
between $^{12}CH_4$ and $^{13}CH_4$.

The Teller-Redlich product rule[27,43,44] for triply-
degenerate fundamentals of tetrahedral XY_4 molecules can
be written in the form

$$\Delta_3 / \omega_3^{(12)} = [\delta - (\Delta_4 / \omega_4^{(12)})] / [1 - (\Delta_4 / \omega_4^{(12)})], \quad (26)$$

where

$$\Delta_i \equiv \omega_i^{(12)} - \omega_i^{(13)} \qquad (27)$$

and

$$\delta \equiv 1 - (m_Y^{(12)}/m_Y^{(13)})[(m_X^{(12)}/m_X^{(13)}) \cdot$$
$$\cdot (m_X^{(13)}+4m_Y^{(13)})/(m_X^{(12)}+4m_Y^{(12)})]^{1/2}.$$
$$(28)$$

In harmonic approximation, Eqs. (26)-(28) predict isotope shifts in $2\nu_3$ and $3\nu_3$ to be two and three times, respectively, the value in ν_3. These systematics have been shown to hold sufficiently well for both bands in question.[45,46]

Our knowledge of the rotational structure in excited vibrational states of octahedral spherical tops in relatively small. A notable exception is ν_3 of SF_6 whose study has been accelerated by virtue of experimental[28] and theoretical[29] efforts at the Los Alamos Scientific Laboratory. Recently, a theoretical model for several excited vibrational states in the ν_3 "ladder" has been constructed[14] in an effort to explain collisionless multiple-photon laser dissociation of SF_6. An important test of such models would be a high-resolution spectroscopic study, both experimental and theoretical, of $3\nu_3$ of SF_6 which seem accessible using state-of-the-art techniques.[12] It would also be helpful to have data and analyses on the $\Delta v_3 = +1$ transitions like $v_3 = 1 \rightarrow 2$, etc.

REFERENCES

1. K. Fox, in Proceedings of the Ohio State University Symposium on Molecular Structure and Spectroscopy, 1971, Abstract K5, unpublished.

2. K. Fox, Phys. Rev. Letters 27, 233 (1971).

3. K. Fox, Phys. Rev. A 6, 907 (1972).

4. T. Oka, in Molecular Spectroscopy: Modern Research II, edited by K. Narahari Rao (Academic Press, New York, 1976).

5. M. R. Aliev, Progress in Phys. Sci. 119, 557 (1976), in Russian.

6. C. W. Holt, M. C. L. Gerry, and I. Ozier, Phys. Rev. Letters 31, 1033 (1973).

7. I. M. Mills, J. K. G. Watson, and W. L. Smith, Molec. Phys. 16, 329 (1969).

8. T. Owen, Astrophys. J. Letters 146, 611 (1966).

9. D. H. Rank, U. Fink, J. V. Foltz, and T. A. Wiggins, Astrophys. J. 140, 366 (1964).

10. R. Goldstein, V. Vali, and K. Fox, Astrophys. J. Letters 180, L129 (1973).

11. V. Vali, R. Goldstein, and K. Fox, Appl. Phys. Letters 22, 391 (1973).

12. N. Fortson and K. Fox, unpublished.

13. K. Fox, in Molecular Spectroscopy: Modern Research I, edited by K. Narahari Rao and C. W. Mathews (Academic Press, New York, 1972).

14. C. D. Cantrell and H. W. Galbraith, Optics Comm. 18, 513 (1976).

15. K. T. Hecht, J. Mol. Spectry. 5, 355, 390 (1960).

16. K. Fox, Ph. D. thesis, University of Michigan, 1961, unpublished.

17. K. Fox, J. Mol. Spectry. 9, 381 (1962).

18. K. Fox, J. Mol. Spectry. 16, 35 (1965).

19. K. Fox, J. Chem. Phys. 43, S25 (1965).

20. J. S. Margolis and K. Fox, J. Chem. Phys. 49, 2451 (1968).

21. K. Fox and I. Ozier, J. Chem. Phys. 52, 5044 (1970).

22. K. Fox, Phys. Rev. A 8, 658 (1973).

23. F. Michelot, J. Moret-Bailly, and K. Fox, J. Chem. Phys. 60, 2606, 2610 (1974).

24. G. Pierre, J. Cadot, R. J. Corice, and K. Fox, Canad. J. Phys., to be published.

25. K. Fox, H. W. Galbraith, B. J. Krohn, and J. D. Louck, Phys. Rev. A, to be published.

26. B. J. Krohn, Los Alamos Scientific Laboratory Report LA-6554-MS (October, 1976).

27. For conventional definitions of spectroscopic quantities used in the present review, the classic reference by G. Herzberg, Infrared and Raman Spectra of Polyatomic Molecules (Van Nostrand, Princeton, N. J., 1945) may be consulted.

28. J. P. Aldridge, H. Filip, H. Flicker, R. F. Holland, R. S. McDowell, N. G. Nereson, and K. Fox, J. Mol. Spectry. 58, 165 (1975).

29. R. S. McDowell, H. W. Galbraith, B. J. Krohn, C. D. Cantrell, and E. D. Hinkley, Optics Comm. 17, 178 (1976).

30. R. S. McDowell, J. P. Aldridge, and R. F. Holland, J. Phys. Chem. 80, 1203 (1976).

31. A. R. Edmonds, Angular Momentum in Quantum Mechanics (Princeton U. Press, 1960).

32. H. H. Nielsen, in Handbuch der Physik, edited by S. Flügge (Springer Verlag, Berlin, 1959), Vol. 37, Pt. 1.

33. J. Moret-Bailly, Cahiers Phys. 15, 237 (1961); J. Mol. Spectry. 15, 344 (1965).

34. E. P. Wigner, Group Theory (Academic Press, New York, 1959).

35. E. B. Wilson, Jr., J. Chem. Phys. 3, 276 (1935).

36. C. D. Cantrell and H. W. Galbraith, J. Mol. Spectry. 58, 158 (1975).

37. E. K. Plyler, E. D. Tidwell, and L. R. Blaine, J. Research NBS 64A, 201 (1960).

38. D. H. Rank, D. P. Eastman, G. Skorinko, and T. A. Wiggins, J. Mol. Spectry. 5, 78 (1960).

39. B. Bobin, J. Phys. (Paris) 33, 345 (1972).

40. K. Fox, K. T. Hecht, R. E. Meredith, and C. W. Peters, J. Chem. Phys. 36, 3135 (1962).

41. J. S. Margolis and K. Fox, J. Chem. Phys. 49, 2451 (1968).

42. K. Fox, unpublished.

43. As quoted in W. R. Angus, C. R. Bailey, J. B. Hale, C. K. Ingold, A. H. Leckie, C. G. Raisin, J. W. Thomson, and C. L. Wilson, J. Chem. Soc. (London), 971 (1936).

44. O. Redlich, Z. Phys. Chem. Abt. B 28, 371 (1935).

45. K. Fox, G. W. Halsey, and D. E. Jennings, J. Chem. Phys. 65, 1590 (1976).

46. L. A. Pugh, T. Owen, and K. Narahari Rao, J. Chem. Phys. 59, 1243 (1973); 60, 708 (1974).

ISOTOPE EFFECTS IN MOLECULAR MULTIQUANTUM AMPLITUDES*

C. K. Rhodes (Presented by C. K. Rhodes)

Stanford Research Institute

Menlo Park, California 94025

and

C. D. Cantrell

University of California

Los Alamos, New Mexico 87545

ABSTRACT

The full isotopic character of multiquantum amplitudes is discussed. Of particular importance are the
roles of enhanced isotopic effects generally characteristic of molecular perturbed spectra and the complete utilization of all the available optical field variables.
Examples involving NH_3, CH_3Br, and SF_6 are examined.

Multiquantum excitation of molecules leading to
dissociation and excitation of highly excited vibrational

---
*Research supported by the United States Energy Research
and Development Administration and ERDA contract AT(04-
3)-115.

states is now a commonly observed phenomenon. These
processes have been ovserved under both collision-free
conditions, as well as those involving considerable
collisional interaction. It is also clear that multiquantum amplitudes can be applied generally to nearly
all classes of molecular systems. This generality has
been explicitly demonstrated in experiments on ammonia,
formaldehyde, sulphur hexafluoride, boron tricholoride,
and osmium tetraoxide, to mention a few of the molecules
with which direct experimental results have been obtained.
Strong isotopic signatures in these excitation mechanisms
have also been observed, with results reported on sulphur
hexafluoride, osmium tetraoxide, formaldehyde, boron
tricholoride, and molybdenum hexafluoride. In this
analysis we will examine the ways in which multiquantum
amplitudes exhibit their isotopic character. We will
emphasize the manner in which the various optical degrees
of freedom under our control can be utilized to optimize
the isotopic differentials expressed by the molecular
system. These considerations involve a mapping of the
optical degrees of freedom onto the relevant motions of
the molecular species. As a prelude to this discussion,
however, we will begin with a review of some simple properties of electromagnetic fields and molecular systems.

Properties of Molecular Systems and Radiation Fields

Molecular systems and the radiation field both contain intrinsic degrees of freedom. We will now examine
those associated with the electromagnetic field. Consider the amplitude describing the coupling of n quanta
to an atomic and molecular system, a process illustrated
graphically in Fig. (1). The amplitude A_n for this process is represented in lowest order by the relation

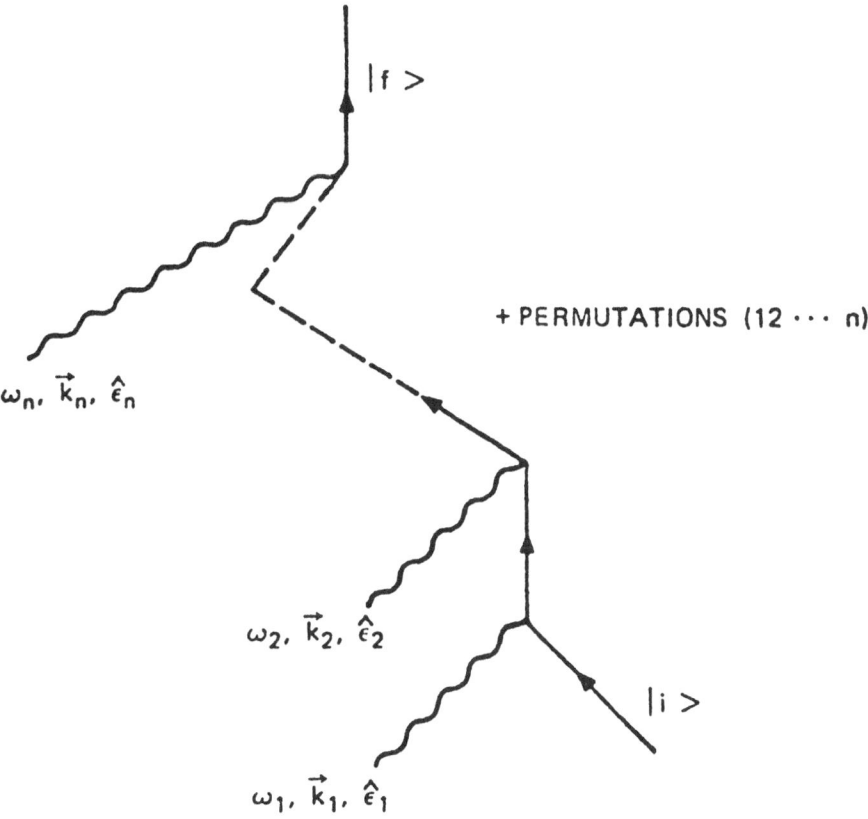

Figure 1.   Diagram of general n^{th} order amplitude.

$$A_n = \sum_{P(123\ldots n)} \sum_{I_{n-1}} \sum_{I_1}$$

$$\frac{\langle f|H_n|I_{n-1}\rangle\langle I_{n-1}|I\rangle\ldots\langle I_1|H_1|I\rangle}{\Delta\omega_{n-1}\cdots\cdots\Delta\omega_1}, \qquad (1)$$

where P(123...n) designates the permutations of (123...n)
which correspond to the additional graphs noted in Fig.
(1). The electromagnetic field variables entering into
the general structure of this amplitude are collected
in Table I. On account of energy conservation, a relationship must govern a linear combination of the frequencies, reducing by one the choices available for the
n frequencies ω_n. Naturally, the intensities of all
waves are independent, contributing n degrees of freedom. On account of the isotropy of space the polariza-
tion $\hat{\epsilon}_n$ and wavevectors $\vec{k}_n$ each involve a reference direction so that they both contribute n-1 degrees of free-
dom. The total number of available degrees of freedom
N is found to be 4n-3, a property which is related linearly to the number of fields n. Therefore, the expression

$$N = 4n-3 \qquad (2)$$

states that each additional field contributes in principle
four control parameters, with the largest relative in-
crement occuring between n = 1 and n = 2. This leads to
the conclusion that the use of a single laser in general
represents the least optimal utilization of the intrinsic

Table 1

PROPERTIES OF ELECTROMAGNETIC VARIABLES
ENTERING INTO A GENERAL nth ORDER
MULTIQUANTUM RADIATIVE AMPLITUDE

Variables	Total Number of Degrees of Freedom	Comments
Frequencies $\omega_1, \omega_2, \ldots, \omega_n$	n-1	One linear combination
Intensities $I_1, I_2, I_3, \ldots, I_n$	n	All independent
Polarizations $\epsilon_1, \epsilon_2, \ldots, \epsilon_n$	n-1	One reference
Wave vectors $\vec{k}_1, \vec{k}_2, \ldots, \vec{k}_n$	n-1	One reference
TOTAL	4n-3	————————

degrees of freedom of the optical fields. This general
advantage in the use of several carefully selected waves
suggests that the development of efficient frequency
conversion processes will play an important role in the
development of optical radiative coupling.

We now examine the intrinsic degrees of freedom
associated with a molecular vibrator. If we neglect
electronic coordinates, a reasonable assumption for an
enormous range of chemical phenomena in which systems
react in their ground electronic states, a molecule
composed of M bodies has a total of 3M degrees of freedom. If we further neglect the overall center of mass
motion, which accounts for three, and the rotational
motions, accounting for an additional three, the remaining total degrees of freedom F is given by[1]

$$F = 3M - 6. \qquad (3)$$

With these assumptions we are explicitly taking into
account only the molecular vibrations, a procedure which
is valid if the neglected degrees of freedom are sufficiently weakly coupled in the process of interest.
In this sense we note, for example, that the neglect of
the electronic degree of freedom becomes invalid at
surface crossings, since the Born-Oppenheimer approximation fails in those regions and a more complicated analysis is required.[2] We further recognize that molecular
symmetry properties vary substantially and for the same
M higher symmetry groups such as O_h generally represent
simpler systems[3] than lower symmetry ensembles such as
C_s. Nevertheless, expression (3) provides us with a
rough measure of the complexity of a given molecular
system.

Field-Molecule Coupling

We now explore the properties associated with the coupling of these two systems, the radiative field and molecular species in a multiquantum process which maximizes the isotopic character of this coupling. In order to explicitly represent the different isotopic influences we write the total Hamiltonian of the molecular system in its matrix form, as shown by equation (4)

$$H = \begin{pmatrix} - & - & - & - & - & - \\ - & - & - & - & - & - \\ - & - & - & - & - & - \\ - & - & - & - & - & - \\ - & - & - & - & - & - \\ - & - & - & - & - & - \end{pmatrix} . \qquad (4)$$

In this way we can explicitly represent two regions of the Hamiltonian, the diagonal elements H_{ii}, representing energy eigenvalues, and the off-diagonal elements H_{ij} ($i \neq j$), corresponding to interaction energies. In general, the former explicitly represent an array of numbers for each vibrational-rotational energy level and in principle include an enumeration of hyperfine states. The off-diagonal elements constitute the transition matrix elements linking states coupled by an external electromagnetic field or possibly an intramolecular interaction characteristic of the molecule. This latter aspect is a feature normally associated with perturbed molecular spectra, a point which will be expanded in the discussion below.

It is immediately recognized from this simple pictorial representation of the Hamiltonian that the diagonal entries constitute a limited one-dimensional world, whereas the off-diagonal matrix elements form a much larger two-dimensional array. This suggests that we do not want

to ignore isotopic effects that arise <u>off</u> the diagonal,
since the greater range of opportunities may result in
instances in which they contribute the main isotopic
signature. As we will show below, the total isotopic
character of a particular amplitude is maximized when
contributions from the <u>whole</u> Hamiltonian, involving
both diagonal and off-diagonal elements are simultaneously utilized. We now specifically discuss these
two contributions.

Diagram Elements

The diagonal elements of the Hamiltonian describe
the energy eigenvalues of the vibrational-rotational
states including the effects of hyperfine structure.
For example, the difference between two of these entries
for an optically allowed transition would represent the
center frequency of that line for a particular hyperfine
component. The linewidth associated with this transition
would generally involve both Doppler and collisional
contributions, the radiative or natural width typically
being negligible in the infrared, the spectral region
dominating our interest.

The isotopic character of these diagonal elements,
and hence, of the transition frequencies arises from
three contributions. Firstly, the isotopic mass difference alters the vibrational frequency. For a diatomic
this factor scales as the square root of the reduced
mass. Polyatomic structures have a more complicated
scaling relationship, depending on the nature of the
normal mode.[1] Secondly, rotational energies are affected
through a change in the molecular moment of inertia.
The change in the rotational constant arises directly from
the alteration of the moment of inertia stemming from
the substitution of a different isotopic mass. This

contribution scales <u>linearly</u> with the reduced mass. In
addition, the term which involves the vibrational-rotational interation will be altered, since the root mean
squared vibrational excursion is now changed by the
isotopic substitution. Thirdly, hyperfine intervals are
also altered. Molecular hyperfine shifts arise from
both electric and magnetic interactions. Generally, in
molecules whose spins are paired in an electronic state
with vanishing orbital angular momentum, the electric
effects dominate. The principle electric hyperfine component is electric quadrupole hyperfine structure, which
depends on the quadrupole moment of the nucleus, the
gradient of the electric field at the nucleus, and hence,
on the character of the bond, and the angular momentum
quantum numbers of the molecular state.[4] Electric hyperfine splittings have been observed in the infrared for
both methyl chloride[5] and osmium tetraoxide.[6]

Off-Diagonal Elements

Off-diagonal matrix elements of the Hamiltonian
govern intramolecular interactions and transition rates
induced between states by external fields. Therefore,
they contain the character of both of the participating
states as well as the operator connecting the levels.
These elements can possess sizeable isotopic variations
and exhibit particularly strong influence in perturbed
spectra. In these perturbed spectra isotopic shifts
arise from intramolecular perturbations of system -- i.e.
interactions of the nearby levels which are not included
in the lowest order Hamiltonian. In this way interactions
among the available modes of excitation characteristic
of an atomic or molecular system can acquire an anamolous
isotopic character. These aspects will be discussed in
detail below with respect to CH_3Br and other examples.

The strength of this interaction depends on several variables, including the nature of the interaction and the magnitude of the coupling. For example, these factors will be influenced by anharmonic terms in the molecular Hamiltonian, Coriolis interactions, and external electromagnetic fields. Normally, these interaction energies are compared to the scale of the energy difference separating the interacting levels in the absence of the perturbing influence. This comparison manifests itself physically in a degree of state mixing given approximately by the ratio of the magnitude of the relevant interaction energy to the energy separating the participating levels. Moreover, these perturbations often have a resonance character, which can be appreciably influenced through isotopic substitution. Therefore, these phenomena can result in greatly enhanced isotopic effects as compared to the "normal" consequences of isotopic substitution.

In our analysis of the isotopic effects of off-diagonal elements, we will concentrate on those associated with perturbed spectra as they represent the most striking manifestation of these effects.

Perturbed Spectra

As noted above, perturbed spectra exhibit enhanced isotopic effects. These phenomena occur generally in atomic and molecular systems. An example of a specific case involving an atom is given by Hg. The partial spectrum of the mercury atom is illustrated in Fig. (2), which shows the allowed $^1S_0 - ^3P_1$ 2537 Å. The former is allowed through the strong spin-orbit interaction characteristic of a heavy atom such as Hg, while the latter is forbidden in all approximations by a one-photon process by angular momentum considerations[7] ($0 \nleftrightarrow 0$). Since

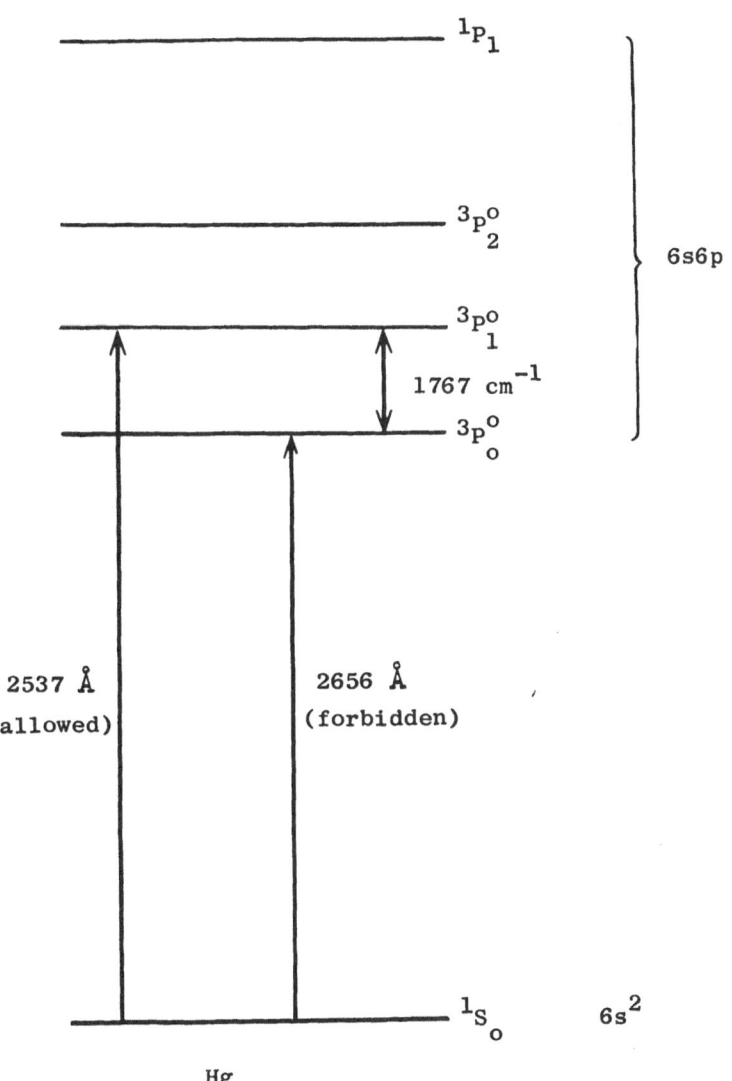

Figure 2. Partial spectrum of the Hg atom illustrating
the lowest excited states of the 6s 6p configuration.

the state parities are appropriate for an E1 transition,
the presence of an additional component of angular momentum coupled to the electronic structure of the Hg
will relax the selection rule preventing the $^1S_0 \rightarrow {}^3P_0$
transition. The necessary angular momentum can be
furnished by collisional interactions with other systems
or by the intrinsic spin of the Hg nucleus. Even a Hg
isotopes have spin zero (I = 0), and hence, cannot provide the additional angular momentum. However, odd A
isotopes such as Ag^{199} and Hg^{201} which have nonzero
values of nuclear spin generate the appropriate state
mixing through the hyperfine term[7] for an E1 amplitude
on the forbidden transition. Therefore, in a mixture
consisting of Hg^{200} and Hg^{201}, the Hg^{200} component would
show an absorption at 2656 Å with the nearest Hg^{200} transition at 2537 Å. This represents an effective isotope
shift[8] of ~ 1767 cm^{-1} as opposed to the normal isotope
shift[9] exhibited by the 2537 Å transition of ~ 0.8 cm^{-1}.
As pointed out by Sobel'man[8], this hyperfine effect results in a cross section of ~ 10^{-19} cm^2.

This simple atomic example illustrates the manner
in which enhanced isotope effects can be produced by the
use of transition moments in addition to energy level
shifts. Similar effects are also present in molecular
systems and are, in fact, a general feature of perturbed
molecular spectra. Such perturbed spectra are well known
in a large number of molecular species.[10] A specific
case for which accurate spectroscopic data are available[11]
is CH_3Br. Other cases involving spherical top systems
are examined in a latter section.

Perturbed Spectra in CH_3Br

Graner has performed an analysis[11] of the Coriolis
interaction of $2\nu_5$ with $\nu_2 + \nu_5$ in both $CH_3{}^{79}Br$ and $CH_3{}^{81}Br$

In these molecules the normal isotope shifts are considerably less than a wave number, but <u>resonant</u> Coriolis mixings couple the $2\nu_5$ and $\nu_2 + \nu_5$ states strongly at differing rotational quantum numbers which correspond to much greater energy differences than the normal mode isotopic shift. This energy splitting between the two isotopic materials is as much as thirty-fold the "normal" vibrational isotopic effect. Indeed, for certain transitions, such as the $^Q Q_0(J)$ lines, the $CH_3^{79}Br$ exhibits appreciable mixing, but $CH_3^{81}Br$ does not show a resonant perturbation. Consequently, at the appropriate energy or corresponding spectral region, transitions in the ^{79}Br species occur with anomalous strength as compared with the ^{81}Br material, with the result that the locally absorbing regions of these two molecules have isotopic gaps (splittings) which are substantially greater than the normal harmonically derived isotopic shifts. Therefore, these perturbations generate complicated, but characteristic isotopics patterns in <u>both</u> level positions and transition moments. These two effects can be simultaneously utilized in a variety of ways, some of which, including the optical Stark effect, will be outlined below.

<u>Isotopic Character of the Optical Stark Effect in Molecular Systems</u>

Off-diagonal elements of the Hamiltonian, whether arising from perturbed spectra or otherwise, can influence the relative spacings of the energy levels through the optical Stark effect. Effects of this nature have been experimentally demonstrated for two-photon couplings, and are generally present in nonlinear amplitudes. The shifts ΔE_n arising from the optical Stark effect are given in the lowest order analysis by the expression[12]

$$\Delta E_n = \tfrac{1}{4} \sum_m \left\{ \frac{|\vec{\mu}_{mn} \cdot \vec{\xi}|^2}{E_n - E_m - h\omega} + \frac{|\vec{\mu}_{mn} \cdot \vec{\xi}|^2}{E_n - E_m + h\omega} \right\} , \quad (5)$$

in which the subscript m designates the intermediate states, ξ is the optical electric field strength, ω is the optical frequency, and μ_{mn} is the electric dipole transition moment. For a two-quantum process, the shift arises from two separate contributions, each corresponding to the pair of diagrams illustrated in Fig. (3). We note that if a particular intermediate state is sufficiently near resonance, the summation is dominated by a single term. This often represents the most interesting case, since it combines a large shift with a sensitivity to the energy level pattern of the molecular system.

We observe that expression (5) above combines the parameters associated with the electromagnetic field with both diagonal and off-diagonal terms of the Hamiltonian, thereby condensing into a single expression all the degrees of freedom at our disposal. The radiation field terms are tools of the experimentalist, while the energy term values and transition moments are characteristic of the molecular system and embody and full isotopic character of that molecule. We will now explore a specific example involving NH_3.

With the definitions of the parameters[13] for NH_3 given in Fig. (4), the equations

$$E_1 - E_2 = -(\omega_1 + \Delta_1), \quad (6)$$

$$E_3 - E_2 = \omega_2 - \Delta_1 + \delta, \quad (7)$$

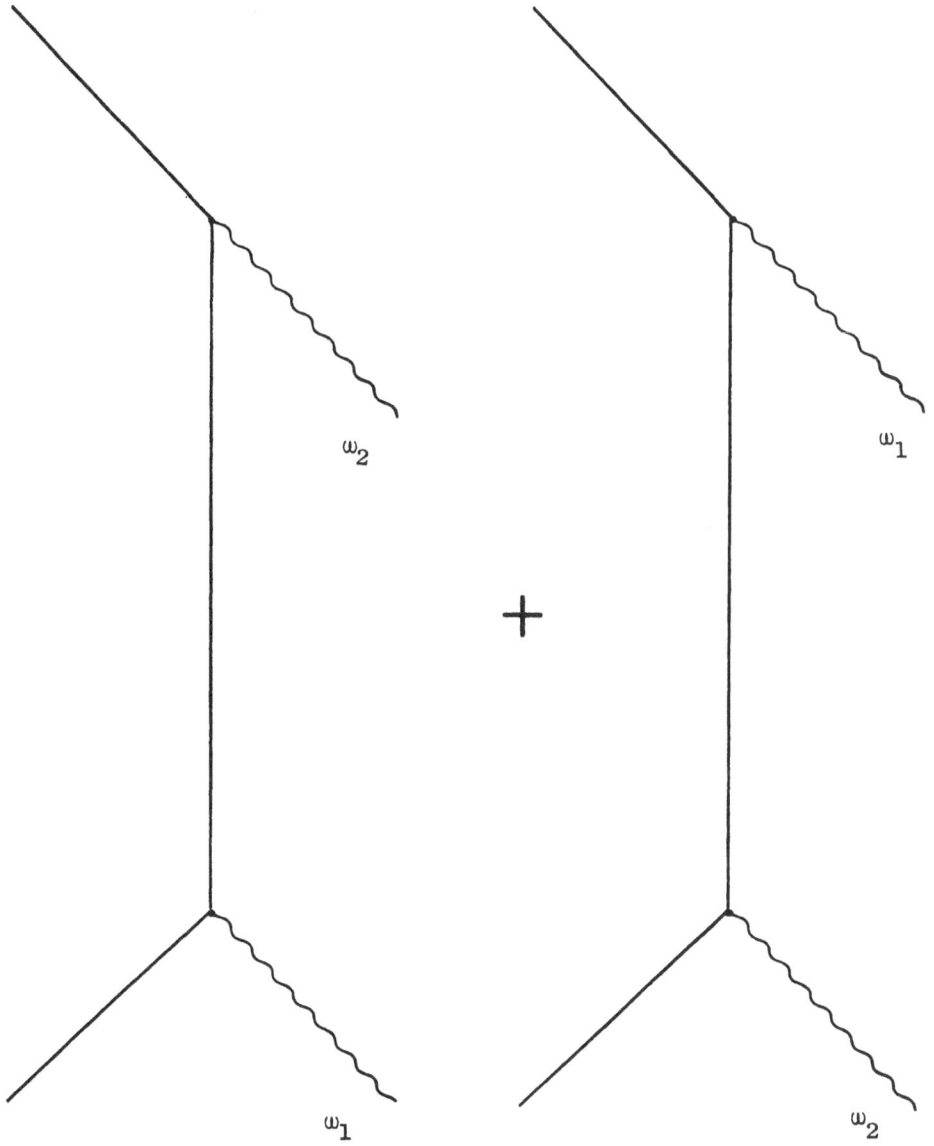

Figure 3.   Diagrams representing two quantum absorption.

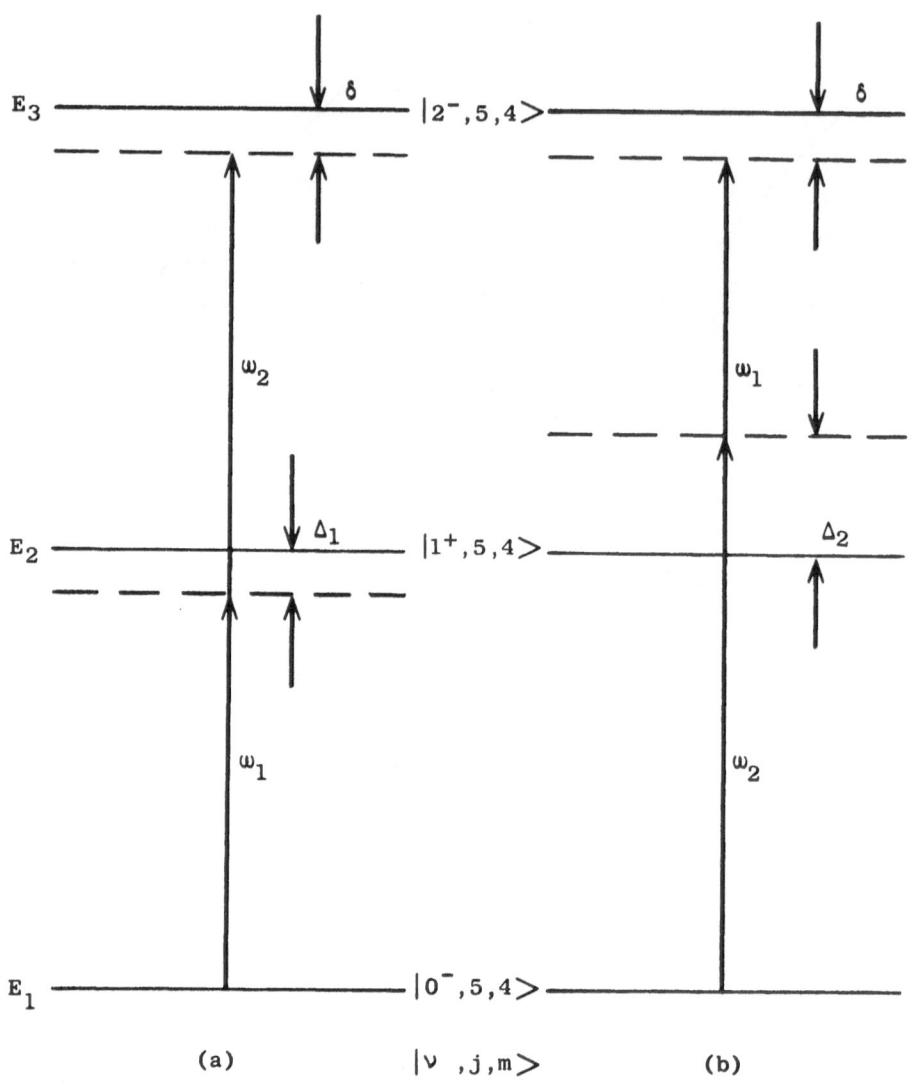

Figure 4.   Diagram of the two terms governing two quantum absorption in NH_3 on the 10^-, 5, 4 > → 12^-, 5, 4 > transition after the work of Bischel, Kelly, and Rhodes cited in Ref. (13).

follow. If we assume that δ/Δ_1 is significantly greater than Δ_1/Δ_2, a valid approximation in this case[13], we can arrive at the simple estimates

$$\Delta E_1 \simeq -\tfrac{1}{4} \frac{(\vec{\mu}_{12} \cdot \vec{\xi}_1)^2}{\Delta_1} , \qquad (8)$$

and

$$\Delta E_3 = -\tfrac{1}{4} \frac{(\vec{\mu}_{23} \cdot \vec{\xi}_2)^2}{\Delta_1 - \delta} . \qquad (9)$$

In expressions (8) and (9) the $\vec{\mu}_{ij}$ are the transition moments connecting their respective levels and $\vec{\xi}_1$ and $\vec{\xi}_2$ are the amplitudes of the two waves. If we now define the total shift S as

$$S = \Delta E_3 - \Delta E_1 , \qquad (10)$$

we can now write

$$S \simeq \frac{2\pi}{c} \left[\frac{\mu_{12}^2}{\Delta_1} I_1 - \frac{\mu_{23}^2}{\Delta_1 - \delta} I_2 \right] , \qquad (11)$$

in which I_1 and I_2 are the intensities of the waves at frequencies ω_1 and ω_2 respectively. This result states that I_2 is linearly related to I_1 for a constant shift S. Rewriting equation (11) we obtain

$$I_2 = \left(\frac{Sc}{2\pi} \right) \left(\frac{\Delta_1 - \delta}{\mu_{23}^2} \right) + \left(\frac{\mu_{12}}{\mu_{23}} \right)^2 \left(\frac{\Delta_1 - \delta}{\Delta_1} \right) I_1 . \qquad (12)$$

We notice that this result contains the spectroscopic
parameters S, Δ_1, δ_1, as well as the matrix elements
μ_{12} and μ_{23}, __all__ of which may be influenced by isotopic
effects. Several special cases can now be examined,
recalling the matrix elements μ_{ij} and the spectroscopic
parameters δ and Δ_1, and S are all influenced by various
intramolecular coupling mechanisms such as Fermi resonance, a resonant anharmonic effect, and Coriolis
coupling.

     Although isotopic effects are not essential in
the example of NH_3, this molecule does serve as a convenient model for an understanding of the effects which
could be exploited in other, presumably more structurally
complex systems. Therefore, the quantitative aspects
of this case are important fiducial marks for scaling
the application of these phenomena to other molecules.

     For NH_3, the transition illustrated in Fig. (3)
has the parameter[13] δ = 294 MHz, Δ_1 = 0.165 cm^{-1}, Δ_2 =
14.9 cm^{-1}, ω_1 = 1033.488 cm^{-1}, and ω_2 = 1048.661 cm^{-1}.
According to expression (11), a shift in the resonance
frequency can be induced by the applied optical fields.
There it is possible to achieve the required resonance
condition by illumination with the appropriate optical
intensities, a fact which has already been utilized in
optical down-conversion.[14] For the parameters given
above, the resonance condition is established at an intensity of approximately 1 MW/cm^2, a value consistent
with unfocused CO_2 TEA lasers.

     The relative polarization of the optical fields is
a parameter which influences both the absorption strength
and the optical Stark shift. This dynamical characteristic, which exhibits itself through the variation of
the direction cosine matrix elements, provides a

considerable range of adjustment of the parameters in
equation (11). We emphasize that the influence of
polarization has been experimentally demonstrated in
previous studies[15] on CH_3F confirming the presence of
these effects.

Enhanced Isotopic Selectivity

Enhanced isotopic selectivity is particularly important in heavy and complex molecular systems. In
those cases selective radiative processes commonly
have two fundamental difficulties; namely,

1) isotopic splittings are often small, and

2) there often exists a high density of states
   leading to broad overlapping absorption
   profiles.

These two factors combine their pernicious effects,
since the high density of states generally overlaps the
small isotopic shifts and consequently reduces or eliminates the isotopic selectivity available by direct absorption in an allowed band. In order to restore isotopically selective excitation, we need to

1) enhance the magnitude of the effective isotopic
   splittings, and

2) reduce the density of competing states coupled
   to the radiation field.

A suitable combination of intramolecular perturbations and nonlinear optical phenomena with emphasis on
full utilization of the optical fields, including intensity, polarization, frequency, and spatial coherence,
can be used to achieve these objectives.

In this context we examined above two specific examples embodying the relevant characteristics, one
illustrating the optical Stark effect noted above with
NH_3 and the second demonstrating enhanced isotope shifts

arising in perturbed spectra of $CH_3^{79}Br$ and $CH_3^{81}Br$.

We assert that the confluence of these effects in multiphoton amplitudes for complex molecules can enable a sizeable enhancement of the isotopic selectivity in these processes.

Multiphoton Dissociation

Considerations of this nature can clearly represent themselves in multiquantum molecular dissociation. This aspect has been discussed[16] in the context of experiments on SF_6. Both anharmonic and Coriolis terms in the Hamiltonian can sufficiently modify the term values of the vibrational manifolds that an appreciable isotopic signature is developed, <u>even though the bands constitute</u> <u>completely overlapping ensembles of transitions in the</u> <u>relevant spectral region</u>. The modification of term values implies on fundamental grounds a corresponding isotopic signature in the transition moments as well.

As noted earlier, collisionless multiple-photon excitation (CMPE)[17-21] has by now been demonstrated in a large number of polyatomic molecules, although not yet in diatomic or small asymmetric-top molecules.[22-24] Indeed, CMPE is unavoidable in some symmetrical polyatomic molecules, such as sulfur hexafluoride. Fig. 5 shows the dependence of the energy absorbed per unit area upon the energy incident per unit area for a sample of sulfur hexafluoride irradiated by CO_2 laser pulses of 1.4 ns duration.[25] The experimental result obtained at the highest incident energy per unit area corresponds to an absorption of approximately twenty photons per molecule. The temptation to give such a remarkable process a scintillating name has been sufficient to evoke the proposed term superexcitation,[26] although this term has been used previously in a related area of molecular physics.[27]

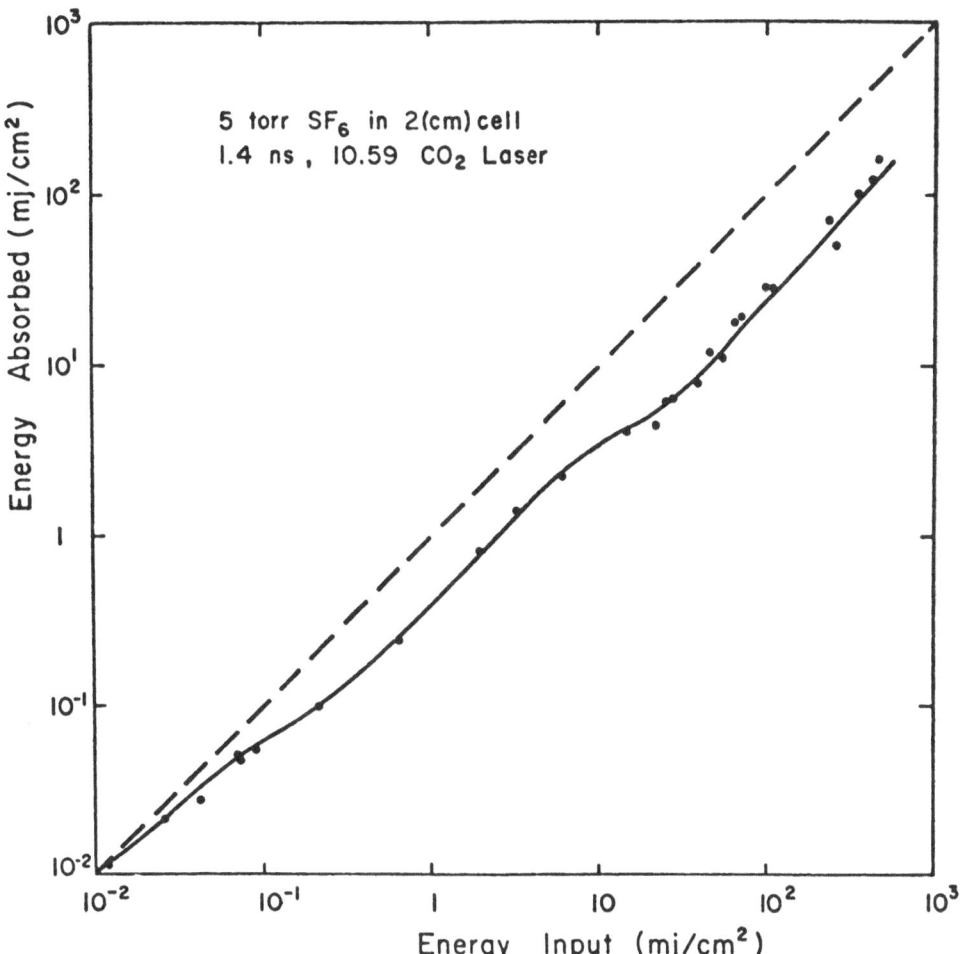

Figure 5. "Absorbed <u>versus</u> incident fluence of 1.4 ns CO_2 laser pulses in 5 torr of SF_6".

It is clear that, whatever the terminology, the possibility of selectively and coherently depositing substantial quantities of energy in molecules on a sub-
collisional time scale is a general phenomenon of considerable importance.

The detailed energy-level structure of the polyatomic molecules in which CMPE has been observed unquestionably plays an important role, analogous to the
effects we have already discussed above for two-photon
processes. For the sake of definiteness we shall concentrate our attention on spherical-top molecules such
as SF_6 and CF_4, although many of our comments will apply
equally well to symmetric-top molecules. In Fig. 6 we
show qualitatively the effects which determine the
structure of the higher energy levels of a spherical-
top molecule.[28-30] The degenerate excited vibrational
levels of a spherical-top molecule are, in general,
split by the anharmonic potential energy into as many
levels as the symmetry of the molecule allows. The details of this splitting are discussed for CH_4 by K.
Fox elsewhere in this volume.[29] The anharmonic splitting
is large compared to the rotational splitting in the
molecules of interest, and probably assists substantially
in the absorption of many infrared quanta by a single
molecule. Although most of the published calculations
of anharmonic splitting assume a power-series expansion
of the anharmonic potential energy and are therefore
limited to low-lying vibrational states, some initial
steps have been taken towards calculating the vibrational
structure of highly excited states.[30]

The exact role of highly excited vibrational states
in the process of CMPE is still a matter of debate. Many
authors argue that CMPE depends upon a "quasi continuum"

of vibrational energy levels, in which essentially any
laser frequency within a broad bandwidth may be resonantly absorbed.[17,20,31,32] Others[16] have suggested
that anharmonic splitting combined with rotational
structure may allow the resonant absorption of a very
large number of quanta without any need to invoke a
quasi continuum. This point is discussed in more detail elsewhere in this volume by H.W. Galbraith and one
of us. In our opinion discrete states are unquestionably
responsible for the absorption of at least five quanta,
and probably many more, in the CMPE of SF_6.

The rotational structure of the higher vibrational
states of a spherical-top molecule is a classic un-
solved problem of molecular spectroscopy, despite strong
attacks upon particular aspects of the problem in the
past.[28,29] However, enough is known that some general
comments may be made. In Fig. 6 the Coriolis interaction energy

$$2B \ \zeta \ \vec{J} \cdot \vec{L} \qquad (13)$$

is somewhat larger than the tensor vibration-rotation
interaction, which arises from the changes of the molecular moment-of-inertia tensor induced by molecular vibration. However, both of these effects can have dramatic
effects upon the energy levels and spectrum of a molecule (such as SF_6) in which very high angular momenta
are excited thermally at the temperatures at which ex-
perimentation is convenient. In a vibrational mode
(such as ν_4 of CH_4)[29] in which the square of the vibrational angular momentum, $\vec{L}^2$, is a good quantum number,
we may denote the states as[16]

$$| \; v \; \ell \; J; \; R \; K_R >, \qquad\qquad (14)$$

where v is the total number of vibrational quanta of
the mode in question; $\ell(\ell+1)$ is the eigenvalue of $\vec{L}^2$;
$J(J+1)$ is the eigenvalue of the square of the total-
angular-momentum operator, $\vec{J}^2$; R is the angular momentum
to which J and ℓ must be coupled[33] in order to diagonalize the Coriolis energy; and K_R is the eigenvalue of
the component of $\vec{R}$ along a molecule-fixed z-axis. The
energy eigenvalues are, if we neglect the tensor vibration-rotation interaction,

$$E(v,\ell,J_R) = v\nu_0 + v(v-1)X + [\ell(\ell+1)-2v]G$$

$$+BJ(J+1) + B\zeta[R(R+1) = J(J+1) - \ell(\ell+1)]. \qquad (15)$$

In this equation, G and X are paramters in the vibrational
anharmonic Hamiltonian, multiplying (respectively)$\vec{L}^2$
and a spherically symmetric contribution derived by a
contact transformation[25] from $(q_x{}^2+q_y{}^2+q_z{}^2)^2$; B is the
familiar geometrical rotation constant; and R is the
purely rotational angular momentum. In the foregoing
equation, each overtone level has been split into $[\frac{v}{2}]$
+ 1 vibrational levels by the Coriolis interaction.

     The effect upon the frequencies of allowed transi-
tions is also striking. The selection rules for the
transitions v → v+1 are: R → R; ℓ → ℓ ± 1 for $\ell \neq 0$
(which we shall denote by a superscript ± in the following); J → J - 1 ($P^{\pm}$ branch), J → J+1 ($Q^{\pm}$ branch), or J
→ J+1 ($R^{\pm}$ branch). Thus, for v>1 there are two, P, Q,
and R branches for the most states with ℓ>0. For v = 1
→v = 2 several branches are missing, as are branches such

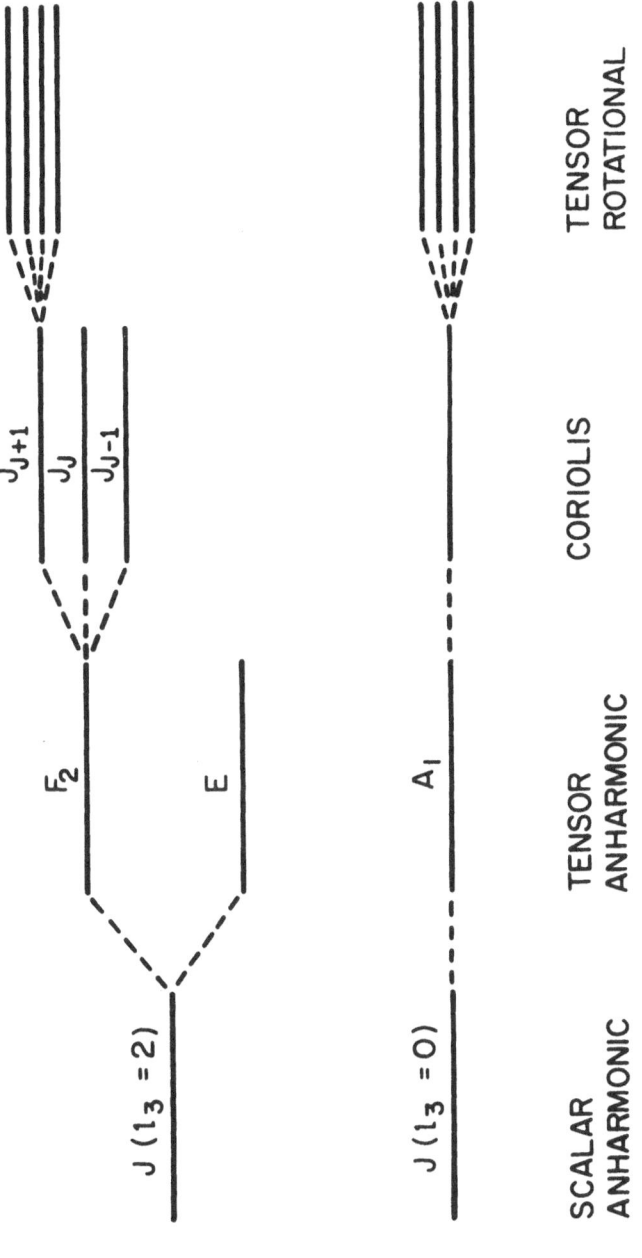

Figure 6. Schematic diagram of the splitting of vibration-rotation energy levels in a spherical-top molecule by anharmonic and vibration-rotation interactions.

as $(J = R + \ell \rightarrow J + 1; \ell \rightarrow \ell-1)$ in the general case.
The transition frequencies are given by the relation

$$\nu(v, \ell^{\pm}, M) = \nu_0 + 2vX \pm (2\ell + 1 \mp 1)G$$

$$+ 2B(1-\zeta)M \mp B\zeta(2\ell + 1 \pm 1), \qquad (16)$$

where the index M takes on the values $- J$ for the $P(J)$
line, zero for the $Q(J)$ line, and $J+1$ for the $R(J)$ line.
The spacing between successive rotational lines in such
a spectrum is the same (within the limit of validity of
the approximations we have made) as the $v = 0 \rightarrow v = 1$
transition, namely, $2B(1-\zeta)$. However, the frequencies
of the multiple P, Q, and R branches are very different
from what one would expect on the basis of simply employing an "effective" value of the rotational constant B,
together with the purely vibrational energy levels.
There is growing evidence [16,34] that the vibrational
Hamiltonian for the ν_3 modes of an octahedral molecule
such as SF_6 is not dominated by the term $G\vec{L}^2$, but rather
by an octahedrally (but not spherically) symmetric contribution derived by contact transfoundation of $(q_x^4 +$
$q_y^4 + q_z^4)$. Physically, this means that the three ν_3
modes become anharmonic in a manner which results in
very little nonlinear interaction among them. If this
proves to be correct, then the picture presented in the
last paragraph must be substantially modified. The appropriate vibrational wavefunctions are the Cartesian
basis $|n_x, n_y, n_z>$ with n_i quanta in the ith component
of the ν_3 mode. The energy levels in this case are
described elsewhere in this volume.[35] The Coriolis interaction gives rise to a different spacing of rotational

energy levels for each of the vibrational levels illustrated in Fig. (2) of the article by Galbraith and
Cantrell.[35] This will lead to rotational perturbations
in some cases and not in others, as we have already described for the simpler example of CH_3Br noted earlier.
Here, too, the existence of rotational perturbations,
or the value of J at which a perturbation starts to
occur, may well be dependent upon the isotopic con-
stitution of the molecule.

In addition to the Coriolis interaction, the
tensor vibration-rotation (VR) interaction (Fig. 6)
plays an important role in the multiple-photon dynamics
of a spherical-top molecule. Its effect is to split
each Coriolis level in Fig. 6 into as many states as
are allowed by the symmetry of the molecule. The consequences of such splittings can be dramatic: for the
SF_6 (v = 0 → v = 1) absorption lines near the CO_2 P(22)
laser line, the tensor VR interaction splits each rotational "line" of SF_6 into components which are spread
over ~ 0.8 cm^{-1}, while the spacing between successive
rotational "lines" is only 0.0555 cm^{-1}.[36] Further, a
strong external optical electric field is capable of
mixing the different symmetry types present in such a
"line", thus making transitions allowed which are forbidden in a vanishingly small electric field. Since
this phenomenon of optical-field-induced symmetry mixing
affects both the upper and lower states of a given transition, it also leads to power broadening in the sense
that the absorption of a second laser frequency to the
same upper state can be affected by the intensity of
the first laser.

Other phenomena already discussed are important for
multiple-photon transitions in spherical-top molecules:

optical Stark shifts, which alter the detuning of a
transition from resonance in a power-dependent way; and
power broadening in the ordinary sense, resulting from
nonlinear saturation of the upper level of a transition.
The power broadening of multiple-photon transitions,
which is somewhat more complex than in the two-level
systems usually considered, has recently been discussed
by Larsen and Bloembergen.[37]   Finally, even in a harmonic
oscillator, which is incapable of saturation in the
usual sense, the interval in frequency over which it is
possible to excite a given (fixed) level is directly
proportional to the incident laser intensity.[23]

Clearly both the selectivity and strength of excitation of a given molecular level can be greatly enhanced by choosing more than one incident laser frequency,
in order to achieve near-resonance with the transitions
depicted in Fig. 1 of Galbraith and Cantrell,[35] while
avoiding the deleterious power-dependent effects just
described. Selectivity in such a process does not depend upon differences in gross properties such as the
isotopic shift of the vibrational frequency of the mode
being excited. The many spectral perturbations to be
expected in spherical-top and symmetric-top molecules
will inevitably lead to shifts of the frequencies and
strengths of transitions which can be exploited for
selective multiple-frequency excitation. Recently, the
isotopically selective laser photo-dissociation of
ethylene has been reported, using an infrared band which
lacks a vibrational isotope shift.[38] Possibly this observation is an experimental illustration of the principles we have discussed. Regardless of the interpretation of this particular observation, we expect that the
purposeful exploitation of these principles will not be
long in coming.

REFERENCES

1. This expression becomes F = 3M-5 for the special case of a linear molecule. See G. Herzberg, Mole- cular Spectra and Molecular Structure II, Infrared and Raman Spectra of Polyatomic Molecules (D. Van Nostrand Co., Inc., Princeton, N. J. 1945).

2. G. E. Zahr, R. K. Preston, and W. H. Miller, J. Chem. Phys. 62, 1127 (1975).

3. For the case M = 7, F = 15, a figure valid for C_s symmetry, while an O_h system has only 6 vibrational fundamentals on account of the degeneracy associated with that highly symmetric point group. These con- siderations will change quantitative comparisons, but not general scaling relationships.

4. C. H. Townes and A. L. Schawlow, Microwave Spectro- scopy (McGraw-Hill, New York, 1955); W. Gordy and R. L. Cook, Microwave Molecular Spectra (Wiley- Interscience, New York, 1970).

5. T. W. Meyer, J. F. Brilando, and C. K. Rhodes, Chem. Phys. Lett. 18, 382 (1973).

6. O. N. Kompanets, A. R. Kukudzhanov, V. S. Letokhov, V. G. Minogin, and E. L. Mikhailov, Zh. Eksp, Teor. Fiz. 69, 32 (1975); Eng. Transl.: [Sov. Phys. - JETP 42, 15 (1976)].

7. R. H. Garstang in Atomic and Molecular Processes, edited by D. R. Bates (Academic Press, New York, 1962) p. 1; C. B. Moore, Accts. Chem. Res. 6, 323 (1973).

8. Ya. B. Zel'dovich and I. I. Sobel'man, JETP Lett. 21, 168 (1975).

9. K. R. Osborn, C. C. McDonald, and H. E. Gunning, J. Chem. Phys. 26, 125 (1957).

10. Y. Y. Kwan and E. A. Cohen, J. Mol. Spectrosc. $\underline{58}$, 54 (1975), (Coriolis ν_t and $2\nu_n$ for C_3v, e.g. NH_3 $\nu_4 = 1$); G. Graner, J. Molec. Spectrosc. $\underline{51}$, 238 (1974), ($2\nu_5(A_1)$ CH_3Br near 2860 cm^{-1}, anharmonic/ Coriolis); C. Betrencourt-Stimemann, G. Graner, and G. Guelachvili, J. Molec. Spectrosc. $\underline{51}$, 216 (1974), (CH_3Br, $\nu_4 \sim 3100$, $\nu_3 + \nu_5 + \nu_6$, both ^{79}Br and ^{81}Br); J. L. Duncan, A. Allan, and D. C. McKean, Mol. Phys. $\underline{18}$, 289 (1970), (CH_3Cl, CH_3Br, CH_3I with ^{12}C and ^{13}C, anharmonic/centrifugal); D. C. McKean, Spectro-chim. Acta $\underline{29A}$, 1559 (1973) (CH_2D spectra and Fermi resonance); N. Brosari-Zizi, C. Alamichel, et C. Amiot, Mol. Phys. $\underline{27}$, 1491 (1974), (CH_3Br $\nu_4 + \nu_6$, ~ 0.005 cm^{-1}, Coriolis); D. R. Anderson and J. Overend, Spectroschim. Acta $\underline{27A}$, 2013 (1971), (CH_3Br ν_3, 0.03 cm^{-1}, rotational distortions); P. K. I. Yin and K. N. Rao, J. Mol Spectrosc. $\underline{51}$, 199 (1974), (PH_3, ν_2 and ν_4, Coriolis); A. G. Maki, J. Molec. Spectrosc. $\underline{57}$, 416 (1975), (O_3, $\nu_1 + \nu_3$ ~ 2100 cm^{-1}, Coriolis); M. S. Child and H. C. Longuet-Higgins, Proc. Roy. Soc. $\underline{A254}$, 259 (1961), (Jahn-Teller); C. DiLauro and I. M. Mills, J. Mol. Spectrosc. $\underline{21}$, 386 (1966), (CH_3F, $\nu_2 - \nu_5$ and $\nu_3 - \nu_6$; CD_3Cl, $\nu_2 - \nu_5$; Coriolis 2nd order); R. S. McDowell, H. W. Galbraith, B. J. Krohn, C. D. Can-trell, and E. D. Hinkley, Optics Commun. $\underline{17}$, 178 (1976).

11. G. Graner, J. Mol. Spectrosc. $\underline{51}$, 238 (1974).

12. A. M. Bonch-Bruevich and V. A. Khodovoi, Usp. Fiz. Nauk $\underline{93}$, 71 (1976) [Sov. Phys.-Usp. $\underline{10}$, 637 (1968)]; P. F. Liao and J. E. Bjorkholm, Phys. Rev. Lett. $\underline{34}$, 1 (1975). This expression is accurate to $\sim 1\%$ for our discussion which follows: see P. F. Liao and J. E. Bjorkholm, Opt. Commun. $\underline{16}$, 392 (1976).

13. W. K. Bischel, P. J. Kelly, and C. K. Rhodes, Phys. Rev. A13, 1829 (1976).

14. R. R. Jacobs, D. Prosnitz, W. K. Bischel, and C. K. Rhodes, Appl. Phys. Lett. 29, 710 (1976).

15. W. K. Bischel, P. J. Kelly, and C. K. Rhodes, Phys. Rev. A13, 1817 (1976).

16. C. D. Cantrell and H. W. Galbraith, Optics Commun. 18, 513 (1976).

17. N. R. Isenor and M. C. Richardson, Appl. Phys. Lett. 18, 224 (1971): N. R. Isenor and M. C. Richardson, Optics Commun. 3, 360 (1971); N. R. Isenor and M. C. Richardson, Proc. Tenth Intern. Conf. on Ionization Phenomena in Gases (Oxford, Donald Parsons, 1971); N. R. Isenor, V. Merchant, R. S. Hallsworth, and M. C. Richardson, Can. J. Phys. 51, 1281 (1973); R. S. Hallsworth and N. R. Isenor, Chem. Phys. Lett. 22, 283 (1973).

18. R. V. Ambartzumian, V. S. Letokhov, E. A. Ryabov and N. V. Chekalin, ZhETF Pis. Red. 20, 597 (1974) [JETP Lett. 20, 273 (1974)]; R. V. Ambartzumian, Yu. A. Gorokhov, V. S. Letokhov and G. N. Makarov, ZhETF Pis. Red. 21, 375 (1975) [JETP Lett. 21, 171 (1975)].

19. J. L. Lyman, R. J. Jensen, J. Rink, C. P. Robinson and S. D. Rockwood, Appl. Phys. Lett. 27, 87 (1975).

20. R. V. Ambartzumian, Yu. A. Gorokhov, V. S. Letokhov, G. N. Makarov, E. A. Ryabov and N. V. Chekalin, pp. 114-121 in Laser Spectroscopy, edited by J. C. Pebay-Peyroula, T. W. Hansch and S. E. Harris (Berlin, Springer-Verlag, 1975).

21. N. G. Basov, V. T. Galochkin, A. N. Oraevsky and N. F. Starodubtsev, ZhETF Pis. 23, 569 (1976) [JETP Lett. 23, 521 (1976)].

22. N. Bloembergen, C. D. Cantrell and D. M. Larsen, pp. 162-176 in <u>Tunable Lasers and Applications</u>, edited by A. Morradian, T. Jaeger and P. Stokseth (Berlin, Springer-Verlag, 1976).

23. J. P. Aldrige, J. H. Birely, C. D. Cantrell and D. C. Cartwright, pp. 57-144 in <u>Laser Photochemistry</u>, <u>Tunable Lasers and Other Topics</u> (Physics of Quantum Electronics, Vol. 4), edited by S. F. Jacobs, M. Sargent III, M. O. Scully and C. T. Walker (Reading, Mass., Addison-Wesley Publishing Co., 1976).

24. V. S. Letokhov and C. B. Moore, Kvant. Elekt. <u>3</u>, 247 and 485 (1976) [Sov. J. Quant. Elect. <u>6</u>, 129 and 259 (1976)].

25. S. Singer, J. J. Hayden and I. Liberman, private communication (1975).

26. T. P. Cotter, W. Fuss, K. L. Kompa and H. Stafast, Optics Commun. <u>18</u>, 220 (1976).

27. R. L. Platzman, Radiation Research <u>17</u>, 419 (1962).

28. W. H. Shaffer, H. H. Nielsen and L. H. Thomas, Phys. Rev. <u>56</u>, 895 and 1097 (1939).

29. K. Fox, J. Mol. Spectrosc. <u>9</u>, 381 (1962).

30. C. D. Cantrell and H. W. Galbraith, Optics Communc. <u>18</u>, 513 (1976).

31. H. W. Galbraith and C. D. Cantrell, J. Mol. Spectrosc. (to be published).

32. N. Bloembergen, Optics Communc. <u>15</u>, 416 (1975).

33. V. S. Letokhov and A. A. Markarov, Optics Commun. <u>17</u>, 250 (1976); <u>Coherent Excitation of Multilevel</u> <u>Molecular Systems in Intense Quasi-Resonant Laser</u> <u>IR Field</u> (Moscow, 1976).

34. K. T. Hecht, J. Mol. Spectrosc. <u>5</u>, 355 (1960).

35. C. C. Jensen, W. B. Person, B. J. Krohn and J. Overend, Optics Commun. (in press)

36. H. W. Galbraith and C. D. Cantrell, This Volume.

37. R. S. McDowell, H. W. Galbraith, B. J. Krohn, C. D.
    Cantrell and E. D. Hinkley, Optics Commun. 17,178
    (1976).

38. D. M. Larsen and N. Bloembergen, Optics Commun. 17,
    254 (1976).

39. V. S. Letokhov and R. V. Ambartzumian, private
    communication (1976).

SPATIO-TEMPORAL STRUCTURATION IN IMMOBILIZED ENZYME SYSTEMS

Kernevez, J.P.; Duban, M.C.; and Joly, C.;

(Presented by Kernevez J.P.)

Université de Technologie de Compiègne

Compiègne Cedex, France

and

Thomas,D.

Laboratoire De Technologie Enzymatique

France

The aim of this paper is to describe the behavior
of immobilized enzyme systems described by nonlinear
partial differential equations. In these artificial
membranes, interaction of diffusion and reaction gives
rise to phenomena of biological significance, such as
transport against the gradient of concentration, hysteresis, oscillations, or pattern formation. Similar qualitative behaviors can be observed in living cells: active
transport, memory, biological clocks, morphogenesis. It
will be shown hereafter that very simple enzyme systems
can exhibit such behaviors.

Since the pioneering work of TURING in 1952 [31],
many authors have studied the appearance of spatio-
temporal structurations in distrubuted biochemical

systems.[1-16,18-20,22,24-28,31,32]   However experimental
realizations of the proposed models are seldom made.
In the following we will present, by their equations,
systems which are effectively built in one of our laboratories.

   The mathematical tools useful for these studies
are the bifurcation and stability theory, and singular
perturbations and asymptotic theory[1-6,8,9,11-14,17,21,
22,29,31].   In the following we shall omit any mathematical
justification of this sort for the phenomena described,
and refer to[33].

1.   THE MODEL CASE

   Enzymes are catalysts of biochemical reactions.
For instance glycose oxidase is an enzyme which is a
specific catalyst of the transformation of glucose into
gluconic acid.   Schematically if we denote E, S, P
respectively: the enzyme, the substrate, and the product,
we have:

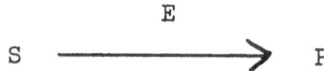

Since MICHAELIS and MENTEN[23] (1913) enzyme reactions
have been studied in homogeneous solutions.   However,
in living cells, enzymes are not in solution, but distributed in heterogeneous structures, where competition
between diffusion and reaction can play an essential role.
That is the reason why it is so interesting to study
enzyme reactions in an artificial membrane, where enzyme
molecules are bound to inactive protein molecules (albumin
for instance).   It is then possible to study, in a well
defined context, enzyme kinetics.

   Such an artificial membrane is produced by a

cocrosslinking method previously described[30]. The thickness of an artificial membrane is between 10 and 50 μ.

The basic experiment is presented in Fig. 1.

The membrane separates 2 reservoirs. In these reservoirs there is a solution of substrate S. S is going to diffuse inside the membrane and react because of the enzyme E. Taking the x - axis perpendicular to the membrane (slab geometry) and choosing the membrane thickness as a unit of length, the "model case" is governed by the equation:

$$s_t - s_{xx} + \sigma s/(1 + s) = 0, \qquad (1.1)$$

where $s = s(x,t)$ = substrate concentration, $0 < x < 1, t > 0$,

$s_t = \dfrac{\partial s}{\partial t}$, $s_{xx} = \dfrac{\partial^2 s}{\partial x^2}$, and $\sigma s/(1+s)$ = MICHAELIS-MENTEN velocity term[18,23].

The DIRICHLET boundary conditions express that in the reservoirs the concentration of substrate is constant:

$$s(0,t) = s_0, \quad s(1,t) = s_1. \qquad (1.2)$$

The initial condition expresses that at the beginning of time, $t = 0$, the membrane is void of substrate,

$$s(x,0) = 0. \qquad (1.3)$$

In Fig. 2 is shown that the evolution of the substrate profile of concentration and it can be observed that there is a stationary state already nearly attained at time $t = 0.7$.

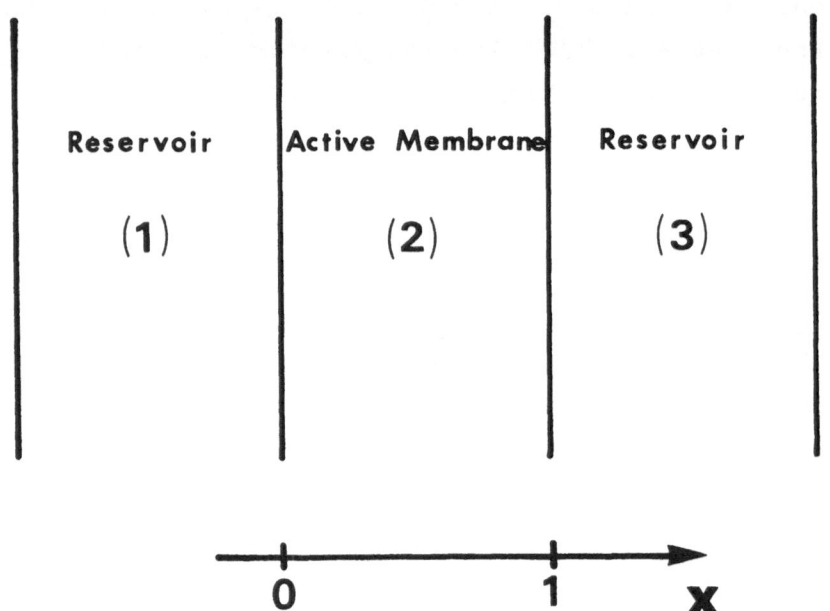

Figure 1 The basic experiment

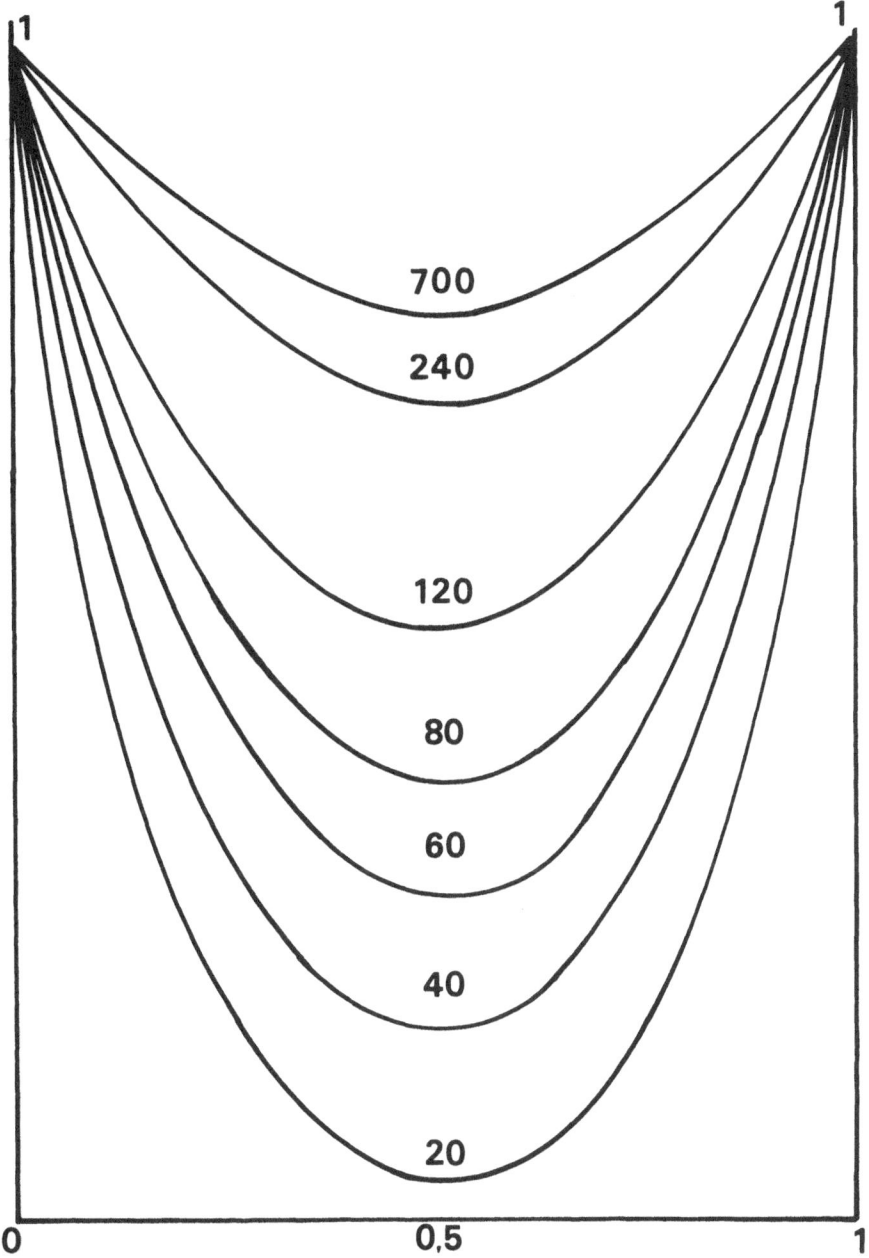

Figure 2 Evolution of the substrate profile of concentration: abscissa=space x; ordinate=concentration s; numbers
beside curves denote time t.

In Fig. 3 is shown a similar transient evolution
of substrate (dotted line) and product (full line)
concentration profiles in the case of an "electrode"
(The enzyme membrane is stuck against an electrode).
Due to the specificity of one enzyme to one substrate,
these electrodes can be used for continuous monitoring
of glucose in blood or saccharose in fermentation processes.

2.   TRANSPORT PHENOMENA

A "glucose pump" is made of 2 layers of different
enzymes, E_1 and E_2, with E_1 transforming S into P in
the first layer, and E_2 transforming P into S in the
second layer. E_1 = hexokinase, E_2 = phosphatase,
S = glucose, P = gluconic acid. On each side of the
membrane, there is a selective layer, impermeable to
P but not to S, and a reservoir containing a homogeneous solution of S. (Fig. 4).

The behavior of this system can be understood by
looking at the substrate S concentration profile in
Fig. 5.: in the first layer there is a consumption of
S, giving a hollow, whereas in the second layer there
is a production of S, whence a hump.

Now if we consider the flux of substrate, which is
proportional to s_x, we see that S is entering into the
membrane at x = 0, and leaving it at x = 1. The net
effect of diffusion and reaction in the bilayer is to
pump glucose from the reservoir x = 0 to the reservoir
x = 1. The equations in this case are:

$$s_t - s_{xx} + F = 0, \quad p_t - p_{xx} - F = 0, \qquad (2.1)$$

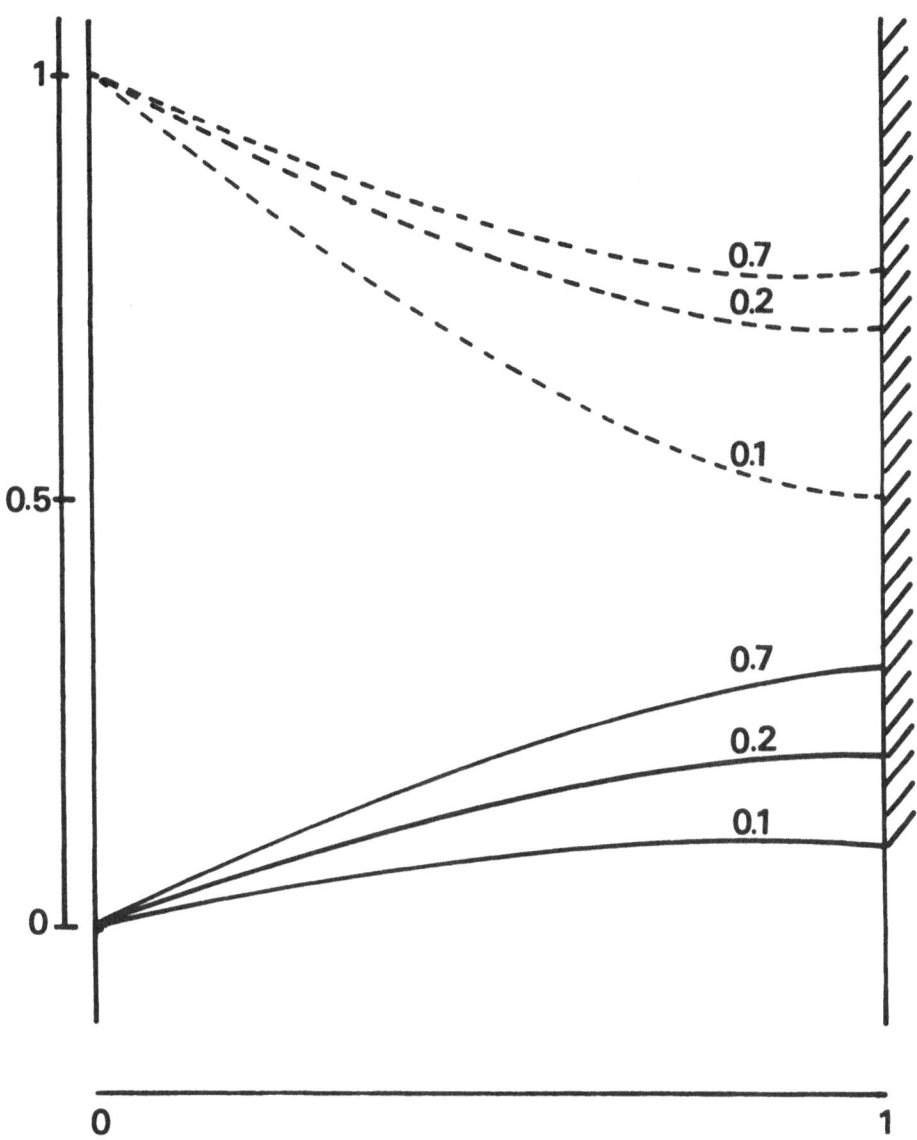

Figure 3 The concentration profiles in a membrane stuck
on an electrode are represented for s (dotted line) and p
(full line) at levels of time t = 0.1, t = 0.2, and
t = 0.7.

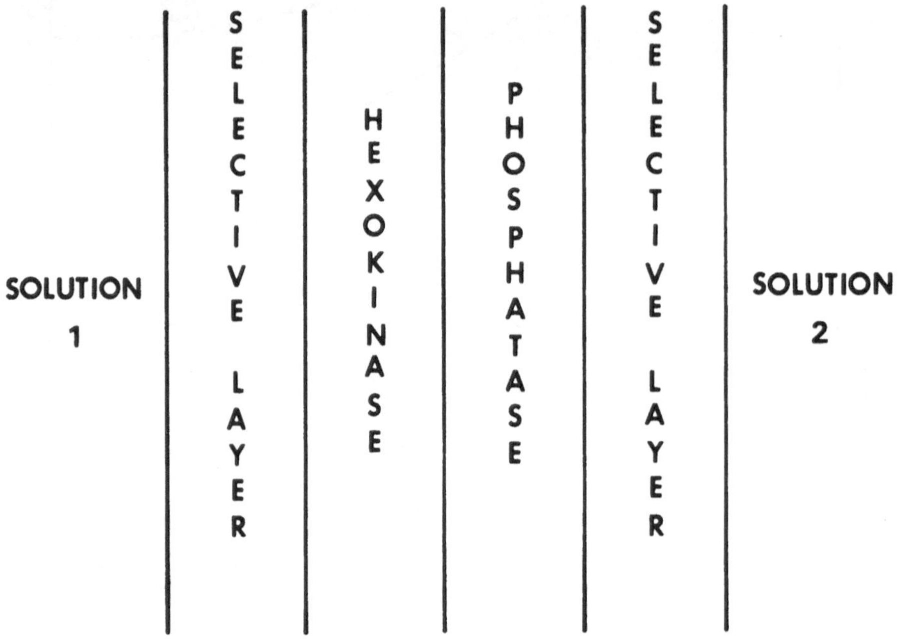

Figure 4 The glucose pump. In the hexokinase ($= E_1$) layer $S \xrightarrow{E_1} P$. In the phosphatase ($=E_2$) layer $P \xrightarrow{E_2} S$.

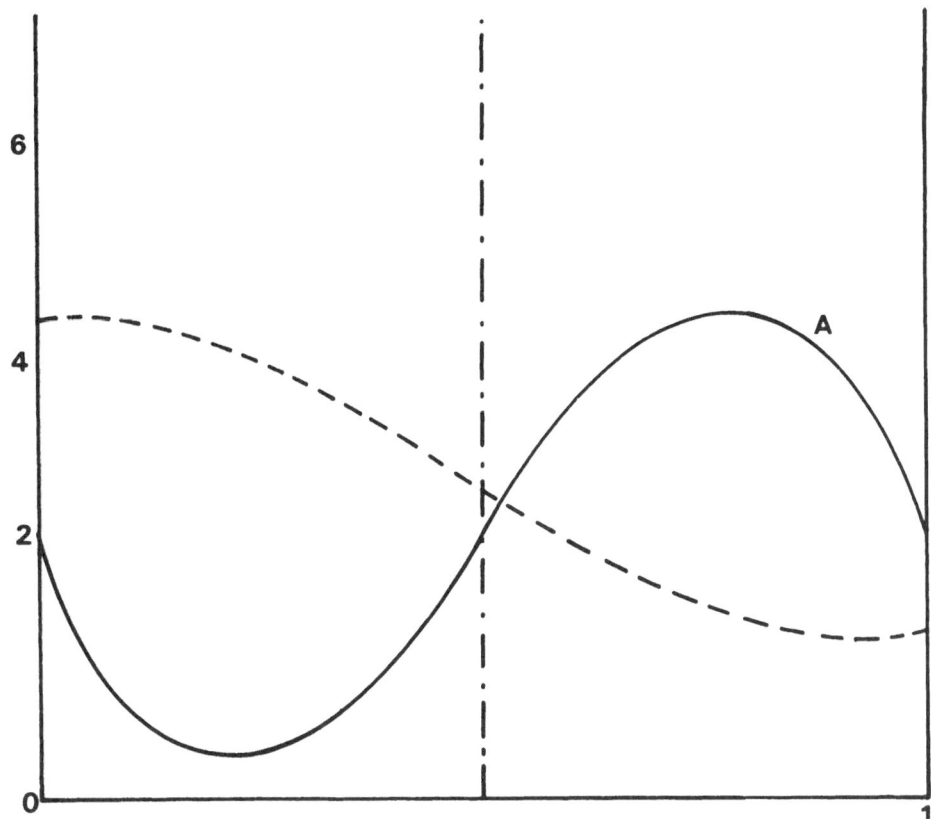

Fig. 5. S (full line) and P (dotted line) concentration
profiles at some time t. Abcissa = space x, ordinate = S
and P concentration.

$$F(x,s,p) = \begin{cases} \sigma \dfrac{s}{1+bp+s} & 0 < x < \dfrac{1}{2} \quad \sigma, b > 0. \\[3mm] & \qquad\qquad\qquad (2.2) \\[1mm] -\sigma \dfrac{p}{1+p} & \dfrac{1}{2} < x < 1 \end{cases}$$

Boundary Conditions

$$s(0,t) = s_0, \quad s(1,t) = s_1, \quad s_0, s_1 \geq 0, \qquad (2.3)$$

$$p_x(0,t) = p_x(1,t) = 0. \qquad (2.4)$$

Initial Conditions

$$s(x,0) = 0, \quad p(x,0) = 0. \qquad (2.5)$$

In fact we can have active transport even for s_1 much larger than s_0.

3. HYSTERESIS

The following simple example will show what we mean by hysteresis: suppose that diffusion and reaction take place in distinct locations (Fig. 6). Then in a stationary state the balance between the flux of substrate coming from an outside reservoir (concentration s_0) to an inside reactor (concentration s) gives the equation:

$$s_0 - s = \rho \, F(s), \qquad (3.1)$$

where ρ is a positive constant and

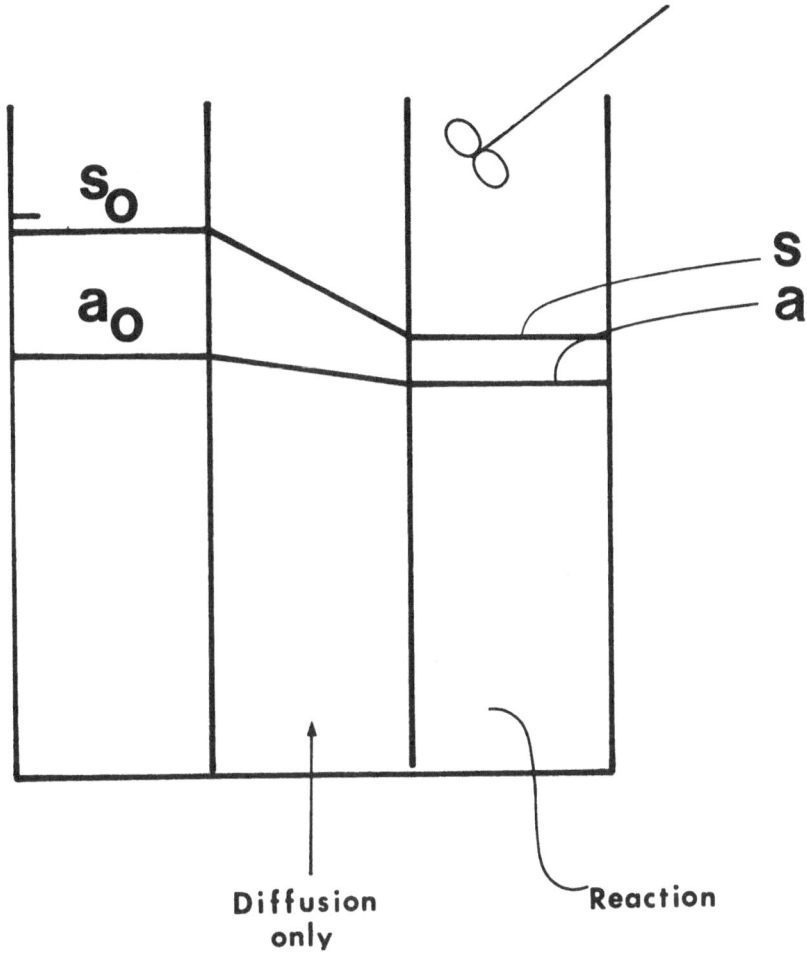

Figure 6 An inactive membrane separates a reservoir
(with substrate S and cosubstrate A concentrations s_0
and a_0) and a reactor (with S and A concentrations s
and a). In III (hysteresis) A is in excess (a = a_0 is
large) and only S is taken into account.

$$F(s) = s/(1+s+ks^2) \qquad\qquad (3.2)$$

(substrate inhibited reaction: for example E = uricase, S = uric acid, P = allantoin).

When the outside concentration s_0 is first increasing, then decreasing, the inside concentration s is varying as indicated in Figures 7 and 8.

Hysteresis in this case is related to the interaction of diffusion s_0-s and of the nonlinear term of reaction ρ F(s). The same qualitative behavior can be observed in the distributed case:

$$-s_{xx} + \sigma \, F(s) = 0, \quad s(0) = s(1) = s_0, \qquad (3.3)$$

F given by (3.2) and s_0 first increasing, then decreasing (Fig. 9).

4.   OSCILLATIONS

To see in what context oscillations can exist in an enzyme system, consider the equations:

$$s_0 - s - \sigma \, aF(s) = 0 \ (\equiv G(s,a)), \qquad (4.1)$$

$$\frac{da}{dt} = \alpha(a_0 - a) - \sigma \, aF(s) \ (\equiv H(s,a)), \qquad (4.2)$$

which represents the case where both substrate s and cosubstrate a react in a homogeneous enzyme reactor, separated from an outside reservoir with concentrations s_0 and a_0 by an inactive membrane with only diffusion (Fig. 6). For example E = uricase, S = uric acid, A = oxygen, P = allantoin.

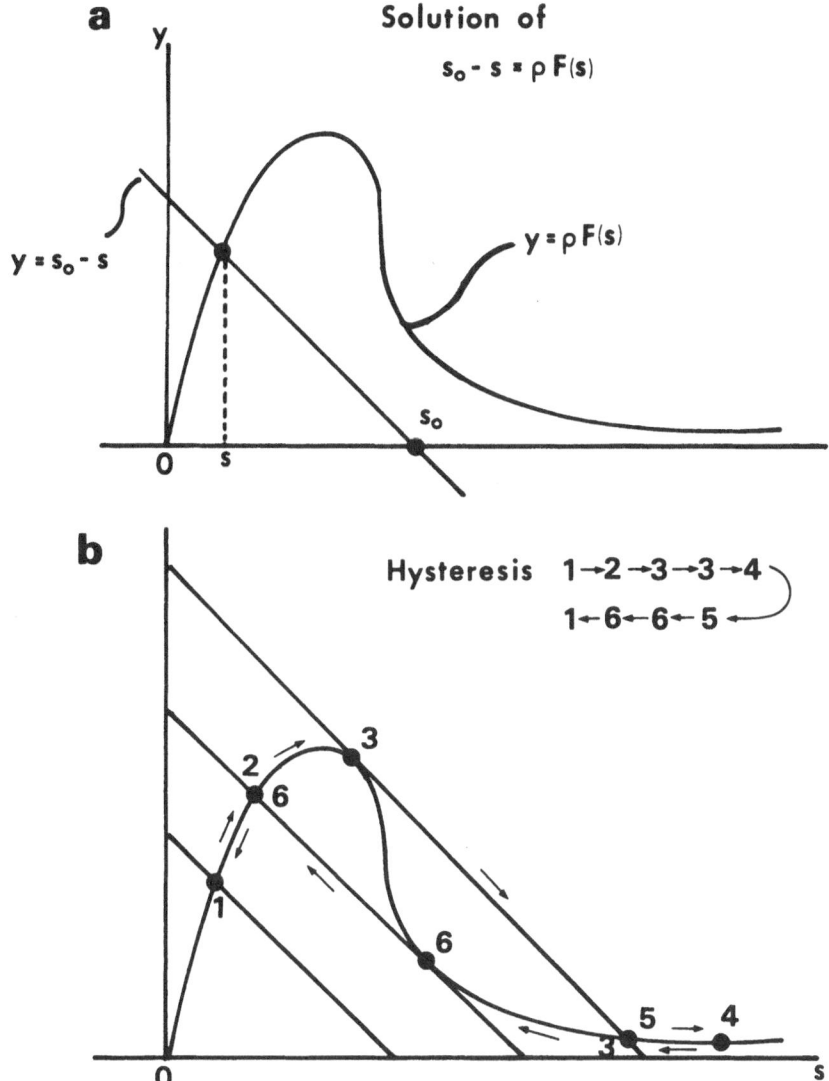

Figure 7 a) for each value of s_0, s is given by the intersection of the straight line $y = s_0-s$ and the curve $y = \rho F(s)$.

b) When s_0 is first increasing, then decreasing, the representative point of the system follows the hysteresis cycle:

1 - 2 - 3 - 3 - 4 - 5 - 6 - 6 - 2 - 1

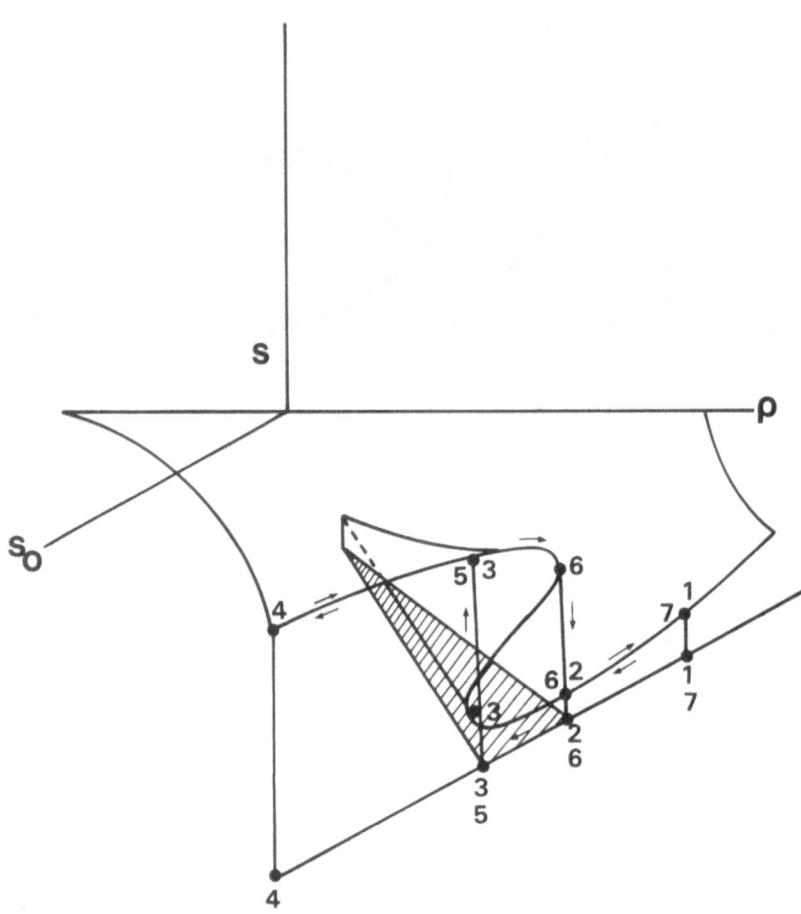

Figure 8 Sketch of the catastrophe surface s = function
of s_0 and ρ defined by (3.1). If ρ = constant, and s_0
varies as indicated in Fig. 7.b., the representative
point follows the path 1 - 2 - 3 - 3 - 4 - 5 - 6 - 6 - 2
- 1 with jumps from 3 to 5 and from 6 to 2.

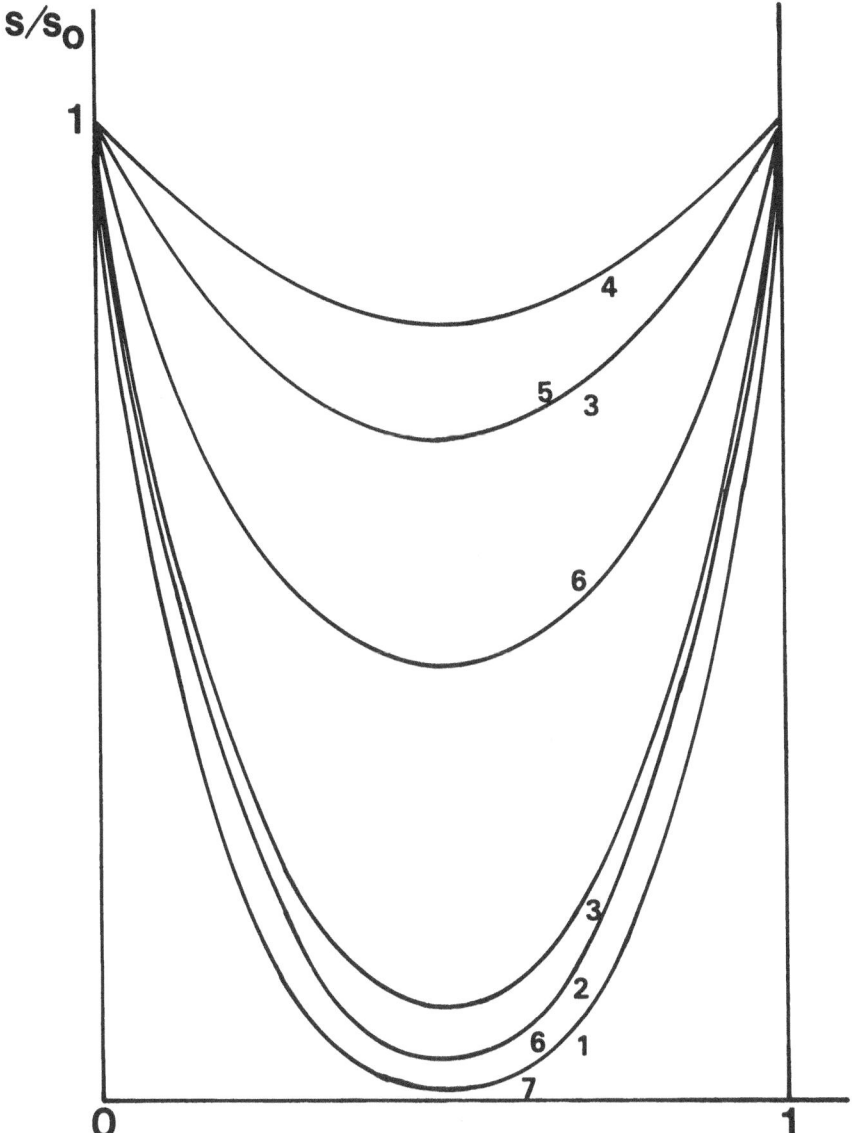

Figure 9 The concentration profiles are solutions of (3.3) for s_0 first increasing, then decreasing. The qualitative behavior is like in Figures 7 and 8, with jumps from 3 to 5 and from 6 to 2. Abscissa = space x. Ordinate = s/s_0 (normalization).

Here the state of the system at any time t is the
point (s(t), a(t)) in the phase plane (s,a). As a
consequence of (4.1) this point always lies on the representative curve [S] of (4.1)(Fig. 10).

It is easily checked that, for suitable values of
the parameters s_0, a_0, ρ, α, the representative curves
(S) and (A) of $G(s,a) = 0$ and $H(s,a) = 0$ are shaped as
indicated in Fig. 10. Moreover (A) separates the first
quadrant in the phase plane into regions where $H(s,a)$
has respectively signs + and -, so that, according to
the case, $\frac{da}{dt}$ will be positive or negative.

Therefore the representative point (s,a) will have
the following periodic motion (Fig. 10): go from I to J,
jump from J to K, go from K to L, jump from L to I, and
so on.

On the castastrophe surface of Fig. 11, where s is
represented as a function of s_0 and $\rho = \sigma a(s_0-s-\rho F(s)=0)$,
if s_0 is given a constant value, ρ will vary like a,
governed by $\frac{da}{dt} = H(s,a)$, and the representative point
(s_0,ρ,s) will describe AB, jump from B to C, describe
CD, jump from D to A, and so on.

We may expect to find an oscillatory behavior for
the corresponding distributed system:

$$s_t-s_{xx}+\sigma F(s,a) = 0,$$

$$a_t-\alpha a_{xx}+\sigma F(s,a) = 0,$$

$$F(s,a) = a \frac{s}{1+s+ks^2} \tag{4.3}$$

$$s(0,t) = s(1,t) = s_0, \quad a(0,t)=a(1,t)=a_0,$$
s and a given at initial time t = 0.

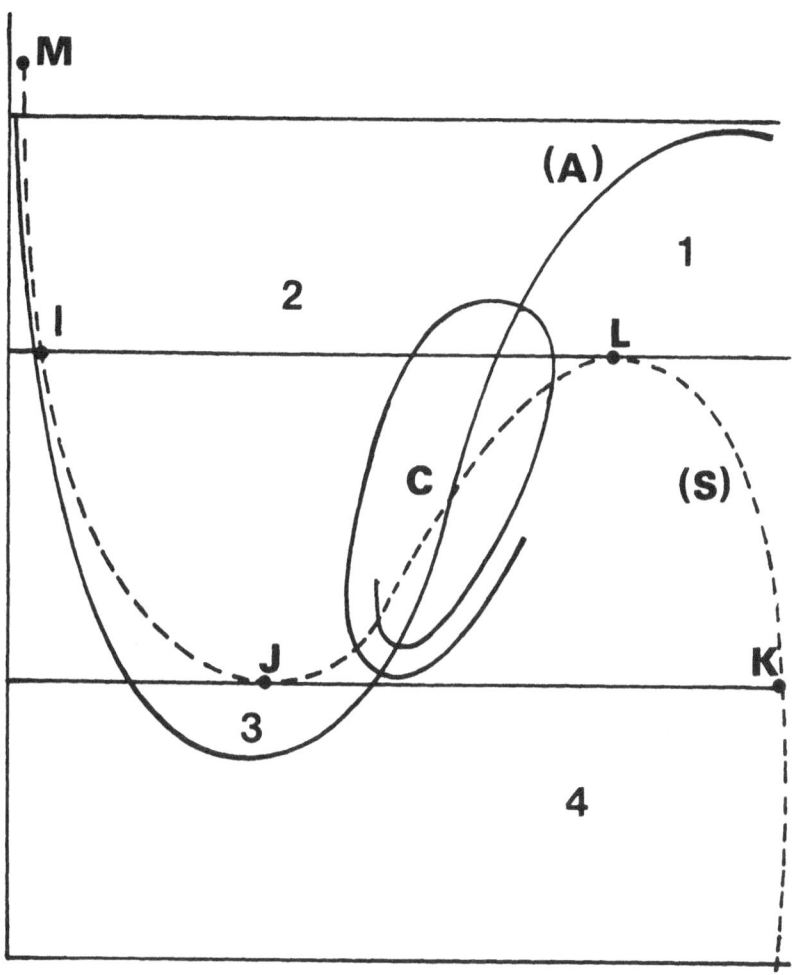

Figure 10 Phase plane representation of s (abscissa) and a (ordinate). In the case of equations (4.1), (4.2) the representative point follows the path J K L I. In the case of $\frac{ds}{dt} = G(s,a)$, $\frac{da}{dt} = H(s,a)$ the point is spiralling towards a limit cycle.

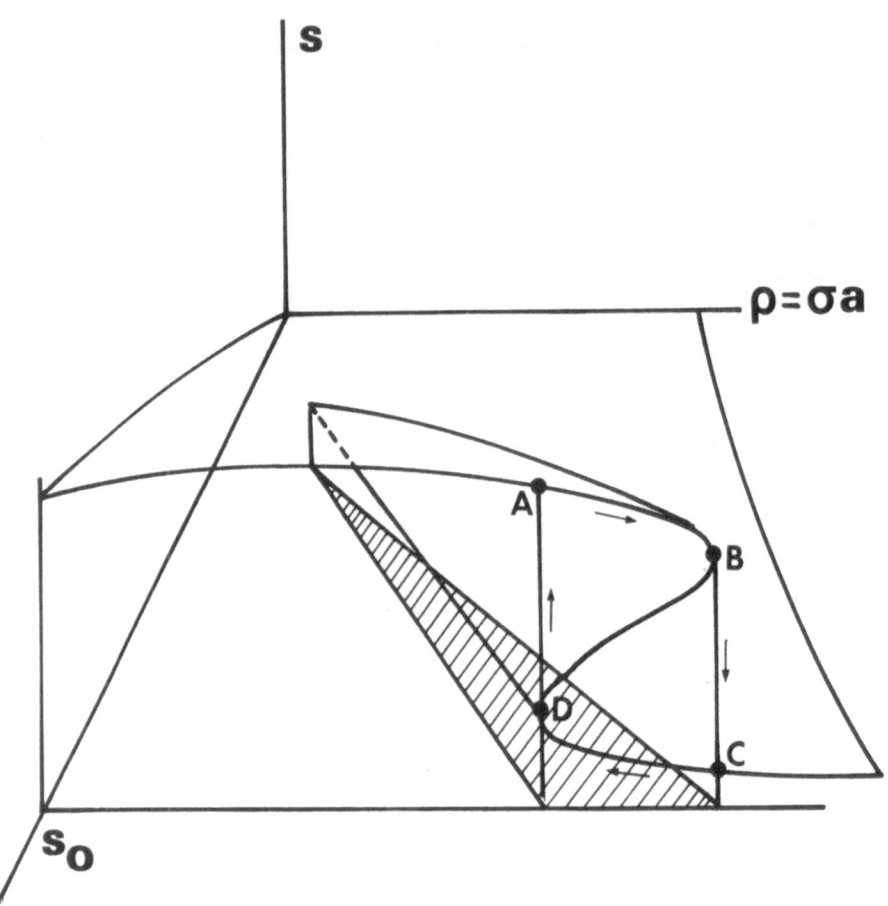

Figure 11 Representation of (4.1), (4.2) on the catastrophe surface of Fig. 9. The representative point follows
AB, jumps from B to C, follows C D, jumps from D to A,
and so on.

Oscillations have actually been found by performing numerical experiments on the computer. If:

$$s_0 - \alpha\, a_0 = 0, \qquad (4.4)$$

then (4.3) has only one stationary state, $(\tilde{s}, \tilde{a})$ given by:

$$-s_{xx} + \sigma\, F(\tilde{s}, \frac{\tilde{s}}{\alpha}) = 0, \quad \tilde{s}(0) = \tilde{s}(1) = s_0, \quad \tilde{a} = \frac{1}{\alpha}\tilde{s}. \qquad (4.5)$$

If we write $s = \tilde{s} + u$, $a = \tilde{a} + v$, and omit terms quadratic or higher order in u and v, the linearized differential equations for u and v are:

$$u_t - u_{xx} + \sigma\, F'_s\, u + \sigma\, F'_a = 0,$$
$$\qquad (4.6)$$
$$v_t - \alpha\, v'_{xx} + \sigma\, F'_s\, u + \sigma\, F'_a\, v = 0,$$

together with the boundary conditions $u(0,t) = u(1,t) = v(0,t) = v(1,t) = 0$.

In (4.6), $F'_s = \frac{\partial F}{\partial s}(\tilde{s}(x), \tilde{a}(x))$, $F'_a = \frac{\partial F}{\partial a}(\tilde{s}(x), \tilde{a}(x))$.

The behavior of (4.6) depends upon the eigenvalues of the operator:

$$\begin{bmatrix} u \\ v \end{bmatrix} \rightarrow \begin{bmatrix} \dfrac{d^2}{dx^2}\,\sigma\, F'_s & -\sigma\, F'_a \\ -\sigma\, F'_s & \alpha\,\dfrac{d^2}{dx^2} - \sigma\, F'_a \end{bmatrix} \begin{bmatrix} u \\ v \end{bmatrix}$$

The plot of the leading eigenvalues as functions of σ is given in Fig. 12. When the (complex conjugate) eigenvalues cross the imaginary axis the trivial stationary state (s,a) loses its stability and it appears as a

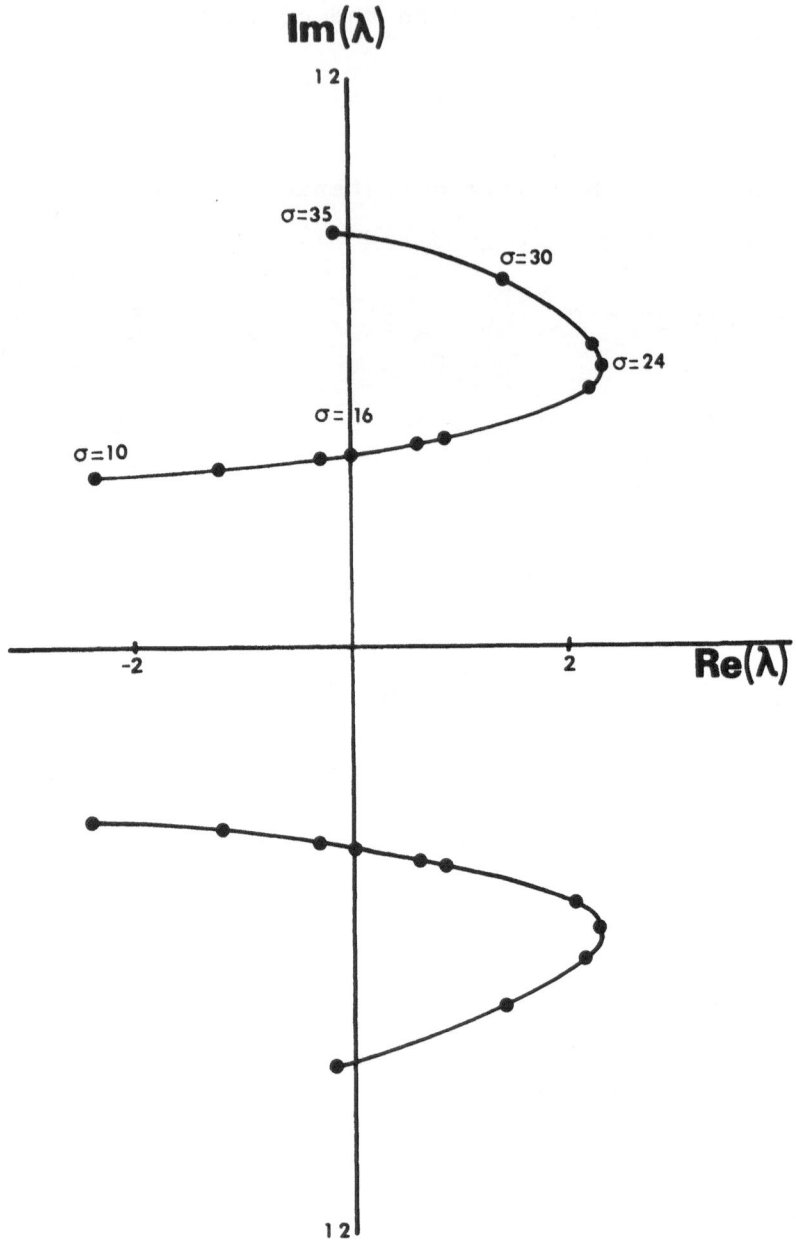

Figure 12 Leading eigenvalues λ as functions of σ.
$s_0 = 100$, $a_0 = 500$, $\alpha = 0.2$.

bifurcated branch of stable oscillatory solutions[33].

5. PATTERN FORMATION

Pattern formation appears when, due to the variation of some parameters, a stable, uniform, stationary
state loses its stability and bifurcates into a stable
stationary state which is structured in space.

This can happen to the system:

$$s_t - s_{xx} + \tau(\rho\ F(s,a) - (s_0 - s)) = 0, \qquad (5.1)$$

$$a_t - \beta\ a_{xx} + \tau(\rho\ F(s,a) - \alpha(a_0 - a)) = 0, \qquad (5.2)$$

$$F(s,a) = a\ \frac{s}{1+s+ks^2}, \qquad (5.3)$$

$$s_x = 0 \text{ and } a_x = 0, \text{ for } x = 0 \text{ and } x = 1. \qquad (5.4)$$

We can chose the parameters α, ρ, s_0, a_0, which are
at our disposal, in order to have only one solution
(s,a) for the system of algebraic equations:

$$\rho\ F(s,a) - (s_0 - s) = 0,$$
$$\qquad (5.5)$$
$$\rho\ F(s,a) - \alpha(a_0 - a) = 0.$$

In that case the functions $s(x,t) \equiv \tilde{s}$ and $a(x,t) = \tilde{a}$ constitute a uniform stationary solution of (5.1),
(5.2), (5.3) and it is the only one.

If we can find the parameters $\tau, \alpha, \rho, s_0, a_0$ so that
this trivial solution is unstable, if the system tends
towards a stable, staionary state, then it will be structured in space. By studying the linearized system of
(5.1)-(5.3) around $(\tilde{s}, \tilde{a})$, it is possible to find conditions

on the parameters in order that the eigenvalues are in
the left half plane, except one, λ_{n_0} , which is positive[33].

In that case the computer experiments indicate that,
whatever the initial values, the nonlinear system (5.1)-
(5.3) tends, when $t \to + \infty$, towards a stable, structured,
stationary state (Fig. 13).

Equations (5.1)-(5.3) modelize the very frequent
situation of an inactive layer separating an active
layer (with enzyme) and a reservoir with a solution
of substrate and cosubstrate at concentrations s_0 and
a_0.

According to this result, "spatial dissipative
structures" are highly probable in biological systems.

CONCLUSION

The nonlinear aspect of the equations previously
given explains the richness of behaviors that these
very simple systems can exhibit. The possibility of
producing the corresponding artificial membranes gives
us a tool to study various phenomena in a well defined
context.

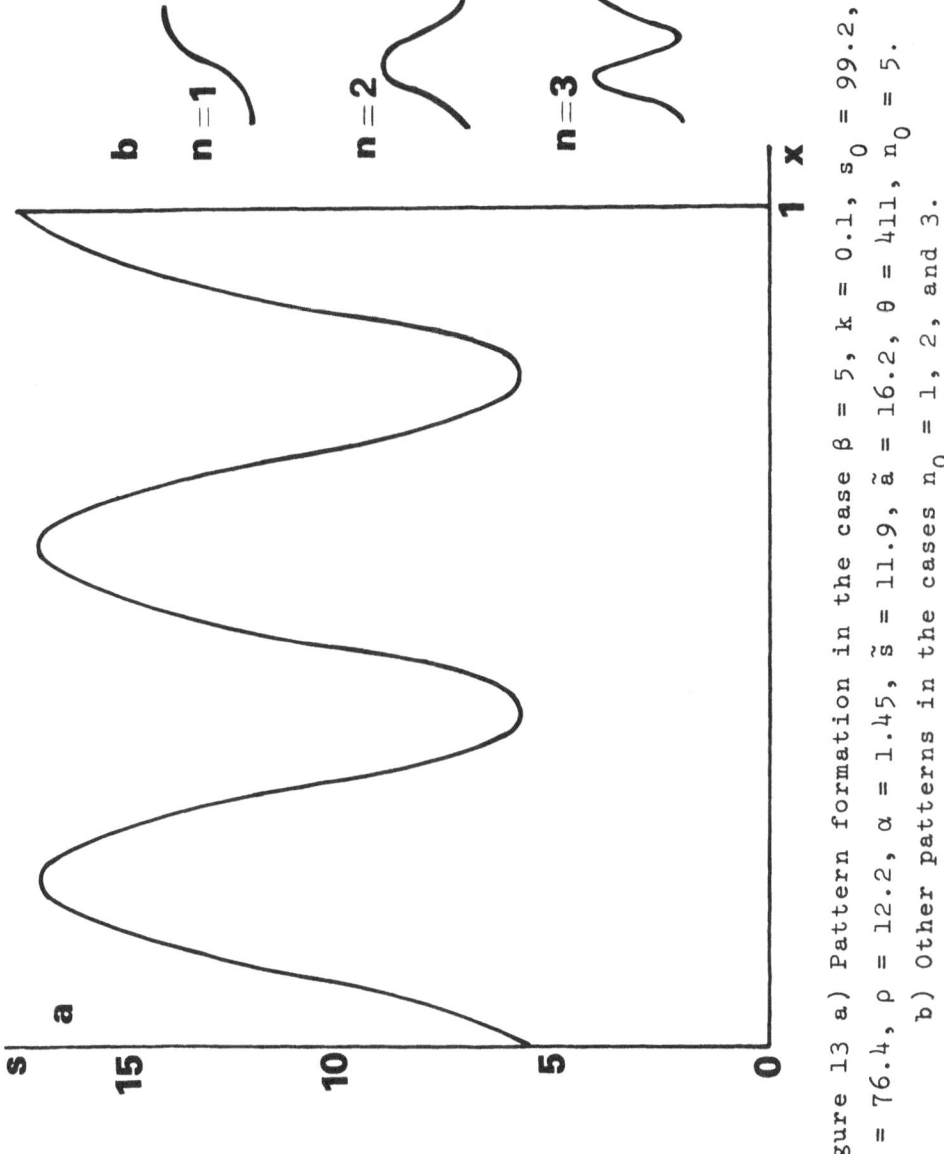

Figure 13 a) Pattern formation in the case $\beta = 5$, $k = 0.1$, $s_0 = 99.2$, $a_0 = 76.4$, $\rho = 12.2$, $\alpha = 1.45$, $\tilde{s} = 11.9$, $\tilde{a} = 16.2$, $\theta = 411$, $n_0 = 5$.

b) Other patterns in the cases $n_0 = 1$, 2, and 3.

REFERENCES

1. Auchmuty J.F.G. and Nicolis G. "Dissipative Struc-
    tures, Catastrophes, and Pattern Formation: A
    Bifurcation Analysis" Proc. Nat. Acad. Sc. USA,
    71, 2748-2751, (1974).

2. Auchmuty J.F.G. and Nicolis G. "Bifurcation Analysis
    of Nonlinear Reaction-Diffusion Equations I:
    Evolution and the Steady State Solutions" Bull. Math.
    Biol. 37, 323-365 (1975).

3. Auchmuty J.F.G. and Nicholis G. "Bifurcation
    Analysis of Nonlinear Reaction-Diffusion Equations
    III: Chemical Oscillations" Bull. Math. Biol. 38,
    325-350 (1976).

4. Boa James A. "A Model Biochemical Reaction" PhD
    Thesis, California Institute of Technology, Pasadena,
    (1974).

5. Boa J.A. and Cohen D.S. "Bifurcation of Localized
    Disturbances in a Model Biochemical Reaction" Siam
    J. Appl. Math. 30, 123-135, N°1, (January 1976).

6. Boa J.A. "Multiple Steady States in a Model Bio-
    chemical Reaction" Studies in Applied Math., Vol.
    LIV, N°1, (March 1975).

7. Caplan S.R., Naparstek A., and Zabusky N.J. Nature,
    245, 364 (1973).

8. Casten R.G. and Holland C.J. "Instability Results
    for Reaction Diffusion Equations with Neumann
    Boundary Conditions" To appear J. Differential
    Equations.

9. Casten R.G. and Holland C.J. "Stability Properties of
    Solutions to Systems of Reaction-Diffusion Equations"
    To appear SIAM. J. on Applied Math.

10. Glansdorf P. and Prigogine I. "Thermodynamic Theory
    of Structure, Stability, and Fluctuations" Wiley
    Interscience, London (1971).

11. Fife P.C. "Pattern Formation in Reacting and Diffusing Systems" J. Chem. Phys. $\underline{64}$, 554-564 (1976).

12. Fife P.C. and Mc Leod J.B. "The approach of Solutions of Nonlinear Diffusion Equations to Travelling and Wave solutions" Bulletin of the American Math. Soc., $\underline{81}$, N°6, 1076-1078 (Nov. 1975).

13. Fife P.C. "Stationary Patterns for Reaction-Diffusion Equations" MRC Technical Summary Report, pp. 1-50 (1976).

14. Fife P.C. "Singular Perturbations and Wave Front Techniques in Reaction Diffusion Problems" SIAM-AMS Proceedings. $\underline{10}$,23-50 (1976).

15. Hardt S., A. Naparstek, L.A. Segel, and S.R. Caplan "Spatio-Temporal Structure Formation and Signal Propagation in a Homogeneous Enzymatic Membrane" pp.9-15 in "Analysis and Control of Immobolized Enzyme Systems" Ed. D. Thomas and J.P. Kernevez, North Holland (1976)

16. Herschkowitz Kaufman M. and Nicolis G. "Localized Spatial Structures and Nonlinear Chemical Waves in Dissipative Systems" J. Chem. Phys. $\underline{56}$, 1890-1895 (1972).

17. Keller J.B. and Kogelman S. "Asymptotic Solutions of Initial Value Problems for Nonlinear Partial Differential Equations" SIAM J. Appl. Math. $\underline{18}$, 748-758, (1970).

18. Kernevez J.P. and Thomas D. "Numerical Analysis and Control of Some Biochemical Systems" Appl. Math. and Opt., $\underline{1}$, N°3, 222-285, (1975).

19. Lefever R. and Prigogine I. "Symmetry Breaking Instabilities in Dissipative Systems II" J. Chem. Phys. $\underline{48}$, 1695-1700, (1968).

20. Lefever R. "Dissipative Structure in Reaction-
    Diffusion Systems" pp.17-39 in "Analysis and Control
    of Immobilized Enzyme Systems" Ed. D. Thomas and
    J.P. Kernevez, North Holland, (1976).

21. Matkowsky B.J. "A Simple Nonlinear Dynamic Stability
    Problem" Bull. Amer. Math. Soc. $\underline{76}$, 620-625,(1970).

22. Meurant G. and Saut J.C. "Bifurcation and Stability
    in a Chemical System" J. Math. Anal. and Appl. -
    to appear

23. Michaelis L., Menten M. "Die Kinetik der
    Invertinwirkung" Biochem. Z., $\underline{49}$, 333-369,(1913).

24. Naparstek A., Romette J.L., Kernevez J.P., and
    Thomas D. "Memory in Enzyme Membranes" Nature,
    $\underline{249}$, 490, (1974).

25. Nicolis G. "Mathematical Problems in Theoretical
    Biology" in "Nonlinear Problems in the Physical
    Sciences and Biology" Ed. I. Stakgold, DD. Joseph,
    D.H. Sattinger, Springer Lecture Notes in Mathematics,
    N°324, (1973).

26. Othmer H.G. and Scriven L.E. "Instability and Dynamic
    Pattern in Cellular Networks" J. Theor. Biol. $\underline{32}$,
    507-537 (1971).

27. Othmer H.G. "Nonlinear Wave Propagation in Reacting
    Systems" Journal of Math. Biol. $\underline{2}$, 133-163, (1975).

28. Prigogine I., Nicolis G. and Babloyantz A. Physics
    Today, $\underline{25}$, 23, (1975).

29. Sattinger D.H. "Topics in Stability and Bifurcation"
    Springer Lecture Notes in Mathematics N° 309, (1972).

30. Thomas D., Boudillon C., Broun G., and Kernevez J.P.
    "Kinetic Behavior of Enzymes in Artificial Membranes"
    Biochemistry, $\underline{13}$, 2995, (1974).

31. Turing A.M. "The Chemical Basis of Morphogenesis"
    Phil. Trans. Roy. Soc., $\underline{B237}$, 37-72, (1952).

32. Zabusky N.J. and Hardin R.H. "Phase Transition, Stability, and Oscillations for an Autocatalytic, First Order Reaction in a Membrane" Phys. Rev. Letters 31, N° 13 (1973).

33. Joly G. "Analyse Mathématique de Structurations Spatio-temporelles dans des Systèmes à Enzymes Immobilisées" Thesis, to appear.

THEORIES AND CONJECTURES ON MEMBRANE-SUPPORTED

WAVES AND PATTERNS

Paul C. Fife

University of Arizona

Tucson, Arizona 85721

1. INTRODUCTION

The work reported on in this symposium by Deem and
Kernevez (see also [2,3,15] and references therein)
suggests an analogy between some biological membrane
processes and packed bad catalytic reactors, well known
and well studied in chemical engineering. The analogy
is quite clear if one envisions the membrane as having
imbedded within it an immobilized enzyme catalyst, and
as allowing other reacting molecules and ions to migrate
through it from one side to the other, or longitudinally.

If one does this, then the versatility of a membrane
in producing chemico-physical effects of some variety is
recognized. For chemical reactors are capable of ex-
hibiting sustained oscillations, multiple steady states,
patterns, and waves, and these phenomena are elicited
merely by chemical reaction processes, together with
forced or diffusive transport (see, for example [11, 14,
17, 18]). They are, moreover, well within the range of
modeling and mathematical analysis. The membrane can
supply all these mechanisms, and many more sophisticated

ones as well. Analytical and experimental analyses of
some of them are reported, as was mentioned, by Kernevez
and Deem in this Proceedings. It should also be mentioned that the possible role of exactly these membrane processes in generating spatial order in organisms was promoted in a paper by Goodwin[12]. In that paper, experimental evidence for the membrane's central role in
spatial order development was mentioned, and a model
put forth to explain gradient formation.

     With this array of research-experimental, numerical,
conjectural, and analytical,- at hand, I feel motivated
to dwell, in this talk, on some of the phenomena theoretically in the vast repertoire of tricks available to
a membrane which can support diffusive or diffusive-like
transport. I emphasize the word "theoretically" because
we are undoubtedly still in the dark as to most of the
chemico-physical phenomena which actually occur. All the
effects I talk about can be modeled by systems of reaction-diffusion equations; and this fact, together with
the availability of analytical methods for studying these
models, is the main lesson I wish to put across. Let me
discuss the various categories of "tricks" in turn.

          2. SIGNAL PROPAGATION ALONG A MEMBRANE
     The best known and most thoroughly studied example
of signal propagation in a biological organism is that
along the membrane of a nerve axon. The signals are
produced in the form of pulses, so that the axon at any
given position eventually returns to its original state
after the signal has passed. The chemico-physical processes involved in this happening, and the main mathematical models for its study, have been well treated in
the talks by H. Cohen and H. Lecar at this symposium, and

anything I could say here would be superfluous. So
let us pass on to the other categories.

3.   PROPAGATING WAVE-FRONTS

Chemical reactors can, under certain circumstances,
support multiple stable stationary states, and it is not
too hard to construct model biochemical reaction schemes
with autocatalysis which do the same[4,16]. Needless to
say, besides multiple chemical states, several physical
states for a membrane are conceivable. For example, the
large molecules therein could exist in various configurations. If we suppose given a membrane which can
exist in two stable states, and the chemical or physical
parameters characterizing the states can somehow diffuse,
then we immediately have the possibility of a progressive
take-over of one of the states by the other. This would
occur in the form of a wave-front, the propagation of the
front along the membrane being due to diffusion and the
hypothesized fact that one of the states is "dominant"
over the other[5].

This take-over phenomenon can certainly happen in
systems governed by reaction-diffusion equations. It
has been extensively analyzed in the case of a scalar
equation with two stable rest states[1,9]. Such an equation
(or rather one for which one of the two rest states is
unstable) was first proposed by Fisher[10] to describe the
propagation of genetic traits in a population. In this
case, the dominance characteristics of one state over
the other can be precisely defined and determined. Wave
fronts have also been analyzed in some types of systems
of reaction-diffusion equations[6,7].

The question may then be asked, is this a once-only
happening, or can the membrane afterwards regain its

original state? Certainly on a longer time scale, the
characteristics of the membrane can change, so that the
state which was originally dominant, and which took over
the other state, now grows weaker, and the second state
becomes the dominant one. Suppose this happens. Then
if, at some point on the membrane, a stimulus is provided, the wave front will propagate back with some
characteristic velocity, restoring the membrane to more
or less its original condition. Then another slow change
in the membrane characteristics will complete the cycle.

   The different membrane states could, again, con-
ceivably involve different permeabilities to certain
molecules or ions. Variable permeability is an important
property of biological membranes; in addition to its more
mundane functions, its possible importance in producing
biological oscillators has been discussed in [13].

   If the two time scales described above are indeed
greatly different, then in relation to the longer one,
the take-over effect is rather sudden, and the membrane
can be considered "switched" on or off, for whatever
purpose that may serve.

4.   PATTERN FORMATION

   Closely associated with front propagation is the
concept of stationary pattern formation. For example,
suppose that the dominance characteristics of the two
states are measured by a parameter, which may be a physical parameter or the concentration of a certain substance, which varies gradually over the extent of the
membrane. Then the velocity of the propagating front
will also vary, and will in fact be zero where the two
states are equally dominant. Then the front will stop
when it reaches that position, resulting in a sharp

division of the membrane into portions characterized by one or the other of the two states. In fact, differentiation of the membrane of a cell, for example, may well be important for the processes of cell division and cell movement.

The process just described is one in which a given nonuniform distribution of some parameter throughout the membrane is amplified to a sharply differentiated pattern. But stable sharply differentiated patterns can also arise when there is no "precursor pattern" available. Let me illustrate this with a model reaction-diffusion system with two components[5,6].

$$\frac{du}{dt} = f(u,r)+\epsilon^2\nabla^2 u, \tag{1}$$

$$\frac{dv}{dt} = g(u,v)+\nabla^2 v. \tag{2}$$

Here the Laplacians are with respect to a coordinate system on the surface of the membrane. We assume that ϵ^2 (the diffusivity of the first substance) is small, and that the function f has the characteristics given in the following diagram, sometimes found for autocatalytic schemes:

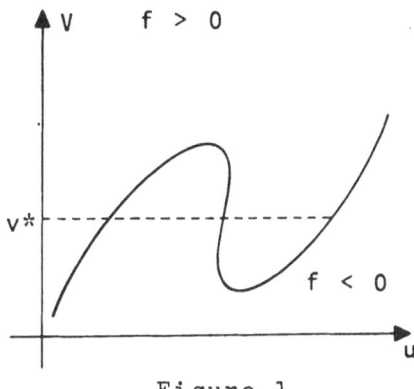

Figure 1

In the stationary state, when ε is small, one expects f = 0 except where the gradient of u is large,
so we are on the S-shaped curve in the phase-plane (Fig.
1). Only the two ascending branches are stable, so we
can assume that we are on one or the other of them.
Call them $h_{\pm}$. At a given point on the membrane, we
shall say it is in "state $h_{\pm}$" according as to whether
the phase point lies on $h_{\pm}$. Thus, in a sense, two stable
states exist. Let us seek a patterned structure in which
the reacting material is differentiated into regions of
state h_{+} and other regions of state h_{-}, and such that
v does not change abruptly from one to the other;
rather it varies gradually. A singular perturbation
analysis can be applied to the transition region; it
tells us that it has width of order ε, and that v = v*,
where v* is the value of v(shown) for which

$$J(v) \equiv \int_{h_{-}(v)}^{h_{+}(v)} f(u,v)du = 0. \qquad (3)$$

This simply says that at the transition, the two states
$h_{\pm}$ are equally dominant, so the front stationary.

Let us first pose the problem on the infinite line[8],
imagining the membrane to be one-dimensional and infinite
in extent. We look for periodic differentiated solutions.
Let $I_{\pm}$ denote the collection of intervals in states $h_{\pm}$.
Then we have the problem: Find a function v(x), satisfying (from (2))

$$\nabla^2 v = \frac{d^2 v}{dx^2} = G(v) \equiv \begin{cases} -g(h_{-}(v),v) & \text{on } I-, \\ -g(h_{+}(v),v) & \text{on } I_{+}, \end{cases}$$

with v = v* at the transition points. Under cerain
sign conditions on g and its derivatives, there will
exist a solution. In fact, there will exist many apparently stable solutions. As shown in [8], there will
commonly be a one-parameter family of them, and they
are such that the wave-length is arbitrary, within limits,
though the relative sizes of the intervals in $I_{\pm}$ remain
fixed.

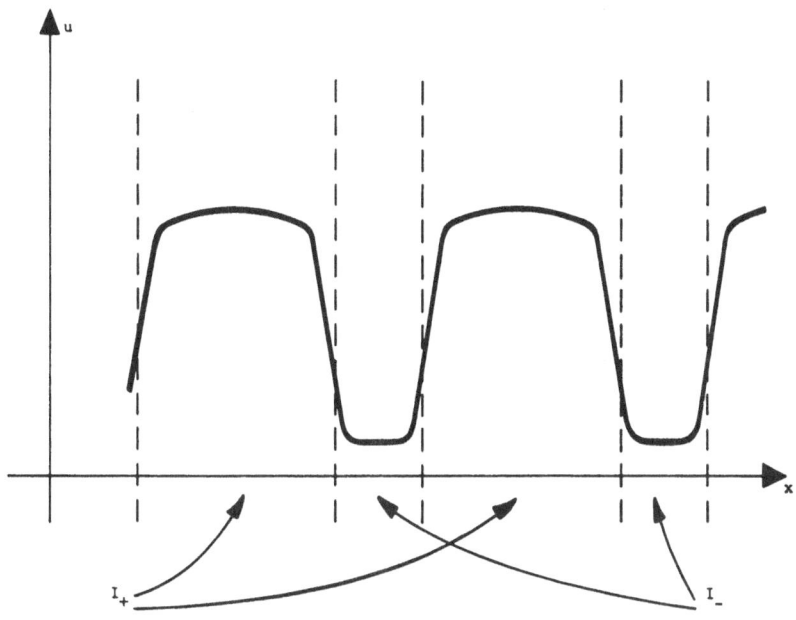

Figure 2

Now let us explore what would happen if we deform
the line into a circle, so that the solution is con-
strained to be periodic in position on the cirlce, with
period equal to the circumference of the circle or to
an integral part of it (Fig. 3).

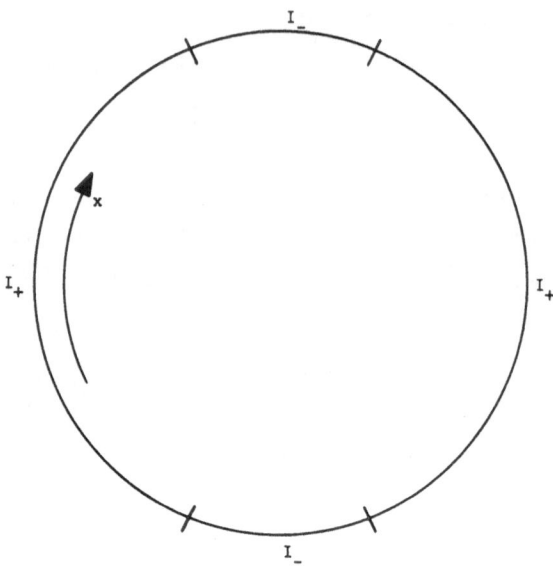

Figure 3

Instead of a one parameter continuous family of
solutions, we now would have a discrete set of solutions.
This type of configuration would be stable to growth,
for if the circle is now gradually stretched or shrunk,
the sizes of the intervals in $I_{\pm}$ will adjust so that the
same pattern will persist, only expanded. This is possible, because of the one degree of freedom we have for
the periodic patterns on the whole line. In this way,
a mechanism for pattern invariance with respect to size
could be established.

5. DIFFUSIVE OSCILLATIONS

The recognized phenomenon of biological oscillations
is in the same category of structured solutions of reaction-diffusion equations with which we have so far been
concerned. Oscillatory wave fronts moving across a

membrane have been discovered and computed by Caplan,
Deem, Naparstek, Thomas, Zabusky, and others[2,3,15]. What
I want to do now is explore the effects of which diffusive
transport can have on an oscillating chemical system.
The method will be asymptotic analysis on an illustrative
example, one of many possible. This example was analyzed
in greater detail in [6].

Imagine a chemical system with the property that, if
well stirred, it exhibits a relaxation oscillation. If
the hypothesized system were to have just two components,
its dynamics would be governed by the equations

$$\epsilon^2 \frac{du}{dt} = f(u,v),$$

$$\frac{dv}{dt} = g(u,v).$$

Here the small parameter ϵ^2 is put in for the purpose of
effecting the two time scales necessary for relaxation
oscillations. A typical such oscillator would be such
that f and g have the characteristics shown in Fig. 4,
where the dotted line is the approximate trajectory of
the periodic solution.

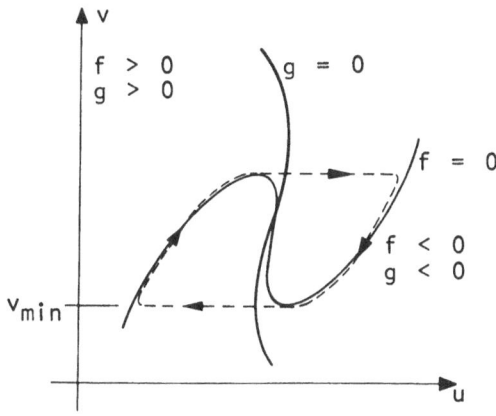

Figure 4

We now suppose there is spatial dependence and
diffusive transport. We assume the diffusivity of v
to be small, and setting it also equal to ε^2, obtain
the equations

$$\varepsilon^2 \frac{du}{dt} = f(u,v) + \varepsilon^2 \nabla^2 u,$$

$$\frac{dv}{dt} = g(u,v) + \varepsilon^2 \nabla^2 v.$$

Whereas the stirred reactor, the whole system change
quite suddenly from one state to another at a certain
stage in the cycle, in the present case, unless the distribution in space is uniform, the change of state occurs
first at some specific location- see the peak in the
diagram of Fig. 5.

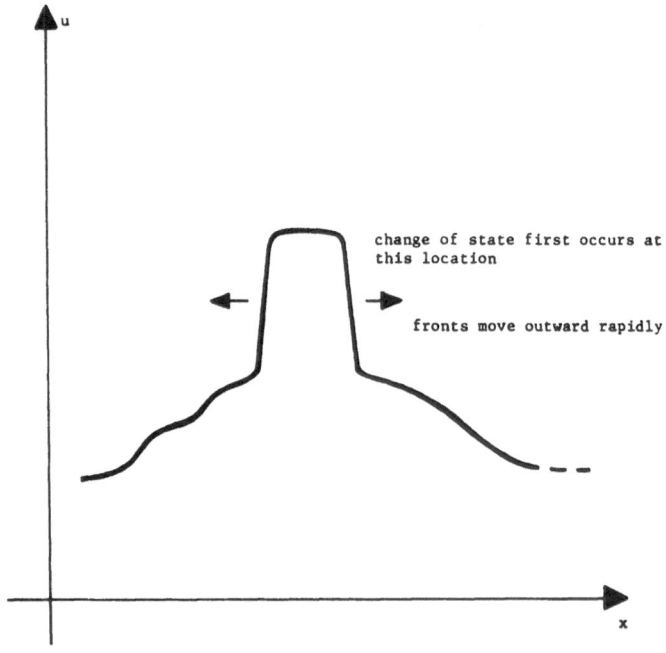

Figure 5

But once this localized transition takes place, a
cooperative phenomenon takes over: the state change at
one location induces a change at neighboring locations,
and so on. This produces two diverging wave fronts
similar to that mentioned before. The front moves
rapidly to a position where v = v* (where (2) is satisfied; see Fig. 1).

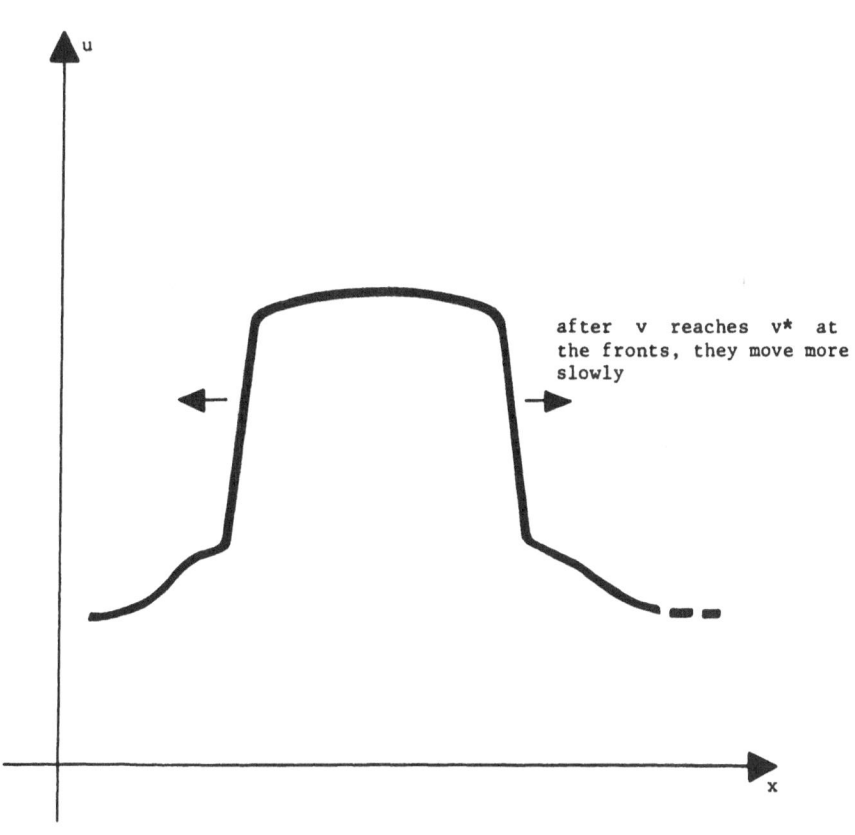

after v reaches v* at
the fronts, they move more
slowly

Figure 6

Once that point is reached, the fronts continue, due to the fact that v is changed through the action of g. This is an "induced" front[6] and it has moderate speed. The motion may continue until another front or a boundary is reached; in other words, until the entire reactor or membrane is in the state h_+. Then according to Fig. 4, v will decrease at each point in the medium. When v_{min} is attained at some position, the medium suddenly jumps to state h_- there, and a sequence of events analogous to the above recurs.

6. CONCLUSION

These phenomena, and many more, can be modeled, and the model analyzed by asymptotic methods. I shall not attempt to interpret any of the things mentioned here as reflecting specific known events in biophysics. I don't know whether they model reality or not. But I do believe that phenomena in some sense akin to those listed here do exist, and in any future definitive study of them, the student should be aware of the possibilities obtainable from simple reaction-diffusion models. The two points I have tried to put across are (1) the bag of tricks theoretically in the possession of biological membranes and other reacting media is indeed full, and furthermore (2) many of these tricks can be modeled and qualitatively analyzed by known mathematical methods.

REFERENCES

1. D.G. Aronson and H.F. Weinberger, Nonlinear diffusion in population genetics, combustion, and nerve propagation, in: Proceedings of the Tulane Program in Partial Differential Equations and Related Topics, Lecture Notes in Mathematics, No. 446, Springer, Berlin, 1975.

2. S.R. Caplan, A. Naparstek, and N.J. Zabusky, Chemical oscillations in a membrane, Nature 245, 364-366, (1973).

3. M.C. Duban, G. Joly, J.P. Kernevez, and D. Thomas, Hysteresis, oscillations, and morphogenesis in immobilized enzyme systems (to appear)

4. B. Edelstein, Biochemical model with multiple steady states and hysteresis, J. Theor. Biol. 29 57-62, (1970).

5. P. Fife, Pattern formation in reacting and diffusing systems, J. Chem. Phys. 64, 554-564, (1976).

6. P. Fife, Singular perturbation and wave front techniques in reaction-diffusion problems, Proc. AMS-SIAM Symposium on Asymptotic Methods and Singular Perturbations, New York, 1976.

7. P. Fife, Asymptotic analysis of reaction-diffusion wave fronts, Rocky Mountain J. of Mathematics, to appear.

8. P. Fife, Stationary patterns for reaction-diffusion equations, in: Papers on Nonlinear Diffusion, Proc. of NSF-CBMS Regional Conference on Nonlinear Diffusion, Research Notes in Mathematics, Pitman, London, to appear.

9. P. Fife and J.B. McLeod, The approach of solutions of nonlinear diffusion equations to travelling front solutions, Arch. Rat. Mech. Anal. to appear.

Announcements in Bull. Amer. Math. Soc. <u>81</u>, 1075-1078, (1975).

10. R.A. Fisher, The advance of advantageous genes, Ann. of Eugenics <u>7</u>, 355-369, (1937).

11. E.D. Gilles, Reactor models, in: Proc. Fourth International Symposium on Chemical Reaction Engineering, Heidelberg, 1976, to appear.

12. B.C. Goodwin, Excitability and spatial order in membranes of developing systems, in: Faraday Symposia of The Chemical Society No.9, 1974.

13. H-S. Hahn, A. Nitzan, P. Ortoleva, and J. Ross, Threshold excitation, relaxation oscillation, and effect of noise in an enzyme reaction, Proc. Nat. Acad. Sci. USA <u>71</u>, 4067-4071, (1974).

14. B. Lübeck, Vergleich von kontinuierlichen Modellen zur Beschreibung des dynamischen Verhaltens von katalystischen Festbettreaktoren, Chem. Eng. Jour. <u>7</u>, 29-40, (1974).

15. A. Naparstek, D. Thomas, and S.R. Caplan, An experimental enzyme-membrane oscillator, Biochimica et Biophysica Acta <u>323</u>, 643-646, (1973).

16. P. Ortoleva and J. Ross, Theory of propagation of discontinuities in kinetic systems with multiple time scales: Fronts, front multiplicities, and pulses, J. Chem. Phys. <u>63</u> 3398-3408, (1975).

17. G. Padberg and E. Wicke, Stabiles und instabiles Verhalten eines adiabatischen Rohrreaktors am Beispiel der katalytischen CO-Oxydation. Chem. Eng. Sci. <u>22</u>, 1035-1051, (1967).

18. W.H. Ray, Bifurcation phenomena in chemically reacting systems, in: Proc. Advanced Seminar on Applications of Bifurcation Theory, Math. Res. Center, U. Wisconsin, 1976, to appear.

PHYSICAL MECHANISMS OF NERVE EXCITABILITY

Harold Lecar

Laboratory of Biophysics, IRP

National Institute of Neurological and Communi-

cative Disorders and Stroke

National Institutes of Health

Bethesda, Maryland 20014

The major physical process underlying transmission of signals in the nervous system is the transient flow of ionic currents across cell membrane in response to various stimuli. All cells are bounded by selectively permeable membranes, which allow ion flow in at least two different ways. There is passive flow through ionic channels, and active "pumped" flow against electrochemical potential gradients, "fueled" by metabolic energy (ATP splitting). We will be concerned with current flow through specific channels which are activated during nerve excitation. Active transport is used to charge the cell batteries but is only of indirect importance in the signalling process.

Transient ionic currents are initiated when a region of the cell membrane separating the interior and exterior electrolyte media, suddenly undergoes a dramatic

change in ionic permeability. The stimuli which induce
permeability change are quite varied, and different cells
of the nervous system contain areas of cell membrane
which are specialized transducers for converting chemical,
mechanical, optical and electrical disturbances into
ionic permeability changes. Once pathways in the membrane have been activated, ion flow is downhill following
an electrochemical gradient. No energy sources other
than the preestablished concentration gradients across
the membrane are needed to drive the currents.

Since the interior of the cell has a high concentration of potassium ions and the external electrolyte
medium is rich in sodium ions, the two phases separated
by the membrane can act as a battery, a concentration
cell. For example, if the membrane were permeable only
to potassium, ions would diffuse until a field were built
up to retard flow down the concentration gradient and the
system would be in equilibrium at the potassium Nernst
potential,

$$V = (kT/e)\ln (\kappa_o/\kappa_i),\qquad\qquad (1)$$

where κ_o and κ_i are the ion concentrations in the two
solutions bounding the membrane. Thus if the membrane
were permeable to potassium only the normal potassium
concentrations would lead to a membrane potential of
-75mV. If the same membrane were permeable to sodium
alone, the potential would be 55mV. Resting nerve cells
have a membrane potential of -60 to -70mV, nearer to the
potassium equilibrium potential. When nerve impulses
are transmitted there is a propagated wave of permeability change, during which the membrane potential undergoes a transient 100 mV swing, as the membrane switches

between states of potassium and sodium permeability.

CURRENT FLOW THROUGH MEMBRANE CHANNELS

Let us briefly consider the current-voltage relation for a cell membrane having selectively permeable
ionic channels, as sketched in Fig. 1. Cell membranes
are of the order of 50-100 Å thick, and we can consider
an idealized channel which allows only a single species
of cation, eg. potassium to permeate. Because the membrane is thin and the pores are sparsely distributed,
space-charge effects can be ignored in the equation for
ion flow driven by a diffusion gradient and an applied
electrical field,

$$J = nA \ (-eD\nabla C + \sigma E + \sigma \nabla U(x)) \tag{2}$$

Here n is the density of channels, A is the effective
area of a single channel, D the diffusion constant of
the pore, σ the conductivity, and U is the potential
energy of any barriers within the channel (such as the
image force for entering the low dielectric membrane).
If the channel were thought to be filled with a viscous
medium, we would have $D = \mu kT$ and $\sigma = ce^2\mu$. Putting
these into Eqn. (2), setting J = 0 and integrating, we
would derive Eqn. 1,

$$V_c = \int_i^o Edx = \frac{kT}{e} \int_i^o \frac{dc}{c} - \int_i^o dU = (\frac{kT}{e})\ln (\frac{c_o}{c_i})$$

since U is zero at both boundaries. The useful constant
kT/e is equal to 25 mV at ordinary room temperatures.
For a mixture of channels or a channel which was partly
permeable to either species of ion, the same reasoning

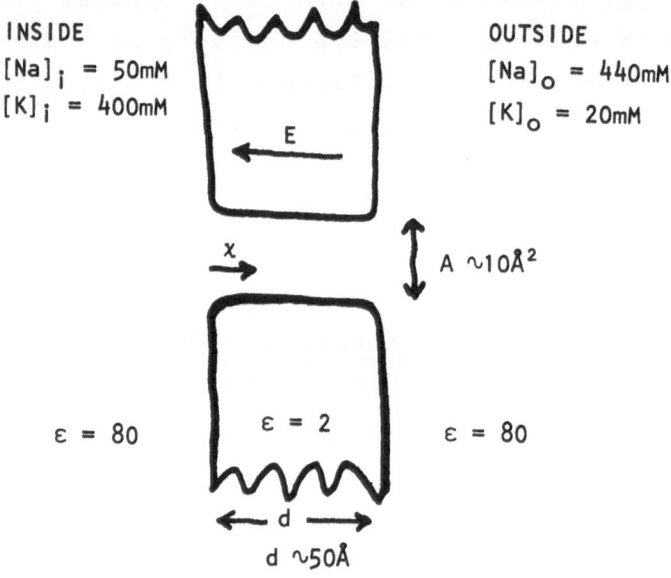

Figure 1. Ion flow through membrane pores.

would lead to

$$V = \frac{kT}{e} \ln \frac{\sum_x (P_x^+(X)_o + P_x^-(X_i)}{\sum P_x^+(X)_i + P_x^-(X_o)} \, , \qquad (3)$$

with $P_{x_{\pm}x}$ equal to μ_x for the simplest case. More generally, $P_x^{\pm}$ is relative permeability for ion species x ($\pm$
being the sign of the monovalent ion). If the P_x were
themselves functions of voltage, membrane potential would
be unstable when a stimulating current is applied.
Equation (2) can be solved for the channel current-voltage
relation. Letting $\phi = \frac{e}{kT} \int^x Edx$ and $U = \frac{e}{kT}U$, we have

$$J = -nA(kTe)\mu[\frac{dc}{dx} + c\frac{d\phi}{dx} + c\frac{dU}{dx}]$$

Multiplying by e^ϕ, and integrating we have

$$J = \frac{nAkTe\mu(C_i e^{\phi i} - C_o e^{\phi o})}{\int_i^o \exp(\phi + U)dx} \qquad (4)$$

Eqn (4) or the sum of such terms for several noninter-
acting species give a reasonable, mildly rectifying
current voltage relation for a thin membrane with ionic
channels.

Other models could be introduced for the current-
voltage relation of an ionic channel; the exact form of
$J(V)$ would depend upon the model, but the qualitative
behavior would be the same. Typically the channel conductance, $\frac{dJ}{dV}$, might increase by about e-fold per 25mV
for a large concentration gradient and is nearly linear
for a small gradient. Observed current-voltage relations
in static situations are of that sort.

In describing nerve excitation we will not be

concerned with the exact form of $J(V)$. In the Hodgkin-
Huxley analysis of ion currents[1] in the squid giant axon,
$J(V)$ was taken as

$$J_K = g_K(V-V_K) \qquad\qquad (5)$$

which appears to be a reasonable approximation under
normal physiological conditions.

SPECIALIZED IONIC CHANNELS

Fig. 2 shows a diagrammatic nerve summarizing some
properties drawn from the tremendous diversity of the
nerve cells. We see that the cells have three functional
regions, an input, where some external disturbance is
translated into a membrane current, a cable, the axon,
capable of propagating an electrical signal over long
distances, and an output end in which the electrical
signals do something to cause the emittance of chemical
transmitters, which are received by the input ends of
the next nerves down the line. The whole system is, of
course, exceedingly complicated with thousands of nerves
sometimes converging on the input connections of one
neuron, and that neuron, in turn, sending out diverging
outputs to many other nerves in the system.

The transient signals at the input and output units
of the neuron cannot travel long distances. The nerve
cell is essentially a leaky submarine cable with a conducting core of electrolyte, having a rather high resistance. Currents initiated at a point along the cable
will decay in a distance of the order of mm.

The elongated process called the axon is used for
long-distance signal transmission. The axon membrane is
capable of voltage-dependent ion permeability changes.

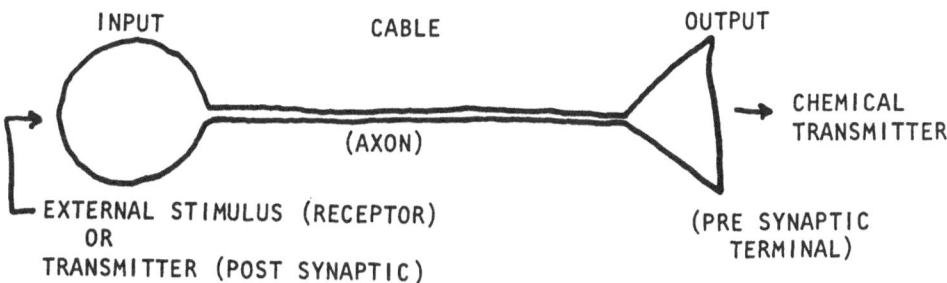

Figure 2.   Schematic neuron.

Thus, current flow through an active region of axon
membrane can stimulate a neighboring region by changing
the membrane potential. This leads to a regenerative
breakdown which can propagate along the cable structure.
Such breakdown events, in which there is a sudden change
of membrane permeability and then a recovery, are called
action potentials.

The action potentials are the units of information
transfer in the nervous system. A typical nerve action
potential is shown in Fig. 3. The voltage swing of
100mV lasts for about 1msec and travels down the cable
at a speed of about 20 m/sec. Action potentials have
a sterotyped shape, once a certain threshold stimulus
has been exceeded. Thus the actual information is coded
in the frequency of nerve firing.

The figure also shows the change in sodium and
potassium conductance which occur during the action potential. Originally these conductance changes were
computed theoretically[1] (Hodgkin & Huxley, 1952) from
equations based on studies of membrane currents under
controlled conditions.

The original work which explained the mechanism of
nerve impulse propagation required a means for separating
the membrane excitation from the cable properties. The
membrane currents were sorted out in the historic experiments of Hodgkin and Huxley[2] performed on the squid
giant axon, an extraordinarily large nerve fiber, which
can have a diamter of 1mm, or more than 20 times the
diameter of the largest vertebrate axons. Giant axons
can be impaled with a wire electrode in order to short-
circuit the resistance of the cable core and establish
an isopotential length of membrane cylinder with purely
radial current flow. Such an axon is called 'spaced

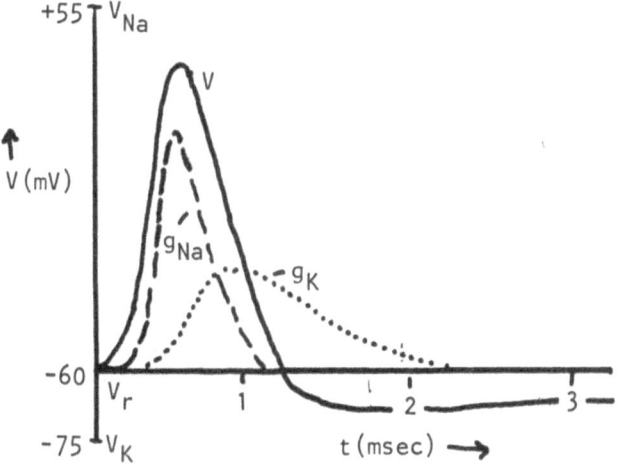

Figure 3. Nerve action potential (after Hodgkin and Huxley, ref. 1).

clamped' and one can study the membrane currents in response to various voltage steps. In order to maintain
a step of voltage across a membrane whose conductance
is changing, the nerve is incorporated as an element in
a feedback system. This method of studying step function
responses is called 'voltage clamp'. The voltage clamp,[3]
first developed by K. S. Cole, is the most important
experimental tool for studying membrane excitability.

The voltage clamp experiments show how the membrane
conductance responds to applied voltage steps. Fig. 4a
illustrates the transient activation of the sodium conductance and the slower relation of the potassium conductance. When the sodium conductance is activated from
rest, its +55mV battery is brought into play and a net
sodium current runs into the axon, counter to the applied
emf which activated the sodium channels. Thus the system
displays a negative dynamic resistance, much like electronic relaxation oscillators.

The main progress in understanding nerve excitation
in recent years has come from the gathering of evidence
that the curves of Fig. 4 represent the activation of
two independent species of ion channels within the membrane. The experimental work characterizing the properties of the channels is discussed in several reviews[4-8].

The voltage clamp method has been extended by in-
genious means to the study of excitable cells which are
either too small or too irregular in their geometry to
be impaled by wires. The picture of membrane excitation
which emerges from voltage clamp studies of different
tissues is quite similar to that derived for the giant
axons, but the details of cable propagation depend upon
cell geometry.[9]

The equivalent electric circuit for a unit area of

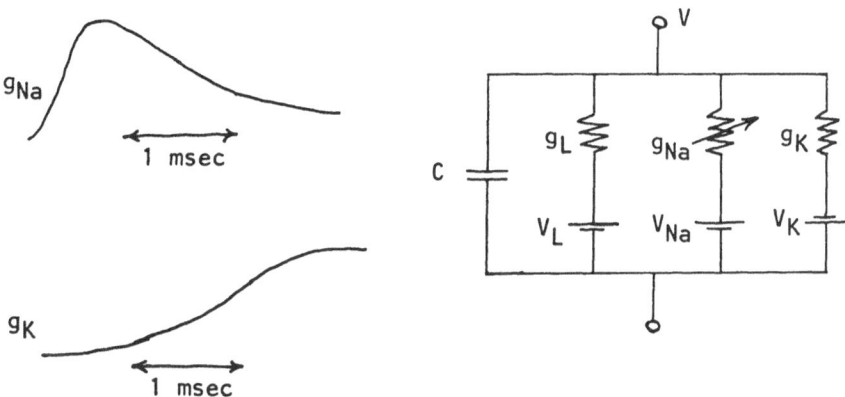

Figure 4. Conductance transients and equivalent circuit
of an axon membrane under space clamp conditions.

membrane is shown in Fig. 4b. There are three conductance
pathways each represented by a resistor and a battery.
The ionic pathways are not linear conductances, but their
short-time properties can be approximated by a linear
relation. Thus, the resistors of Fig. 4b are supposed
to represent ionic diffusion pathways, with emfs coming
from the concentration gradients across the membrane.
The resistors with arrows are voltage dependent. The
major voltage dependence comes from the activation of
pathways which can undergo conductance transitions.

The ionic currents obey the equations

$$J = C\frac{dV}{dt} + J_i \tag{7}$$

$$J_i = g_L(V-V_L) + g_{Na}(V-V_{Na}) + g_K(V-V_K) \tag{8}$$

Here the first term of Eqn (8) is just the observed leak
current through the membrane and the other two terms are
the currents flowing through the sodium and potassium
pathways. In order to describe the transient activation
of the conducting channels, we must have equations giving
the response of g_{Na} and g_K to changes in membrane potential. Hodgkin and Huxley did this by using auxiliary
variables extracted from the voltage clamp data. Thus

$$g_{Na} = \bar{g}_{Na}\, m^3 h \tag{9}$$

$$\frac{dm}{dt} = \frac{m_\infty(V)-m}{\tau_m(V)} \tag{10}$$

$$\frac{dh}{dt} = \frac{h_\infty(V)-h}{\tau_h(V)} \tag{11}$$

$$g_K = \bar{g}_K \; n^4 \tag{12}$$

$$\frac{dn}{dt} = \frac{n_\infty(V) - n}{\tau_n(V)} \tag{13}$$

Here $\bar{g}_{Na}$ and $\bar{g}_K$ represent the maximum conductances per unit area, and are hence equal to the density of channels, N, times the conductance of a unit channel, γ. Later we shall discuss the experimental means of finding N and γ.

The equations are completely specified in terms of six empirical functions of applied potential, the steady state values and relaxation times of the three auxiliary conductance variable, m, h and n. These functions are plotted in Fig. 5 and all appear to have similar shapes. The steady state functions, m_∞, h_∞, and n_∞ have sigmoid shapes much like the Langevin function for orienting dipoles. The relaxation time related to the process is resonant with voltage. At first blush, one doesn't usually associate a voltage-dependent relaxation time with dipole relaxation. This is because dipole energies in ordinary electric fields are much less than kT. For a cell membrane, however, the fields are quite large. A 100mV potential across a 50 Å membrane gives 200,000 v/cm, near breakdown for many materials. The energy gained by moving a single electronic charge across this potential drop is of the order of 4kT so these characteristic voltage dependent functions are a clue to the structural changes brought on by potential change.

Eqns 7-13 form a fourth-order nonlinear system for predicting all of the phenomena associated with the generation of action potentials in space-clamped axons. The equations predict thresholds, oscillations, latencies, refractory periods. They also provide a means of

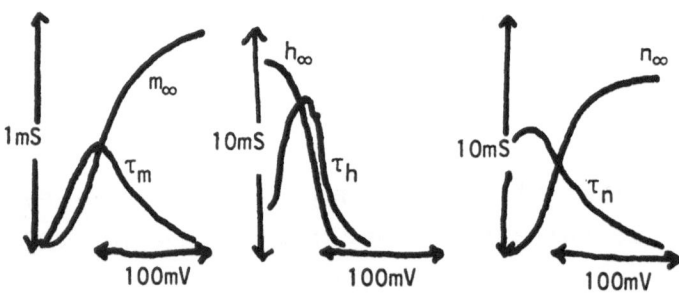

Figure 5.   Conductance parameters in Hodgkin-Huxley equations.

predicting how temperature, ionic composition and various
drugs modify excitation when one knows how the voltage-
clamp data varies in the presence of these influences.

Perhaps, the most spectacular success of the Hodgkin-
Huxley theory was the prediction of the propagated action
potential by combining Eqns 7-13 with the membrane cable
equations,

$$\frac{a}{2R} \frac{\partial^2 V}{\partial x^2} = J = C\frac{\partial V}{\partial t} + J_i(V,m,n,h) \tag{14}$$

where a is diameter, R is the resistivity of the core,
and J_i contains all the membrane conductance dynamics.

Propagation can be described as a nonlinear eigenvalue problem. There is a constant-speed solution to
the full partial differential equation, and the speed
that fits agrees with experiment. That is, one sub-
stitutes a trial solution,

$$V = V(x-\theta t),$$

so that Eqn (14) reduces to

$$\frac{a}{2R} \theta^2 \frac{d^2 V}{dt^2} + C\frac{dV}{dt} = J_i(V,m,n,h) \tag{15}$$

The value of θ, for which Eqn (15) has a solution is in
good agreement with observation[1].

One interesting application of the full set of
equations is the description of propagation in excitable
cells which have membrane properties following a system
like Eqns 7-13 but with different cable structures. The
most important example is propagation in myelinated
nerve[10]. Myelinated nerves of vertebrates do not form
a continuously conducting membrane cylinder, but rather

have small lengths called nodes of Ranvier which conduct
according to Eqns 7-13. The nodes are separated by mm-
long insulating lengths. Conduction occurs from excitable node to node and the partial differential equation is replaced by a set of ordinary differential equations describing the saltatory conduction. Saltatory
conducting is a way of getting higher speeds from small
nerves. Another example of Eqns 7-13 imbedded in a more
complicated cable structure is the system for describing
the conduction of action potentials through striated
muscle[11], in which the outer cylinder has excitable tu-
bules running into the interior of the cell. Thus there
is a wide class of phenomena where Eqns 7-13 imbedded in
different cable structures provides a complete description.

There are also excitable tissues with different
species of channels. Heart muscle appears to operate
by the same principle, but with more complicated dynamics.
A number of excitable cells contain calcium channels in
addition to the sodium and potassium channels. Thus
there is considerable variety, though the Hodgkin-Huxley
model is a good prototype of all electrical excitation.

It must be remembered that the Hodgkin-Huxley equations are an empirical description. The underlying parameters are not uniquely measurable, and a number of
alternative dynamical schemes have been proposed which
fit the voltage clamp data and also describe excitability.
Despite the nonuniqueness, the equations provide a convenient framework both for discussion of the physical
mechanisms underlying excitation and for the mathematical
modeling of excitation phenomena.

The Hodgkin-Huxley equations have been the basis of
much modeling of nerve excitation and transmission. In

addition to the numerical studies of the full set of
equations, there have been a number of studies of simpler
systems of equations related to the Hodgkin-Huxley model
and capable of describing some significant features of
excitation. Readers interested in the mathematical
theories of excitation and propagation may read the reviews of FitzHugh[12], Scott[13] and Cohen[14]. In order to
give the flavor of these studies, we shall briefly mention two problems - the excitation threshold and the
constant speed of propagation.

The Hodgkin-Huxley equations exhibit a sharp
threshold for input stimuli. Physically this is to be
expected from a circuit such as that of Fig. 4b, in
which the two large conductances, g_{Na} and g_K have op-
positely biased batteries. Since a voltage positive
with respect to rest tends to increase g_{Na}, leading to
further regenerative voltage increase, a sodium current,
J_{Na}, larger than the sum of $J_K + J_L$ will cause the system
to move rapidly away from the resting state as long as
g_{Na} is activated. Any current value less than the critical amount will cause a disturbance which will eventually dissipate by current flow through the potassium
and leakage pathways.

One approximation in which the threshold properties
of the equation can be seen clearly is the V, m-reduced
approximation of FitzHugh[12]. Since the variables, V
and m have much faster response times than n and h, we
might look at the equations with only V and m varying
in time and n and h, the recovery variables, frozen in
their tracks. The resultant second order system can be
studied in a V, m phase plane.

$$\frac{dV}{dt} = C^{-1}[J - J_i(V,m)]$$

$$\frac{dm}{dt} = \frac{m_\infty(V) - m}{\tau_m(V)} \tag{16}$$

$$J_i(V,m) = [\bar{g}_{Na}h(0)]m^3(V - V_{Na}) + (g_K(0) + g_L)(V - \frac{g_K(0)V_K + g_L V_L}{g_K(0) + g_L})$$

As shown in Fig. 6, the V, m-reduced equations have
three singular points. The two stable singular points,
A and C, correspond to the resting state and the peak
of the activity (the latter point would not be stable
if the recovery variables were operating). The singular
point B is a saddle point corresponding to the threshold;
the dashed line in the figure is the threshold separatrix,
which divides the two classes of trajectories. The V,
m model gives a good account of many of the threshold
properties when a nerve is stimulated by sufficiently
short duration pulses so that the recovery variables
aren't brought into play. More complete descriptions
of reduced phase spaces and their relation to the threshold properties of the complete equations are given in
the papers of FitzHugh[12].

A second example is the prediction of the constant-
speed propagation for an even more drastically reduced
model. If in addition to freezing the recovery variables,
the sodium conductance is speeded up, so that $\tau_m \to 0$ and
J_{Na} is always at its steady state value,

$$J_{Na} \cong \bar{g}_{Na} h(0) m_\infty^3(V)(V - V_{Na}) = J_i(V)$$

the excitable system is now represented by an instan-
taneously responding negative resistance element. Now

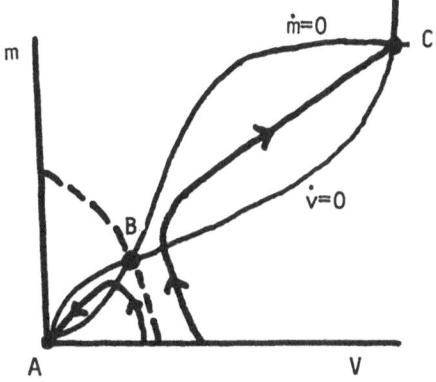

Figure 6.   V, m reduced phase plane (after FitzHugh, ref. 12)

Eqn (14), for the propagated wavefront of activity becomes a nonlinear diffusion equation of a type first
studied by Kolomogorov (cf. Scott, 1975). Such equations have constant-speed solutions. We can further
approximate Eqn (14) by letting the current voltage relation be represented by a cubic.

$$J_i(V) \cong (gV_D^{-3})(V-V_A)(V-V_B)(V-V_C) \qquad (17)$$

Then Eqn (15) can be solved by writing the equation for
$\frac{d\dot{V}}{dV}$ and searching in the V, $\dot{V}$ phase plane for the solution curve extending from V_A to V_C. Problems of this
sort have been discussed in a recent review by Hunter,
McNaughton and Noble[15]. The original problem as stated
here was first solved by Huxley for the case $V_A \cong V_B =$
0 and $V_D = V_C$.

He obtained an expression for the conduction speed

$$\theta = \frac{1}{2}(ag/RC^2)^{1/2} \qquad (18)$$

We shall not discuss other mathematical models based
on the Hodgkin-Huxley model, but rather refer the reader
to the reviews cited. Perhaps the most flexible and useful of the simplified models is the FitzHugh-Naguomo
model[12,14] which uses a cubic approximation for the
sodium current, but also includes a recovery variable
which roughly corresponds to the potassium current. This
model shares many of the properties of the Hodgkin-Huxley
equations in a phase space of reduced dimensions.

GATED CHANNELS IN NATURAL AND SYNTHETIC MEMBRANES

The first definite indications that conductance

occurs at discrete sparesely distributed sites comes
from studies of the action of the puffer fish poison,
tetrodotoxin. The number of tetrodotoxin molecules ad-
sorbed to the membrane for blockage of the sodium conductance can be measured[16]. Assuming one molecule per
conduction site, leads to a density of about 500 sites
per square micron of squid axon membrane.

Thus there are discrete conductance sites, separated
by ~1000Å. One would like to know whether these sites
are identical channels or conducting domains, whether
each microscopic conductance region is a microscopic
negative resistance element, or whether the electric
field across the membrane can somehow switch ion channels
on and off. Some insight into the nature of the conducting sites which give excitable membranes steep voltage
sensitive conductance can be obtained from experiments
on a class of synthetic membranes which resemble cell
membranes in their major physical properties.

Synthetic lipid bilayer membranes[17] can be made by
spreading a solution of lipids over a thin hole in a
partition separating two electrolyte solutions. Bimolecular films can form spontaneously as solvent and
lipid flow out to the periphery of the film. The stable
bimolecular membranes are exact counterparts of cell
membranes. Lipid bilayers, like cell membranes, are 40
to 70 A thick, and have capacitance of about 1 mF/cm^2.
This is what you would expect for a thin layer of oily
material with dielectric constant of 2. The breakdown
voltage for such bilayers is about 200 mV. Bilayers even
show the same water permeability as cell membranes.

One would expect a slab of material with dielectric
constant 2 to be impermeable to ions coming from solution,
since the image force to be overcome is of the order of

46 kcal/mole. In fact unmodified bilayers are good in-
sultators, with conductances of about 10^{-8} mho/cm^2.
This is to be compared to 10^{-4} to 10^{-3} mho/cm^2-for a
resting cell membrane and 10^{-} mho/cm^2 for an axon membrane at peak activity.

The conductance of a bilayer can be dramatically
increased by the addition of a number of different substances. These substances act either as carriers for
ions or as pores through the membrane. Carriers are
substances which combine with the ion at the membrane
surface to form a dielectric jacket; the ion carrier
complex is effectively a larger ion which then doesn't
have as big an electrostatic self energy to surmount,
when entering the lipid phase.

Certain additives can induce a voltage dependent
permeability in bilayers[18]. The voltage dependent conductance and voltage dependence relaxation times induced in synthetic membranes by a bacterial protein called
EIM resemble the empirical functions of Fig. 5.

EIM excitability can be studied one pore at a time[19]
Discrete conductance steps can be observed when EIM is
first added to the solutions bathing the membrane. These
steps represent conductance units of about 4×10^{-10} mho
or ion fluxes of 10^{8} ions per sec in a 100 mV potential.
Fig. 7a shows the results of an event in which a single
EIM conductance channel remained active in the membrane
for a long time. Trains of current humps can be observed
for this isolated channel, and one can ask how these
fluctuations relate to the macroscopic voltage dependent
permeability.

The frequency and amplitude of the conductance jumps
are measured for different potentials. The jump amplitudes are linear in potential through the negative

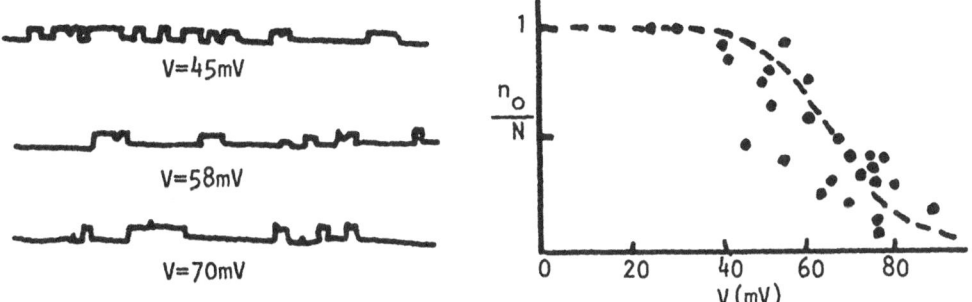

Figure 7.   EIM unit conductance jumps and relative
frequency of channels in open state (after Lecar, et
al, Ref. 18).

resistance region. The jump frequencies, on the other
hand, vary exponentially with applied potential. Fig.
7b shows the fraction of time that a single channel stays
open as a function of membrane potential. The numbers
next to the points in the figure indicate the number of
channels in a particular few-channel experiment. The
dashed curve in the figure is the normalized voltage
dependent conductance as determined from a many-channel
membrane. It is also the theoretical curve derived from
a very simple theory. Consider ion channels capable of
two structural states, one conducting, the other non-
conducting. The free energy difference between the two
states has an intrinsic conformational term and an
electrical field term,

$$\Delta W = \Delta W_c + QV = Q(V-\bar{V}) \tag{19}$$

then in equilibrium at any voltage, the Boltzmann distribution of open and closed channels would lead to the
desired sigmoid function for the steady state conductance.
Referring to the free energy diagram, we expect

$$n_{open}/n_{closed} = \exp(\Delta W/\kappa T) \tag{20}$$

and $n_0 + n_c = N$, the total number of channels. Thus

$$g_\infty(V) = n_0(V)\gamma = \gamma N(1+\exp\frac{Q}{\kappa T}(V-\bar{V}))^{-1} \tag{21}$$

where γ is the conductance of a single channel. Furthermore, the simplest picture of transitions over a single
barrier leads to the proper bell-shaped relaxation time.
     If we think of the conformational part of the free
energy barrier itself as relatively unchanged by the

electric field, then the relaxation rates for the open-
to-close transition are

$$\alpha, \beta = \lambda \exp[\pm(Q/2\kappa T)(V-\bar{V})] \qquad (22)$$

and

$$\tau(V) = \frac{1}{\alpha+\beta} = \frac{1}{2\lambda} \operatorname{sech} \left[\frac{Q}{2\kappa T}(V-V)\right] \qquad (23)$$

The functions of $g_\infty(V)$ and $\tau(V)$ of Equations (21) and
(23) have exactly the appearance of the voltage dependent parameters in the Hodgkin-Huxley theory. To
see the full range of the curves of Eqns 21 and 23 would
require considerable charge motion or dipole reorienta-
tion. Even for fields as large as those within the membrane it would take the motion of several electronic
charges per channel across the entire membrane potential
to account for the degree of saturation we see here.
For example, EIM requires 3.3e per channel.

For EIM, the transition really is a Poisson process,
as if some charged group sat at the bottom of a deep
free energy well and waited a long time till it could
receive enough thermal energy to flip across the membrane.
For the axon, the kinetics of conductance change follow
more complicated laws. If the Hodgkin-Huxley equations
were literally true there might be several groups which
must act cooperatively to open a channel.

CHANNELS IN NATURE: GATING NOISE AND GATING CURRENTS

The observation of single channel current fluctuations
in bilayers depended on a number of fortunate experimental
circumstances. These may not necessarily occur for natural
membranes because you can't easily control the number of

channels, there is more background noise, and the channels may have smaller conductances. However, the same
information can be inferred from the electrical noise
caused by many channels opening and closing at random.

Consider the N channels of the simple switching
model. In steady state at a given voltage each channel
generates a random telegraph signal. The channels are
all independent and the N channels opening and closing
generate a mean current given by

$$\bar{I} = n_0(V)\gamma(V-\bar{V}) \tag{24}$$

and a fluctuating current having the power spectrum

$$\rho_I(\omega) = \frac{(\alpha/\pi)\gamma^2 n_0(V)(V-\bar{V})^2}{1+\omega^2\tau^2(V)} \tag{25}$$

Here we note that the ratios of variance to mean of the
ratio of zero-frequency power to mean gives the current
of a single channel. The spectrum of Eqn (25) is
Lorentzian and such spectra have been studied now for a
number of excitable membranes including the axon, post
synaptic membrane, and photoreceptors. From these
measurements typical values ranging from 10^{-12} - 10^{-10}
mho were found.

For a channel obeying the HH kinetics the statistics
are not simply Poisson and the spectrum is given by a
weighted sum of Lorentzians. The sodium and potassium
channel noise spectra have been observed and studied under
a variety of conditions. Noise measurement has become
the prime means of estimating the elementary channel conductance[20].

Since the opening of the channels is preceded by the

motion of a considerable amount of charge, 3 to 6 electron
charges per channel, this charge motion might be ob-
servable as a displacement current. The displacement
current was predicted by Hodgkin and Huxley, but never
observed, because it is swamped by the early surge of
ionic current. It is now possible to unmask the gating
currents[6,16] in systems in which all the ionic currents
are blocked by channel-clogging drugs.

The gating currents appear as asymmetrical and
saturable displacement currents more or less following
the kinetics of a hypothetical m subunit. The effective
charge of a gating particle, as estimated by Keynes and
Rojas is 1.3e, close to 1/3 the charge needed to activate
the channel. The kinetics of the gating follow the m
kinetics in some respects. However there is not an exact
correspondence. Under certain circumstances the gating
currents are inactivated by depolarizing pulses, thus
showing that activation and inactivation are coupled.
No gating currents have yet been observed which correspond
to h and n.

The gating current measurements lead to an independent
measure of the channel density. The number obtained,
$\sim 500/cm^2$ is in good agreement with estimates from tetrodotoxin binding and noise. One interesting application
of the approximate theories of propagation which relates
some of the applied mathematics with the membrane mechanism
studies was the recognition by A. L. Hodgkin[21] that the
gating currents introduce a source of dielectric loss
proportional to the sodium channel density. This requires a reconsideration of the approximate theory of
conduction speed. In various simple theories of conduction speed, it can be shown that θ is proportional to
different powers of C_m and g_{Na} (cf. Eqn 18). Since both

of these quantities are proportional to the density of
Na channels, Hodgkin was able to show that conduction
speed had a maximum value for a particular channel density. For the conditions of the squid giant axon, the
optimum density is 575 per square micron in agreement
with all the experimental estimates. This is a rather
nice example of the blending of the mathematical and
physical considerations.

In summary, signal transmission in nerve axons is
well described by an empirical set of differential equations, the Hodgkin-Huxley equations. Although the molecular basis of these equations has not been determined,
it is now clear that they represent the dynamics of the
opening of discrete ion-selective channels in the cell
membrane. Other membrane transduction processes in nerve
cells, such as synaptic transmission or photoreception
also involve specialized ionic channels, and the focus
of much of present research is on the inference of the
molecular structures and transformations of these channels

REFERENCES

1. Hodgkin, A. L. and Huxley, A. F. (1952). J. Physiol.
   117, 500-544. "A quantitative description of membrane current and its application to conduction and
   excitation in nerve."

2. Hodgkin, A. L. (1964) "The Conduction of the Nervous
   Impulse". Liverpool U. Press, Liverpool.

3. Cole, K. S. (1968). "Membranes, Ions and Impulses".
   University of California Press, Berkeley.

4. Hille, B. (1970). Progs. Biophys. Mol. Biol. 21:
   1-32. "Ionic Channels in Nerve Membranes."

5. Ehrenstein, G. and Lecar, H. (1972). "The mechanism
   of signal transmission in nerve axons." Annl Review
   of Biophysics and Bioengineering 1: 347-368.

6. Armstrong, C.M. (1974). "Ionic pores, gates and
   gating currents." Quarterly Reviews of Biophysics
   7, 179-210.

7. Hille, B. (1975). "Ionic selectivity of Na and K
   channels of nerve membranes." In "Membranes: A
   series of advances". G. Eisenman, ed. Marcel Dekker,
   Inc., N.Y.

8. Ehrenstein, G. (1976). "Ionic channels in nerve
   membranes." Physics Today 29, No. 10.

9. Jack, J.J.B., Noble, D. and Tsien, R.W. (1975).
   "Electric Current Flow in Excitable Cells".
   Clarendon Press, Oxford.

10. FitzHugh, R. (1962). "Computation of impulse initia-
    tion and saltatory conduction in a myelinated nerve
    fiber." Biophys. J. 2: 11-21.

11. Adrian, R. H. and Peachey, L. D. (1973). "Recon-
    struction of the Action Potential of Frog Sartorius
    Muscle." J. Physiol. 235, 103-131.

12. FitzHugh, R. (1969). "Mathematical models of ex-
    citation and propagation in nerve." In "Biological
    Engineering", H. P. Schwann, Ed. McGraw-Hill, Inc.,
    N.Y.

13. Scott, A.C. (1975) "The electrophysics of a nerve
    fiber." Reviews of Modern Physics $\underline{47}$, 487-533.

14. Cohen, H. (1976). "Mathematical Developments in
    Hodgkin-Huxley Theory and its approximations."
    Lectures on Mathematics in the Life Sciences.

15. Hunter, P.J., McNaughton, P.A. and Noble, D. (1975).
    "Analytical Models of propagation in excitable cells."
    Prog. Biophys. & Mol. Biol. $\underline{30}$, 99-144.

16. Keynes, R. D. (1975). "The ionic channels in ex-
    citable membranes." In Energy Transformation in
    Biological Systems. Ciba Foundation Symposium 31,
    191-203.

17. Mueller, P. and Rudin, D.O. (1969). "Translocators
    in Biomolecular Lipid Membranes." Current Topics
    in Bioenergetics $\underline{3}$: 157-

18. Ehrenstein, G. and Lecar, H. (1977) "Electrically
    Gated ionic channels in lipid bilayers." Quarterly
    Reviews of Biophysics $\underline{10}$ (In press).

19. Lecar, H., Ehrenstein, G. and Latorre, R. (1975).
    "Mechanism for channel gating in excitable bilayers."
    Annals of the N.Y. Acad. Sci. $\underline{264}$: 304-313.

20. Conti. F. and Wanke, E. (1975). "Channel noise in
    nerve membranes and lipid bilayers." Quarterly
    Reviews of Biophysics $\underline{8}$, 451-506.

21. Hodgkin, A. L. (1975). "The optimum density of
    sodium channels in an unmyelinated nerve." Phil.
    Trans. R. Soc. Lond. B. $\underline{270}$: 297-300.

OF TRIALS AND ERRORS

Feza Gürsey

Yale University

New Haven, Connecticut 06510

ACCEPTANCE SPEECH FOR THE 1977 J. ROBERT OPPENHEIMER
MEMORIAL PRIZE, JANUARY 19, 1977

Ladies and Gentlemen,

I feel slightly bewildered and deeply grateful on accepting this award. Since I was not able to decipher the citation I am still not quite sure about what I have done to deserve the honor. But, as the Turkish saying goes, one should eat the grapes and not ask about the vineyard where they were picked. On the other hand, the committee has shown wisdom in being imprecise, as in the perspective of history most specific contributions will appear to be as many trials and errors, with the errors outweighing the trials.

In any case I should be less than sincere if I do not express my happiness about many aspects of this ceremony.

To begin with, it is the first time that the prize has been split. The singlet prize has turned into a doublet with up and down components, with me on my way down and my younger colleague on his way up. Being one

of two certainly alleviates any loneliness I might have
felt otherwise. I am also particularly proud to share
this honor with Sheldon Glashow. It reminds me of the
time, fifteen years ago, when we were together in the
first summer school devoted to Group Theoretical Methods
in Physics for which I was responsible in Istanbul. I
had asked Shelly to give (in parallel with David Speiser)
the first summer courses on the brand new SU(3) symmetry
of Gell-Mann and Neeman. I reread his lecture notes the
other day. His colorful words ring surprisingly new
and true. He talks about the eightfold way, the de-
couplet, the omega minus before its discovery (remember
it was 1962), about gauge theories, spontaneous symmetry
breakdown and ways of synthesizing strong, weak and
electromagnetic interactions. In fact, the only major
thing missing from the lecture, but supplied abundantly
by his personality was charm, which he proposed with
Bjorken only a year later.

     Another source of excitement for me was to see
Behram Kursunoglu's name conclude the telegram I received
a few weeks ago. We have now been friends for thirty
seven years. Since I have been long familiar with his
great sense of humor I could not be certain whether he
really meant the words in the telegram or intended it as
a private joke. Now, like many others in the present
company, I feel grateful to him for creating this most
stimulating and enjoyable series of conferences and de-
dicating the prize to the memory of J. Robert Oppenheimer.

     Oppenheimer's name in this connection revives some
of my most precious memories. I have had the privilege
of being always encouraged by him in my efforts to do
physics and bridge two cultures. He studied me with
affectionate irony, telling me I was an unusual case, not

young in age, but in many respects like a young physicist
because of my late reentry in Physics. My place in a
no man's land, between the U.S. and Turkey, somewhere
between orthodoxy and crackpotism, between old age and
young age both amused and interested him. My gratitude
to J. Robert Oppenheimer is only matched by the deep
affection I have for his memory.

        After these random thoughts connected with the award
let me add some personal notes about my career as a
physicist. I heard my friend Shelly Glashow some weeks
ago jokingly refer to his thoroughly immodest odyssey in
a talk he gave at Yale. Since a modest odyssey would be
self contradictory, I prefer to give you some examples
of my early failures and - to use Dyson's phrase - missed
opportunities for which, I feel, various subtle punishments should be devised to accompany the awards of prizes,
the way percussion accompanies the sound of flutes in
Eastern music.

        The first in my series of errors occurred when I
went to England together with Behram in 1945 to do graduate
work in Physics. The British Council had decided that
we should study in Edinburgh. Behram did go there, but
I stayed in London which seemed to offer greater op-
portunity for fun and bohemian intellectual life. Max
Born who was king of physics in Edinburgh was puzzled by
my choice and wanted to meet me during one of the trips
he made to London to attend the meetings of the Royal
Society. He asked me what I wanted to work on and whom
I wanted to work with in London, naturally assuming that
I had a definite program of study in Physics all worked
out. Of course I had no idea. Then, pacing up and down
in the courtyard of the Royal Society he said, "I see,
you have no idea. If you want a suggestion, I can tell

you what I would do if I were your age. I would learn
group theory, especially the continuous groups and their
representations. I have the feeling that they will gain
increasing importance in Modern Physics. You should be
prepared when the time comes."

When the time came thirteen years later I was not
prepared. First, I had not understood what Born was
talking about. Second, in the intervening years I had
learned about the Lorentz group and the de Sitter group
through my interest in General Relativity, but I knew
practically nothing about unitary representations or
dimensions of multiplets or the kind of groups that are
not rotation groups, like unitary or symplectic groups,
let alone exceptional groups. The year was 1958 and I
was having a brief association with Wolfgang Pauli a few
months before his death. For him I was the Turk from
Brookhaven where I had been working for a year. He was
lecturing in Berkeley and had me invited there. He was
deeply disillusioned with the outcome of his collaboration with Heisenberg on the unified non-linear field
theory for particles which involved only one fundamental
spinor field. But he was convinced that some underlying
symmetry existed between hadrons and leptons, mainly because of the Pauli group for neutrinos which I had shown
to represent also the isospin rotations of a left handed
isotopic doublet. This group shared by leptons and
hadrons had impressed Pauli who asked me to find out the
minimum number of fundamental fields for various possible
schemes that would generate all particles. Among others
I found one with three spinor fields. But since they did
not correspond to known particles I assumed that they did
not obey the asymptotic condition. Pauli liked the idea
of unseen constituents. But nobody else did. At the

same time we were talking about the baryons. I asked
him if strange baryons could have a different parity.
He said to me, "The eight baryons of the Gell-Mann-
Nishijima scheme look like members of the same multiplet.
They should have the same parity. There must be some
group which corresponds to these baryons". Thus we had
3 constituents and 8 baryons. But because I had not
heeded Born's advice I could not put 3 and 8 together.
As you know that was done by Gell-Mann and Neeman three
years later. I deserved a good punishment for this fit
of blindness and got it when Pauli asked me to give a
talk on this subject in Brookhaven. The moral of the
story is this: if young people in the audience expect
some advice from their older colleagues, it is quite
useless. The advice never works.

Other punishments I deserve would be for the failure
of going to Birmingham in 1947 when Peierls was assembling
a group to work on Quantum Field Theory and wanted to
include me in it. Among the people who joined the group
were Dalitz, Salpeter and Dyson. I had to relearn Field
Theory the hard way much later. Again I had not appreciated the importance of that activity and showed a
poor sense of timing.

Yet another hiding should be administered for my
failure to understand the physical implications of the
theory after construction a non-linear Lagrangian with
chiral symmetry in 1958. This was done admirably by
Gell-Mann, Nambu and Weinberg later on.

Such examples should be sufficient to illustrate the
unending comedy of errors. When a few things manage to
work out in the end it is a source of constant amazement
to me. Some of our breed have the Midas touch. Like
Professor Dirac or Professor Gell-Mann, anything they

touch turns to gold. For most of us, however, some
ideas may work out in spite of ourselves, the way they
do for Inspector Clouzeau in the famed Pink Panther
series. The bumbling detective survives just because
he is too innocent and too dumb to realize the dangers
involved. Does the bumbling method work out of sheer
luck? Do errors have a way of correcting themselves,
if not for each physicist, at least for the world of
physicists? I have the sneaking suspicion that all roads
eventually lead to Rome in Physics.

The ludicrous story of the way $SU(6)$ symmetry has
survived after many disguises like $SU_w(6)$, current algebra, the Melosh transformation and partial justification through asymptotic freedom is a pure Pink Panther
episode.

The deep reason for spontaneous healing properties
of injured or sick theories may be that there are only
certain possible structures that are both consistent and
growing mathematical organisms. Both our imagination
and Nature's imagination are limited. Thus, if we follow
any lead, even a false lead, if we start with any model,
even a bad model we are condemned to land in the end on
the lap of the collection of possible structures and
proceed towards Nature through a random walk process that
works like successive approximations.

Now, besides the guidance provided by experiment,
can we say anything a priori about these structures? We
have two guides from eminent practitioners of our art.
Dirac says that (as it is written in the first page of
our program), "the world is described by beautiful mathematics". From Wigner we learned that there are certain
possible structures associated with invariance principles,
namely the irreducible realizations of the corresponding

symmetry group. According to this view Nature loves
irreducible representations. Thus, as a working hypothesis we can inspect the possible structures and select
the most beautiful, deep and far reaching ones as candi-
dates for the fundamental structures associated with
fundamental laws.

The most inspiring guidelines for selecting structures have been those which lead to unification on one
hand and geometrization on the other. We find an aes-
thetic value in unified theories and feel most satisfied if they exhibit a geometrical structure. Both of
course are Einsteinian themes. The Riemann geometry for
gravitation and attempts at generalization to non-
Riemannian geometries have been around for sixty years.
Weyl introduced a geometry with gauge invariance for
the unification of gravitation and electromagnetism.
Modern gauge theories found an interpretation in terms
of fiber bundle geometry. An even more general geometry
is now being worked out for gauge super-symmetry and
supergravity. Thus, from Riemann spaces to fiber bundles
involving Grassmann numbers we witness beautiful geometrical structures unifying an increasing number of different
parts of physics that are associated with different fundamental interactions.

There is another line of development of geometrical
structures less well known to physicists. It is the
geometrization of Quantum Mechanics due to Weyl, Jordan,
Wigner, von Neumann and Birkhoff. These physicists and
mathematicians have showed that there is a one to one
correspondence between finite Hilbert spaces and projective geometries. The projection operators in Quantum
Mechanics correspond to geometric objects like points,
lines, etc. Projection operators are hermitian,

idempotent and represent states. By extending complex
numbers to quaternions and octonions one obtains new
kinds of geometries where some projective properties like
the Pappus theorem or Desargues theorem no longer hold.
The latter has to do with properties associated with
the projection of a figure in a space on a lower dimensional subspace of the same kind. Hence, by en-
larging the algebra of the underlying numbers in the
Hilbert space, we obtain new models both for Quantum
Mechanical Spaces and Projective geometries. In Modern
Mathematics these new projective geometries have an importance comparable to that of non-Euclidean and Riemannian geometries in the nineteenth century.

The first example of the octonionic generalization
of a finite Hilbert space was given by Jordan in 1933,
Jordan,von Neumann and Wigner in 1934. It was a possible
structure. They had no interpretation for the degrees
of freedom it represented and did not know the group of
invariance of probability amplitudes in such a space.
Mathematicians later found its geometric interpretation
in terms of new non-Desarguesian projective geometries
and determined the invariance groups to be the exceptional
groups. It is F_4 for the particular Hilbert space of
Jordan, von Neumann and Wigner, E_6, E_7 and E_8 for more
general geometries of the octonionic kind. In all these
models the group $SU(3) \times SU^c(3)$ arises naturally, one
$SU(3)$ being a candidate for color and the other for unitary symmetry.

It is now tempting to connect these possible geometries with the Hilbert space of internal symmetries.
In one such model there are six quarks and ten leptons.
The results up to now are consistent with phenomenology
and the scheme will become more plausible if new heavy

leptons and heavy quarks are discovered.

It will be quite startling if Nature takes advantage
of all the geometrical possibilities for unifying various
fundamental interactions. Finally, is there one single
geometry that unifies supergravity with the non-Desarguesian geometries? So far no such geometry is known, but
if there is such a possible structure we must not despair
if it is not discovered by the Einstein-Dirac method.
One day the tortoise will catch Achilles and the theory
will be discovered by the bumbling detective method.
Thank you again.

FEZA GÜRSEY RECEIVING THE
1977 J. ROBERT OPPENHEIMER MEMORIAL PRIZE
FROM FIRST RECIPIENT, P. A. M. DIRAC
January 19, 1977

PARTICIPANTS

Kazuo Abe
University of Michigan

Richard Arnowitt
Northeastern University

Marshall Baker
University of Washington

James Ball
University of Utah

Michael Barnett
Stanford Linear Accelerator
  Center

Asim Barut
University of Colorado

Mirza Bég
Rockefeller University

C. Bouchiat
University of Paris, South

Richard Brandt
New York University

Laurie Brown
Northwestern University

Arthur Broyles
University of Florida

David Campbell
Los Alamos Scientific
  Laboratory

Cyrus Cantrell
Los Alamos Scientific
  Laboratory

Peter Carruthers
Los Alamos Scientific
  Laboratory

Kenneth Case
Rockefeller University

Owen Chamberlain
University of California

M. S. Chen
Center for Theoretical Studies

Ta-Pei Cheng
University of Missouri

Hirsh Cohen
IBM

Charles Conley
University of Wisconsin

John Cornwall
University of California

Stanley Deans
University of Florida

Gary Deem
Bell Laboratories

Robert Diebold
Argonne National Laboratory

P.A.M. Dirac
Florida State University

Bernice Durand
University of Wisconsin

Loyal Durand
University of Wisconsin

Glennys Farrar
California Institute of
  Technology

Richard Feynman
California Institute of
  Technology

Thomas Fields
Argonne National Laboratory

Paul Fife
University of Arizona

Paul Fishbane
University of Virginia

Hermann Flaschka
University of Arizona

Kenneth Fox
University of Tennessee

Paul Frampton
University of California

Daniel Freedman
State University of New York
  at Stony Brook

Richard Friedberg
Columbia University

Harald Fritzsch
CERN

Harold Galbraith
Los Alamos Scientific
  Laboratory

Murray Gell-Mann
California Institute of
  Technology

Howard Georgi
Harvard University

Fred Gilman
Stanford Linear Accelerator
  Center

Sheldon Glashow
Harvard University

O. W. Greenberg
University of Maryland

Marcus T. Grisaru
Brandeis University

David Gross
Princeton University

Eugene Gross
Brandeis University

Feza Gürsey
Yale University

C. R. Hagen
University of Rochester

Francis Halzen
University of Wisconsin

Brosl Hasslacher
California Institute of
  Technology

A. W. Hendry
Indiana University

Jarmo Hietarinta
Ohio State University

Bernard Hildebrand
Energy Research and
  Development Administration

Robert Hofstadter
Stanford University

Joseph Hubbard
Center for Theoretical Studies

Antal Jevicki
Princeton University

Kenneth Johnson
Massachusetts Institute of
  Technology

Gordon Kane
University of Michigan

Stuart Kasden
Princeton University

David Kaup
Clarkson College of
 Technology

Boris Kayser
National Science
 Foundation

J. P. Kernevez
Universite de Technologie
 de Compiegne

John Klauder
Bell Laboratories

Abraham Klein
University of Pennsylvania

A. D. Krisch
University of Michigan

Behram Kursunoglu
Center for Theoretical
 Studies

Kenneth Lane
Stanford Linear Accelerator
 Center

Herbert Lashinsky
University of Maryland

Harold Lecar
National Institutes of
 Health

Don Lichtenberg
Indiana University

Harry Lipkin
Fermi Laboratory

Marvin Marshak
University of Minnesota

David McLaughlin
University of Arizona

Sydney Meshkov
National Bureau of Standards

Himrich Meyer
University of Wuppertal

Michel Mille
Center for Theoretical Studies

Peter Minkowski
University of Bern

John Moffat
University of Toronto

Pran Nath
Northeastern University

André Neveu
Institute for Advanced Study

Patrick O'Donnell
University of Toronto

Reinhard Oehme
University of Chicago

Harold O'Gren
Indiana University

Martin Olsson
University of Wisconsin

Heinz Pagels
Aspen Center for Physics

Sandip Pakvasa
University of Hawaii

William Palmer
Ohio State University

Michael Parkinson
University of Florida

Emmanuel Paschos
Brookhaven National
  Laboratory

J. Patera
University of Montreal

Earl Peterson
University of Minnesota

Arnold Perlmutter
Center for Theoretical
  Studies

Chris Quigg
Fermi Laboratory

Pierre Ramond
California Institute of
  Technology

Mario Rasetti
Center for Theoretical
  Studies

L. G. Ratner
Argonne National Laboratory

Charles Rhodes
Stanford Research Institute

Jabus Roberts
Rice University

Rudolf Rodenberg
III. Physikalisches Institut
  der Technischen Hochschule

Joseph Romig
Boulder, Colorado

S. Peter Rosen
Energy Research and
  Development Administration

Carl Rosenzweig
Syracuse University

Ralph Roskies
University of Pittsburgh

Jonathan Rosner
Institute for Advanced
  Study

Hanno Rund
University of Arizona

Robert Sachs
Argonne National Laboratory

Abdus Salam
International Centre for
  Theoretical Physics

Howard Schnitzer
Brandeis University

Jonathan Schonfeld
California Institute of
  Technology

Julian Schwinger
University of California

Alwyn Scott
University of Wisconsin

Gino Segré
University of Pennsylvania

L. M. Simmons, Jr.
Los Alamos Scientific
  Laboratory

Richard Slansky
Los Alamos Scientific
  Laboratory

Charles Sommerfield
Yale University

Vigdor Teplitz
Virginia Polytechnic Institute
  and State University

G. 't Hooft
Stanford Linear Accelerator
  Center

Gerald Thomas
Argonne National Laboratory

Yukio Tomozawa
University of Michigan

T. L. Trueman
Brookhaven National
  Laboratory

Hung-Sheng Tsao
Rockefeller University

P. Van Nieuwenhuizen
State University of New
  York at Stony Brook

Kameshwar Wali
Syracuse University

Jill Wright
Bedford College

Eli Yablonovitch
Harvard University

York-Peng Yao
University of Michigan

G. B. Yodh
University of Maryland

Akihiko Yokosawa
Argonne National Laboratory

Henry Yuen
TRW Systems

Norman Zabusky
University of Pittsburgh

Fredrik Zachariasen
California Institute of
  Technology

Bruno Zumino
CERN

PROGRAM

ORBIS SCIENTIAE 1977

MONDAY, JANUARY 17, 1977

LAW SCHOOL AUDITORIUM

OPENING ADDRESS OF WELCOME

SESSION I -

"POLARIZED PARTICLES AND SPIN EFFECTS IN HIGH ENERGY
PHYSICS"

Moderator:
Alan Krisch, University of Michigan

Dissertators:
P.A.M. Dirac, Florida State University
    "DYNAMICAL METHODS FOR STREAMS OF MATTER"(30 min.)

Behram Kursunoglu, University of Miami
    "ORIGIN OF SPIN" (30 min).

F. Halzen, University of Wisconsin
    "POLARIZATION EXPERIMENTS - A THEORETICAL REVIEW"
    (40 min.)

J. B. Roberts, Rice University
    "MEASUREMENT OF SPIN DEPENDENCE OF σ_{tot}(pp)"(20 min.)

A. Yokosawa, Argonne
    "pp SCATTERING - AMPLITUDES MEASUREMENTS AND A
    POSSIBLE DIRECT-CHANNEL RESONANCE IN pp SYSTEM"
    (20 min.)
K. Abe, University of Michigan
    "LARGE $P_\perp^2$ SPIN DEPENDENCE OF P-P ELASTIC
    SCATTERING(20 min.)

SESSION II -

"POLARIZED PARTICLES AND SPIN EFFECTS IN HIGH ENERGY
PHYSICS"

Moderator:
Alan Krisch, University of Michigan

Dissertators:
L. Dick, CERN
    "POLARIZATION EXPERIMENTS IN EUROPE"(40 min.)

R. E. Diebold, Argone
    "ELASTIC AND INELASTIC POLARIZATION EFFECTS OBSERVED
    WITH THE ARGONNE EFFECTIVE MASS SPECTROMETER"(20 min.)

414

E. A. Peterson, University Minnesota
"INCLUSIVE ASYMMETRIES" (20 min.)

Harold Ogren, Indiana University
"PRELIMINARY POLARIZATION RESULTS AT FERMILAB
ENERGIES"(20 min.)

Owen Chamberlain
"POLARIZED TARGET EXPERIMENT AT FERMILAB"(20 min.)

TUESDAY, JANUARY 18, 1977

SESSION III -

"SUPERSYMMETRY AND SUPERGRAVITY"

Moderator:
Daniel Z. Freedman, State University of New York
at Stony Brook

Dissertators:
Richard Arnowitt, Northeastern University
"LOCAL SUPERSYMMETRY AND INTERACTIONS"(30 min.)

Daniel Z. Freedman, SUNY at Stony Brook
"SUPERGRAVITY FIELD THEORIES AND THE ART OF
CONSTRUCTING THEM"(30 min.)

Peter van Nieuwenhuizen, SUNY at Stony Brook
"RENORMALIZABILITY OF SUPERGRAVITY"(30 min.)

Bruno Zumino, CERN
"TOPICS IN SUPERGRAVITY AND SUPERSYMMETRY"(30 min.)

SESSION IV -

"FLAVOR INTERACTIONS"

Moderator:
H. Fritzsch, CERN

Dissertators:
H. Fritzsch, CERN
"QUANTUM FLAVORDYNAMICS"(30 min.)

F. Gilman, SLAC
"PRODUCTION OF NEW PARTICLES IN ELECTRON POSITRON
ANNIHILATION"(30 min.)

H. Meyer, Wuppertal University, Germany
"RECENT RESULTS ON e^+e^- ANNIHILATION FROM PLUTO
AT DORIS"(30 min.)

M. Barnett, SLAC
"THE SEARCH FOR HEAVY PARTICLES"(30 min.)

C. Bouchiat, Ecole Normale Superieure, Paris
    "PARITY VIOLATION IN ATOMIC PHYSICS AND
    NEUTRAL CURRENTS"(30 min.)

WEDNESDAY, JANUARY 19, 1977

SESSION V -

"HADRON PHENOMENOLOGY"

Moderator:
Sydney Meshkov, National Bureau of Standards,
                    Washington, D. C.

Dissertators:
Richard P. Feynman, California Institute of Technology
    "CORRELATIONS IN HADRON COLLISIONS AT HIGH
    TRANSVERSE MOMENTUM"(30 min.)

Jonathan L. Rosner, Institute for Advanced Study,
                    Princeton
    "FINAL STATES IN CHARMED PARTICLE DECAYS"(30 min.)

Howard Georgi, Harvard University
    "THE WINNER OF THE VECTOR-MODEL LOOK-ALIKE
    CONTEST"(30 min.)

Harry J. Lipkin, Fermi Laboratory and Argonne
    "THE ALEXANDER...ZWEIG RULES AND WHAT IS WRONG
    WITH PSEUDOSCALAR MESONS"(30 min.)

SESSION VI -

"ATTEMPTS TO SOLVE QUANTUM CHROMODYNAMICS"

Moderator:
Murray Gell-Mann, California Institute of Technology

Dissertators:
David Gross, Princeton University
    "A MECHANISM FOR QUARK CONFINEMENT"(30 min.)

H. David Politzer, Harvard University
    "THE STATUS OF ξ-SCALING"(30 min.)

Ta-Pei Cheng, University of Missouri
    "MUON NUMBER NONCONSERVATION IN GAUGE THEORIES"(30 min.

J. M. Cornwall, UCLA
    "SEMICLASSICAL PHYSICS AND CONFINEMENT"(30 min.)

G. t'Hooft, SLAC and the University of Utrecht
    "SOME OBSERVATIONS IN QUANTUM CHROMODYNAMICS"(30 min.)

THURSDAY, JANUARY 20, 1977

SESSION VII -

"SOLITONS AND NONLINEAR PARTICLE THEORY"

Moderator:
David K. Campbell, Los Alamos Scientific Laboratory

Dissertators:
Andre Neveu, Institute for Advanced Study, Princeton
    "SOME RECENT DEVELOPMENTS ON SOLITONS IN
    TWO-DIMENSIONAL FIELD THEORIES"(45 min.)

Antal Jevicki, Institute for Advanced Study, Princeton
    "PATH INTEGRAL QUANTIZATION OF SOLITONS"(30 min.)

Richard Friedberg, Barnard College
    "NONTOPOLOGICAL SOLITONS"(30 min.)

Paul Frampton, University of California
    "VACUUM BUBBLE INSTANTONS"(20 min.)

SESSION VIII -

"TOPICS IN NONLINEAR MATHEMATICS WITH APPLICATIONS TO
PHYSICS"

Moderator:
Hermann Flaschka, University of Arizona

Dissertators:
Henry Yuen, TRW, Redondo Beach, California
    "NONLINEAR DEEP WATER WAVES: A PHYSICAL
    TESTING GROUND FOR SOLITONS AND RECURRENCE"(30 min.)

David Kaup, Clarkson College, Potsdam, N.Y.
    "SOLITONS AS PARTICLES, AND THE EFFECTS OF
    PERTURBATIONS"(30 min.)

N. J. Zabusky, University of Pittsburgh
    "COHERENT STRUCTURES IN FLUID DYNAMICS"(30 min.)

FRIDAY, JANUARY 21, 1977

SESSION IX -

"NONLINEAR MOLECULAR PROCESSES"

Moderator:
Cyrus Cantrell, Los Alamos Scientific Laboratory

Dissertators:
Eli Yablonovitch, Harvard University
    "COLLISIONLESS MULTIPHOTON DISSOCIATION OF SF_6:
    A STATISTICAL THERMODYNAMIC PROCESS"(30 min.)

H. W. Galbraith, Los Alamos Scientific Laboratory
    "STRUCTURE OF THE VIBRATIONAL STATES IN THE ν_3-
    FUNDAMENTAL AND ITS OVERTONES IN SF_6 AND
    MULTIPHOTON ABSORPTION EFFECTS"(30 min.)

Kenneth Fox, University of Tennessee
    "REVIEW OF ROTATIONAL STRUCTURE IN EXCITED
    VIBRATIONAL STATES OF SPHERICAL-TOP MOLECULES"
    (30 min.)

C. K. Rhodes, Stanford Research Institute
    "ISOTOPE EFFECTS IN MOLECULAR MULTIQUANTUM"(30 min.)

SESSION X:

"NONLINEAR STRUCTURES AND OSCILLATIONS IN BIOCHEMICAL
MEDIA"

Moderator:
Norman J. Zabusky, University of Pittsburgh

Dissertators:
J. P. Kernevez, Universite de Technologie de Compiegne
    "SPATIO-TEMPORAL STRUCTURATION IN IMMOBOLIZED
    ENZYME SYSTEMS"(30 min.)

Paul C. Fife, University of Arizona
    "THEORIES AND CONJECTURES ON MEMBRANE-SUPPORTED
    WAVES AND PATTERNS"(30 min.)

Harold Lecar, National Institutes of Health
    "PHYSICAL MECHANISMS OF NERVE EXCITABILITY"(30 min.)

Hirsh Cohen, International Business Machines Corp.
    "MATHEMATICAL DEVELOPMENT IN HODGKIN-HUXLEY
    THEORY AND ITS APPROXIMATION"(30 min.)